Advanced Materials and Methods for WATER TREATMENT

Advanced Materials and Methods for
WATER TREATMENT

SK Nataraj
Professor
Centre for Nano and Material Sciences (CNMS)
Jain University
Kadakolla-Bengaluru, India

TECHSAR™
NEW DELHI

Published and Printed by
TechSar Pvt. Ltd.
4435-36/7, Ansari Road, Daryaganj
New Delhi-110 002 (India)
E-mail: tech_info@techsarworld.com
Website: www.techsarworld.com

ISBN: 978-81-959662-5-7

Reprint 2025

Dedicated to Parents

Veeramma & Sharanappa

To Loving Family:

Ashwini, Anvita and Vrishank

Preface

Demand for safe and clean water is ever-increasing and on the other hand, efforts to recover wasted resources, in particular, water are also gaining significant importance. Researchers, scientists, innovators, and policymakers throughout the world are investing their time and efforts to build effective and sustainable infrastructure to manage and recover resources from discarded wastes of various states and nature. Nevertheless, the increasing complexities of waste and the identification of new and emerging pollutants in various discarded sources pose a serious challenge to conventional technologies. Therefore, constant monitoring of water management systems and wastewater treatment infrastructure is essential and it continues to be a challenging task as the newly emerging class of pollutants in water resist the conventional treatment methods. Therefore, conventional and existing water treatment materials, methods, systems, and technologies becoming out of date.

To address future challenges associated with water and wastewater treatment issues, researchers, innovators, and technologists are developing new materials stock through innovative approaches to make water treatment technologies more sustainable and effective. These efforts also include the development and packaging of advanced materials and technologies or processes more economical, have higher durability, and have less to no maintenance costs. Among many such efforts, nanoscience and nanotechnology-assisted approaches are the primary set of solutions that have shown great potential in the effective abetment of pollution-causing constituents as well as the scale-up activities to be useful for real field applications.

Even though every existing and conventional water and wastewater treatment technology has improved over the years, however, these systems can fail or prove ineffective against an advanced class of pollutants and contaminants which can create an ecological and health disaster. In fact, domestic wastewater, industrial discharges, and municipal wastewater treatment systems that fail to treat contaminated water and wastewater properly can potentially harm human health, ecological systems, and the environment at large.

Therefore, this book gives a comprehensive account of recent developments in materials and process developments targeting future water and wastewater treatment issues. This book would serve as a guide to various stakeholders such as researchers, technologists, policymakers as well as student community on the various materials stock and methods developed in recent years to address complex pollutants that are difficult to treat or remove with conventional as well as existing water treatment methods. Therefore, in this book broad as well as specific case studies have been used to contextualize the importance of future water and wastewater-related problems and in each case, effective materials and methods developed to address the issues have been discussed in detail.

Furthermore, this book discusses the significance of each material selection, the importance of methods adopted and factors that influence the performance in pollutant treatment and removal in each case of application have been highlighted. Specifically, the first chapter gives an overview of current methods and their limitations which further highlights a brief note on recent advances in each technology segment. Chapter 2 gives detailed advanced methods and techniques followed in wastewater recycling efforts involving both modified versions of conventional technologies and innovative new approaches. This chapter in brief also gives an account of advances in waste management, disposal, and economical aspects, and most importantly water recovery and reuse in non-drinking water sectors that are identified as water-intensive industries or entities. Further chapters give detailed information on the specific material and technologies including novel membrane materials, functionalized polymers, block copolymers, and polyelectrolyte complexes. In addition to conventional pressure-driven membrane processes, advanced forward osmosis which is emerging as an effective alternative has been discussed in detail in a subsequent chapter. Another important segment of modern and advanced water treatment efforts comes from the use of functional materials comprising biopolymers, carbon materials, and nanocomposite materials that have shown great potential in both water and wastewater treatment applications.

Nonetheless, a large segment of water sources and wastewater constituents has been classified into micro-scale categories which can be effectively treated with ultra-porous filters with low-energy inputs. In this category, advanced electrospun fibers, aerogels, and ceramic filters have been discussed. The last two chapters focus on enlisting the advantages and limitations of modern methods which further account for detailed future challenges in developing effective end-use modules and systems. The use of nanoscience and nanotechnology principles have been widely explored in developing various functional nanomaterials that have been found highly effective in the treatment of water and wastewater sources. The last chapter specifically gives a detailed account of nanomaterial-based water treatment solutions. In particular, functional nanomaterials are developed using biopolymers, activated ceramic or inorganic nanomaterials, functional carbon-based micro-cleaners, and so on.

Therefore, advanced and future water treatment materials and technologies hold far-reaching potential in addressing water scarcity, and mitigating water contamination that may directly enter the ecosystem and environment and these efforts promote responsible water management practices. Nevertheless, future water treatment technologies will focus on utilizing the raw materials that are abundant in nature which utilize minimum energy inputs to convert them into active water treatment media. Also, future technologies will also be integrated with energy source systems that make overall water treatment systems sustainable and point operating processes. For instance, future water treatment technologies might include advanced treatment media integrated with solar technology, high-tech filtration, and nanotechnology approaches are expected to revolutionize in minimizing material utilization.

I express my gratitude to my research scholars at Sustainable Materials and Processes Lab, Centre for Nano and Material Sciences, Jain University for their constant queries which contributed to the betterment of the manuscript. Also, my sincere thanks go to some very important people for making this book happen: my parents (Veeramma, Sharanappa), in-laws (Yashodha, Mallikarjun) and brothers (Girish, Santosh, and Harish), my family, especially my wife Ashwini, and my children Anvita and Vrishank for providing me with the right atmosphere, sacrificing their time and allowing me to work on this book without disturbance.

S.K. Nataraj

Contents

Chapter

1 Introduction and Overview of Current Water Treatment Technologies

— Abstract —

This chapter overviews the current status of water treatment technologies used across all the sectors. This chapter covers a broader spectrum of separation technologies for broader water treatment applications. In this chapter, details of advances in modern water treatment techniques, list of physical methods, brief summary of chemical methods, combinations of physico-chemical methods, biological methods, thermal methods, and advanced physical methods like membrane-based separation processes are discussed. Further, this chapter includes brief comparative study and sustainability aspects of each water treatment method currently in use.

1.1 Introduction

The world is heading towards a severe water crisis due to rapid urbanization, changing weather patterns, rising consumer demands, rapid industrial boom and increasing contamination of freshwater sources as well as largely looming severe climate changes. On the other hand, the water treatment industry is set to embrace several innovative approaches to address the impending crisis. Interestingly, future water treatment technology approaches will not only be restricted to medium-to large-scale water purification and wastewater treatment units but also to be more decentralized and portable scale components. This would enhance the reach of water treatment technologies at remote areas. On the other hand, increased clean energy concepts are forcing the separation and purification industry to adopt cleaner and greener approaches to make water treatment processes increasingly viable and realistic. This is being realized by adopting low power consumption components into separation technologies during water treatment and pollution treatment plant design and integration at treatment sites. However, there are several challenges associated with adopting compatible and competitive components into integrated treatment systems. Further, these alterations would likely present industries and manufacturers a multifaceted challenge that could be an opportunity researcher, and technologists to innovate new materials and methods to stay competitive within the industry. These new opportunities may also come up with better solutions with high performance material stock and technology components that benefit manufacturers, end users and the environment at large. [1]

The world water treatment, desalination and wastewater treatment technologies market are segmented into materials or accessories, membrane components, engineering equipment, pumps and maintenance materials. This market for the previously mentioned separation process and technology related sectors is projected to grow at an average over 10% compound annual growth rate (CAGR) every year for water and wastewater treatment technology market segments, in particular. [2] The separation and purification technology market is segmented by equipment type and end-user industry. The material stocks are further divided into separation media, chemicals, engineering materials, and equipment materials. The treatment equipment have been further divided into desalination process, organic separation, oil/water separation, emerging pollutants separation or removal, physical separation techniques like suspended solids removal, dissolved solids removal, biological treatment like anaerobic and aerobic treatment, nutrient and precious metals recovery, disinfection processes like ozonation, oxidation, advanced oxidation process (AOP) and other treatment apparatuses. [3] At the consumer end, the market is divided mainly into beverages, food processing, municipal water treatment, healthcare processes, poultry, agriculture and so on. Each category can be divided into separate segments based on their market size and potential growth based on current market revenue, namely:

(a) in the medium span, growing demand for domestic use of purified water and swiftly waning freshwater resources and growing awareness for domestic wastewater treatment are prospective to drive the demand treatment technologies,

(b) currently practiced high performance and high capital cost separation and purification technologies like membrane separation processes are expected to hinder the growth of the market, and

(c) materials innovations in separation and purification technologies are anticipated to act as an opportunity for new emerging market.

In addition to this, there are several related factors which may directly or indirectly affect the overall advancement in the water treatment technologies. This may also open new avenues for innovating new and advanced materials and approaches to realize them in field applications. Among several opportunities, some can be summarized as aspects of water that are likely to have an overall impact on the advancement in water treatment technologies at industrial scale in the near future.

Worldwide scientists, researchers and policy makers are worried about the ever-looming climate change pattern which is disrupting weather patterns, increasing earth's water temperature, and worsening greenhouse effect. These changing weather patterns are leading to dangerous weather events, unpredictable rainfalls, uneven water distribution, intensifying water scarcity and their availability. These aggravated conditions in combination with increasing water pollution would catalyze the climate change process. Water pollution originates from various sources that may be single source or/and multiple sources or nonpoint sources. If the water pollution occurs from a single source like accidental discharge or disaster is known as point-source water pollution. On the other hand, non-point-source water pollution occurs from many sources or various sources. However, water pollution largely occurs due to intentional or unintentional spills or leaks from cargo transportation carrying oil or chemical containers, domestic discharges and industrial effluent discharges directly into surface water resources or drains. On the other hand, increased industrialization utilizes large quantities of freshwater resources and most of it results in wastewater generation at the end of the process. [4]

Among air pollution, soil pollution, water pollution and sound pollution, water severely affects the immediate ecosystem and environment at large. In case of large quantity discharges, water pollution spreads across larger landmass or large aquatic areas affecting the hundreds of kilometers away from the source distressing millions of living beings. This uncontrolled spread of pollution, broadly known as trans-boundary pollution, causes the severe destruction of lakes, reservoirs, rivers, seas and larger biodiversity. Further, polluted water streams deplete aquatic environments and trigger uncontrolled spread of phytoplankton in lakes and ponds eutrophication as shown in Figure 1.1. Even though this process of large collection of wastewaters in lakes or ponds avoids fouling drains and waterways, it accumulates large quantities of organic and inorganic pollutants which increase the possibilities of chemical oxygen demand (COD) and biological oxygen demand (BOD) in the watercourses. In this condition, a surge in water body temperature can result in the death of many aquatic beings and disrupt many aquatic habitats such as mass killing of fishes, coral bleaching and eutrophication. [5]

Figure 1.1. Process of formation of eutrophication in lakes which is caused due to the water pollution and growth of algae bloom.

Ultimately, all polluted water streams contribute to the altercation of the food chain and rise in the temperature of water bodies. Further, this situation can drastically affect the quantity and quality of potable and usable water availability to survive a large dependent population. On the other hand, severe water pollution majorly contributes to climate change leading to:

(a) extreme weather changes and deviations in water cycle patterns,
(b) frequency and intensity of extreme weather changes are only expected to increase with climate change,
(c) high or extremely high-water vulnerability and scarcity to meet their everyday needs,
(d) entry of contaminated water into potable water resources and their supply chain leads to waterborne diseases like typhoid and cholera for which larger population is vulnerable,
(e) rising temperatures in water bodies can lead to growth of deadly pathogens in lake or pond sources,
(f) polluted water poses an enormous threat to soil health,
(g) these climate change induced weather patterns aggravate water stress areas which limits the accessible water resources thereby leading to water distribution conflicts, and overall, climate changes are known to cause rising sea levels putting fresh water to vulnerable to turn salty and contribute to sea level rise. It is estimated that by 2050, the rise in sea level and potential extreme storm surges would affect more than half of the world's inhabitants living within 200-250 km of the coastline. Therefore, largely climate change has roots in water pollution in one or the other way. [6]

1.2 Advances in Conventional Water Treatment Techniques

As per United Nation's (UN) estimation, by 2050 the world's population is projected to reach more than 9 billion. To meet the food requirement, it is estimated that global food production will be augmented by 70% from now. Nevertheless, the world food demand is likely to experience a scenario of over near 10,000 million tons in 2030 and over 14,000 million tons in 2050. [7] On the supply chain, certified food production is projected to cross 10,000 million tons in 2030 and over 15,000 million tons in 2050. On the other hand, the energy-water-food (EWF) nexus is known as the significant interconnectedness among all three vital resources for living beings and human societies, in particular. However, nexus of all three vital resources is central to sustainable world development. However, the inextricable relationships among these vital resources require a suitably integrated clean and sustainable approach to safeguarding water, energy, and food security (Figure 1.2).

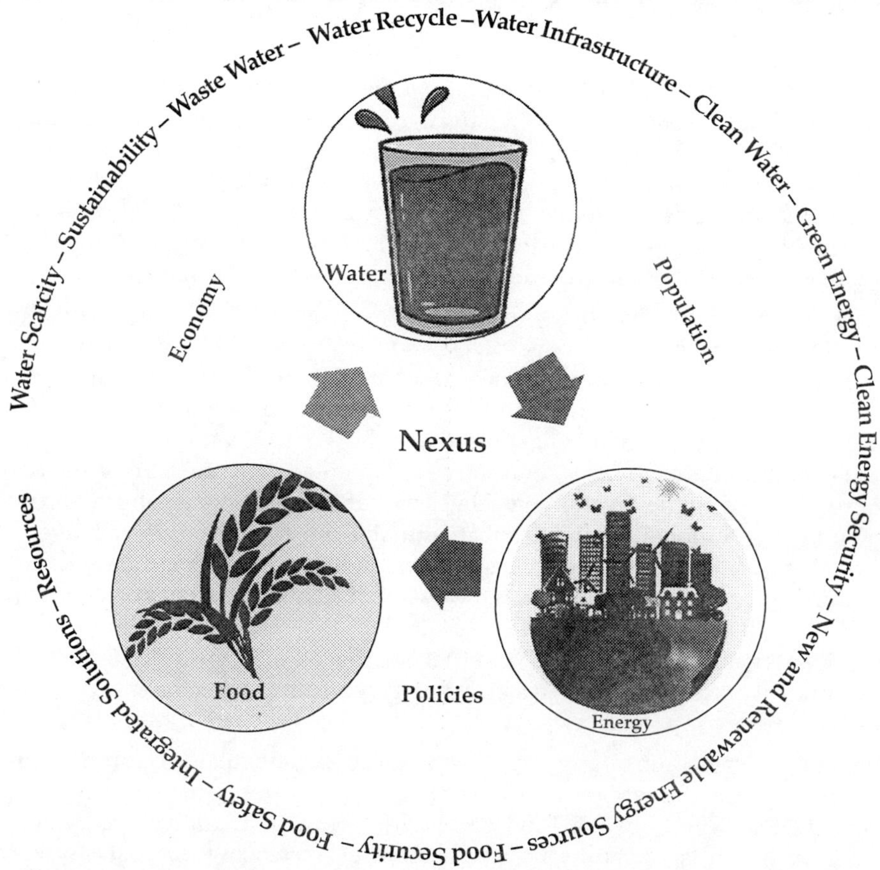

Figure 1.2. Water, energy, and food nexus and importance of correlating factors.

Sustainable energy utilization drastically reduces water consumption in many industrial processes. This reduces the challenges associated with water resources and their accessibility throughout the ecosystems. Absence of sustainable water supply impacts food production and agricultural yield. Increasing reforestation efforts and growing high yield biomass for commercial purposes are exploring utilization of wastewater for irrigation. Recovered water from wastewater resources has been used for maximizing the number of crops produced on wastelands. The use of wastewater for growing fruits, vegetables, floriculture and livestock may contaminate the harvest; however, cultivating commercial crops would be standardized in near future. On the other hand, agricultural wastewater is mainly the surplus water that runs off the field at the low end of furrows, boundary strips, basins, and swamped areas during flood irrigation. The surplus wastewater is also known as agricultural or irrigation tail water. In flood irrigation, a certain amount of tail water runoff is necessary to ensure adequate wetting and penetration of top layer of soil along the length of the trough or border strip being flooded and wetted to achieve an optimal irrigation productivity. On the other hand, another source of agricultural surplus water is effluent from plants processing crops harvested and washed from the field and contaminated water generated during processed food, larger food manufacturing facilities. These larger processing units generate substantial volumes of agro-product processing wastewater typically comprising high concentrations of organic matter. Often, food processing wastewater from agricultural product harvesting and processing plants is sent to a nearby municipal effluent treatment plant. Their typical contamination and concentration level demand special handling and treatment protocols. Therefore, with suitable sample analysis, protocols, and treatment techniques both agriculture surplus flooded water and processing wastewater can be reclaimed and reused for sustainable farming.

In 2015 United Nations member states adopted the 2030 agenda for sustainable development goals (SDGs) which provides a largely consented blueprint and shared outline for peace, prosperity for people, greener planet, and economic growth for now and into the future. For implementation of 17 adopted sustainable goals and objectives, the United Nations plays a vital role through member states and intergovernmental structures. To implement these 17 SDGs to the level of expected level of success, systematic coherence in strategies at global, regional, sub-regional, country and provincial levels is crucial. At the core SDGs are pressing issues which determine the safety and sustainability of the earth through the action of all member states on economy, environment, education, energy sustainability, and most importantly, clean water and sanitization. Nonetheless, the whole world recognizes the importance of clean energy and safe water thereby influencing the energy, water, and food nexus. [8]

The energy-water-food nexus is central to sustainable development and larger economic targets for many developing countries. Greener energy and clean water influence the food production both quantitatively as well as qualitatively. It is driven by the changing diet, increasing population, rapid urbanization, and economic growth. Historically, agriculture sectors have been the largest consumer of the global freshwater resources and more than one-quarter of the energy used worldwide is consumed by food production, harvesting and supply chain. However, the complex connections between these vital domains require a properly integrated and cohesive approach for ensuring water and energy security.

The major concern associated with water is due to its finite resource but serving exponentially more population and activities. Therefore, to safeguard everyone in a billion population who have access to a reliable supply is crucial to human survival, industrial growth, agricultural activities and sustainable economic progress. On the other hand, as water is extracted more, these finite resources turn out to be more strained and the other two dependentsectors like energy and food suffer more. Interestingly, all sectors listed in 17 SDGs depend on water and decision-makers and policy makers in all sectors are now intensely focusing on water resource management, environment protection, water supply and sanitation as part of policy and practice. Among many considerations, the finite and essential nature of freshwater comes to the center to make decisions in addressing the water-energy-food nexus thereby consequences associated with climate change. [9]

1.3 Conventional Water Treatment Techniques

From ancient times, civilizations have established and thrived over centuries around water resources. Large and clean water sources have always been pillars for both traditional and modern societies. Other than domestic use, water has been used for agricultural irrigation, industrial production, as coolant in coal-based electricity and heat generation, and so on. The most immediate effect of overexploitation of water in all possible sectors is generation of wastewater which contributes enormously to the contamination level in the environment. According to the new definition, wastewater is the contaminated or polluted form of freshwater generated from rainwater runoff, industrial activities, and human activities. All the wasted water from human activities is also called sewage water. Traditionally, wastewater is categorized by the methods through which it is produced, namely, industrial wastewater, domestic sewage, agriculture wastewater, storm sewage and waste-waters with emerging pollutants. The major contributor to domestic waste wateris water from kitchen sinks, showers, toilets, baths, laundries, and personal care activities. It is estimated that domestic households generate an average of 200–300 L of wastewater per person every day. This contains an average of 99% of it as recyclable water and remaining one percent solutes or pollutants dissolved or dispersed in it.

The major concern is that worldwide more than 80% of discharged or disposed water after usage returned to the environment is untreated. However, if properly treated, all can be reused directly. Unfortunately, at present just one percent of total water used across the world is recycled. Of the billions of liters of wastewater discharged every day, the majority of wastewater is discharged into rivers, estuaries and oceans which leads to adulteration, pollution, and destruction of natural and aquatic habitats. Moreover, these contaminated water resources surrounding the ecosystems also expose the wildlife and humans in those habitats toto harmful effects from disposed of chemicals. On the other hand, feces and urine from both animals and humans carry numerous disease-causing contaminants and organisms. Wastewater from industry may contain harmful chemicals like semi-volatile organics, synthetic substances, heavy metals and other pollutants known to cause numerous health and environmental problems. Another major concern for the future is concern for emerging pollutants in wastewater. The emerging pollutants in wastewater are majorly classified as pharmaceutical and personal care products containing human waste, soaps, shampoos, detergents, nitrogen, phosphorus, pesticides, drug molecules and excessive pharmaceutical ingredients. According to several set norms, wastewater needs to be released into aquatic systems or natural habitats once it is completely or partially treated meeting set standards by state and local regulatory authorities. [10]

Rapid growth of metropolises is forcing overexploitation of freshwater resources for domestic use and discharge of large quantities of wastewater in the same proportion. As per estimates nearly 80% of total water supplied for domestic use largely turns into wastewater. Among these, city (tier-I) and town (tier-II), it is estimated that ~70% of used water enters the environment untreated. This is due to most municipalities do not have minimum required wastewater treatment infrastructure. In such a situation, it is expected that by 2050, production of wastewater would be multi-fold from now on, thereby further widening the treated and untreated water gap. Thus, overall detailed estimation and analysis of water resources usage or overexploitation indicates that in the coming years, there will be a twin-edged problem to deal with drastically decreasing fresh water availability and increased wastewater generation due to augmented population and industrialization. [11]

Apart from domestic wastewater, millions of liters of wastewater is generated by manufacturing industries of which only 60% is treated. Amongst, small-scale industries make large numbers which may not afford the cost of wastewater treatment infrastructure. Among medium- to large-scale industries which have well established water treatment plants take care partially or complete treatment of wastewater following set standards. Traditionally, several water treatment methods have been adapted including primary, secondary and tertiary wastewater treatment techniques involving physical, chemical and biological methods. In recent times, membrane-based processes, advanced oxidation process, dissolved air floatation, nanomaterial-based filters, dual media filter, photo-catalytic degradation, activated carbon filter, ion-exchange methods, sand filtration and tank stabilization, flash mixer, clari-flocculator, sludge drying beds secondary clarifiers so on as shown in Figure 1.3. [12]

1.4 Advances in Physical Methods

Wastewater treatment processes are essential to protect, preserve and conserve the health of various diverse ecosystems on the earth. If wastewater is effectively and properly treated, it can be used as a new source of water for several applications and purposes. Efficient wastewater treatment allows recovery of a maximum amount of clean or reusable water after segregating all suspended or dissolved solid waste. Even though wastewater is characterized for its physical, chemical, biological, and complex emerging characteristics, fundamentally basic parameters like pH, odor, color and temperature playan important role in inducing complexities to wastewater. Most importantly, high temperature, extreme pH conditions of wastewater impact piping, storage and transportation materials and processes like septic tanks and drainages, respectively. [13]

Increased discharge of wastewater to natural freshwater resources or environment causes increased concern and this necessitates the adaptation of usable water treatment techniques. Climate change and increasing concern over emerging pollutants are diverting global interest to adopt appropriate wastewater treatment techniques as industries, communities, and individuals recognize ways to keep crucial resources accessible and suitable for use. Therefore, advanced wastewater treatment is essential as the world needs to mitigate the impacts of depleting freshwater sources in competition with increased population, urbanization, and industrialization. However, wastewater treatment cannot be always achieved efficiently due to the

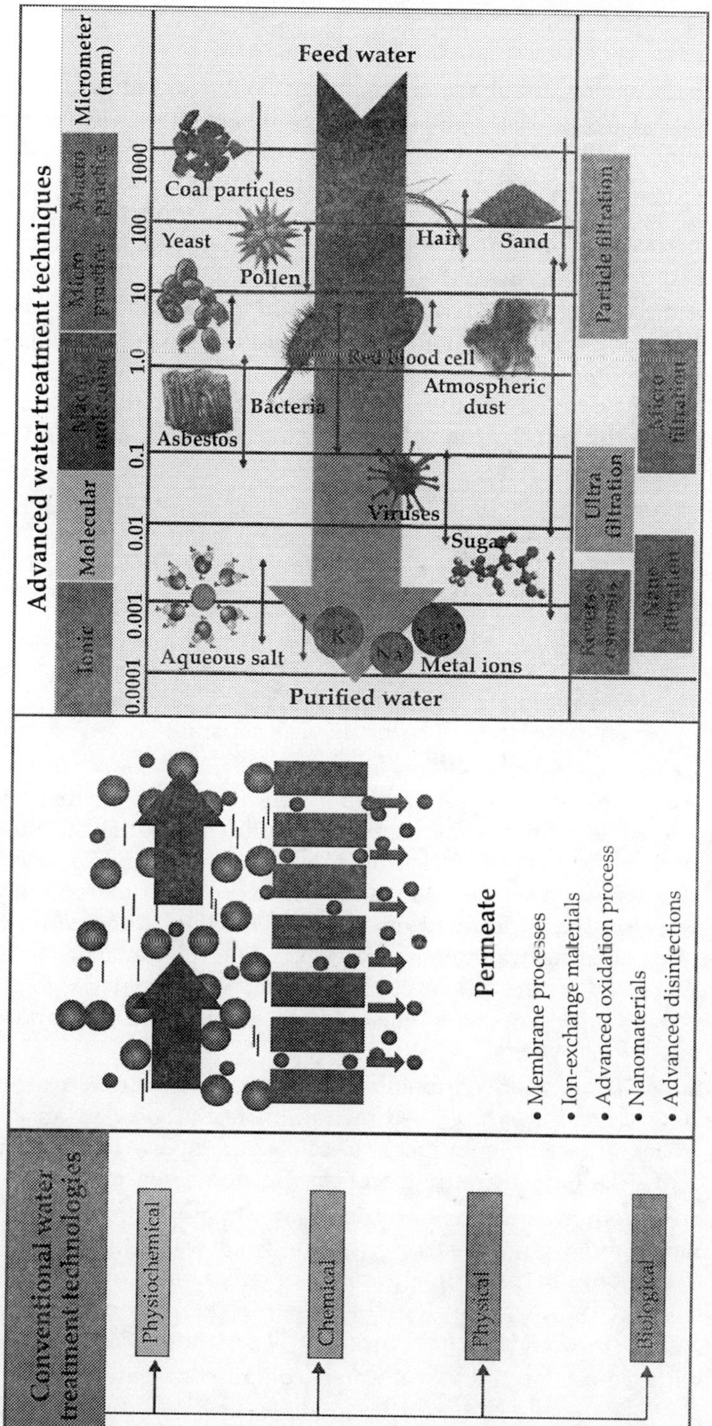

Figure 1.3 List of conventional and advanced water treatment techniques.

nature of unknown pollutants such as emerging contaminants and their mysterious consequences. Therefore, to keep up with these emerging challenges, researchers, innovators and technologists are working on new techniques and methods of water treatment. With the adaptation of these treatment technologies, it is likely to further enhance the quality of recovered water beyond the limitations and restrictions of conventional technologies to realize the goal of resource recovery or resource conservation. In this direction, advanced wastewater treatment infrastructure may directly and indirectly enhance the possibilities and availability of recycled water reclamation and supply for future applications.

Among three main water treatment stages, the primary water treatment step involves mainly physical methods of removing suspended solids through various means either direct segregation or by allowing them to settle from wastewater. In this method, the waste or undesired suspended solutes are processed through a physical technique with the help equipment and filtration or separation media such as:

(a) *Screening*: In this step, incoming effluent is passed through screening apparatus where objects such as plastics, wood fragments, rags and grease are removed. Further in this process, all screened material separated, washed, pressed, and disposed of as landfill and filtered liquid component is then pumped to the next step.

(b) *Grit removal:* In this step, heavier but fine substances such as sand, aggregated suspended solids and gravel are removed from the pre-screened wastewater. In this step also, all separated fine solid pollutants are disposed of in a landfill.

(c) *Primary settling:* In this step, lighter pollutant materials that take more time to settle than heavier suspended solids are passed through large circular tanks called clarifiers to remove from clarified treated water. Further, clarified water from settled finer pollutant material known as primary sludge is pumped off from the clarifier to the exit tank from the top. Floating fragments and debris like microplastics and grease are skimmed off the top and sent with the settled material to digesters. Chemicals and reagents are also added to eliminate phosphorus from clarified wastewater.

(d) *Sedimentation*: Conventionally, larger particles are removed from wastewater using simple sedimentation techniques. In an advanced approach, the sedimentation process is combined with a filtration step to treat simultaneous settling and clarification of wastewater.

(e) *Aeration*: Another efficient physical water treatment method is aeration in which air is circulated through the wastewater to provide oxygen to it. Aeration helps in floating oily or grease and lighter suspension on the surface of wastewater which are then easily separated out either manually or with the help of mechanical separators. [14, 15]

Generally, most of the wastewater contains suspended particles of different particlse or solid sizes which can be classified as dissolved solids with size < 0.08 μm, colloidal suspensions with particle size ranging from 0.08 to

1 μm, supra-colloidal suspensions with solute size >100-100 mm and settleable particles having particle size >100 μm. For such characteristic wastewater a treatment selected greatly depends on the size of suspended particles present in the wastewater. In practice, primary water treatment steps mainly involving physical methods are efficiently used for the treatment. Suspended solids with size that are visible to the bare eye can be separated either by subsiding under the influence of gravity or by flotation, based on their relative densities of suspended solids in water. In addition, these suspended solids may also be easily segregated via various mesh filters. However, very fine suspended particles of a colloidal nature (size < 1 μm) which are generally characterized as highly stable suspensions cause prolonged contamination. This is due to the stability of these particles possessing electrostatic surface charges or zeta of the same nature (usually negative). This leads to the existence of constant and steady repulsive forces between them, which prevents colluding and aggregation which thereby results in settling under gravitational influence. Therefore, colloidal systems are generally regarded as highly stable suspensions that are impossible to segregate either by settling or flotation. Also, due to their size limitation it is not possible to separate colloidal solids by mesh filtration since colloidal systems pass through easily. However, separation of fine suspended particles and colloidal systems can be achieved following physiochemical treatments methods. [16]

1.5 Advances in Physiochemical Methods

As discussed in the previous section , the larger particles are removed from wastewater using various physical methods. Majorly, physical methods categorized as primary treatment involving the physical removal of bulky as well as small suspended particles through mesh, cloth, screen filtration, and sedimentation. This suspended solid separation is carried out in many stages. Initially, fluctuating and floating debris are removed using screening or the grit filters and on the other hand soil, sand particles and small pebbles are removed by sedimentation. Further, advanced physical methods of treatments include flotation, sedimentation, and physical adsorption creating barriers such as screens, bar racks, deep bed filters, and membrane filters. On the other hand, physicochemical wastewater treatment methods primarily focus on the separation of pollutants in colloidal range. This is accomplished through the addition of chemicals or substances which helps in breaking of colloidal suspension known as coagulants and flocculants. The colloidal systems having sable and similar surface charges allow them to remain in colloidal state indefinitely and therefore form into suspension or flocs with settling properties upon breaking the charge balance. Prior to this stage in physical methods like screening, skimming and sedimentation are employed to remove coarse solids. However, in these steps no chemicals are involved in the solid separation process. The insoluble material is made to separate in the screening process and sedimentation settles down at the bottom and subsequently skimming is used to float the grease or oily suspension from wastewater. [17]

On the other hand, the physicochemical methods mainly involve individual or combined steps of:

(a) **Coagulation:** The term coagulation, derived from the Latin (coagulare), means driving together and the process coagulation means destabilization or neutralization of negative zeta or charges on colloidal systems in the wastewater with the help of added destabilizing agents called coagulants. This is a rapid coagulant mixing process which is applied to separate colloidal pollutants in a rapid mixing in a short contact time. Generally, the quantity of coagulant added to wastewater for the process of coagulation to take place depends on the quality and level of contamination which is mainly characterized by domestic or industrial wastewater. In the case of domestic or municipal wastewater, usually < 50 mg/L dose of coagulant is added, while coagulant dose for industrial wastewater varies based on concentration. In advanced approaches nano-sized or appropriately functionalized coagulants are most used such as aluminum sulfate, sodium aluminate mixtures of organic ferric chloride, ferric sulfate, aluminum polychloride, inorganic compounds, lime and the more lately developed formulation including iron polychloride. [18]

(b) **Flocculation:** The term flocculation is derived from the Latin word *floculare,* meaning formation of flocs or bridges. To initiate complete sedimentation of colloidal pollutants come together in this process forming flocs assembly together. With time, flocks grow in volume and attain high density which ultimately experiences gravitational influence to undergo sedimentation. In advance approaches flocculation processes are enhanced via rapid mixing known as perikinetic or natural convection. This helps in enhancing constant points in a colloidal mixture between flocculant and pollutant colloid via agitating the wastewater and coagulant mixture in which mechanism is also known as orthokinetic or forced convection flocculation. Overall, in the flocculation step, flocculants added to assist flocculation allow flocculation to come together and adhere, snowballing their size, aggregation thereby leading to increased density. Flocculants can be categorized by their nature, origin, and type like mineral, organic flocculant, synthetic or natural or their surface zeta or electric charge like cationic, anionic or non-ionic. Recently, advanced approaches involving greener aspects greatly consider organic flocculants extracted from natural origin or from biomaterials such as cellulose derivatives, alginates from seaweed extract and starches from plant grain extracts. However, effectiveness of biopolymer-based flocculants is relatively inferior at present, however advanced approaches use functionalized biopolymers to achieve high rates. Most popularly used flocculants originate from long chain macromolecules which readily undergo soluble in water. Once long chain macromolecules added to wastewater mixtures these readily interact with colloidal pollutants as they are simply made of synthetic monomer carrying electric charges and or ionisable groups on their backbones. Flocculants generally impart charge imbalance to attract each other for this reason they are known as polyelectrolytes. Most of the time inorganic-based flocculants are highly efficient agents and usually recommended in low concentrations for solid ranging from 0.05%-0.1%, for liquid dispersion between 0.1% - 0.2% and for liquids in solution 0.5%-1.0%. It is important to note the use of excessive polyelectrolytes in wastewater for separation of colloidal pollutants from wastewater may harm the whole flocculation process. [19]

(c) ***Advanced sedimentation:*** However, these methods are applied in configurations in which techniques are applied individually or all steps are carried out in the same unit. [20]

There are several features like physicochemical properties of pollutants in the wastewater play an important role in the removal efficiency of any applied treatment technique. Traditionally, these basic characteristics have been analyzed and characteristically determined by conventional laboratory jar methods or by applying RoboJar systems. In this method, the testing procedure contains six jars of identical size with varying doses of coagulant added at the same time. The system proceeds with a rapid mixing sequence for a prearranged time followed by slow mixing for another set time ultimately resulting in pollutant/coagulant mixture settling sequence. Once the solid mass is settled, the remaining supernatant of clarified water is separated from the tank. These, traditional tests provide a good estimate of the definite treatment process required, parameters to be set for sedimentation, the rate of rapid or slow mixing, dose of coagulant/flocculant, characteristics of wastewater and treatment plant. Throughout the process concentration and the flocculation efficiency of the process is constantly monitored from the beginning to the end. These intermediary monitoring steps are carried out by measuring conventional basic wastewater characteristics like pH, turbidity, conductivity, total dissolved solids (TSS), chemical oxygen demand (COD), alkalinity, biological oxygen demand (BOD), nutrients levels like nitrogen and phosphate. Therefore, these constant measurements give the specific treatment efficiency treatment in a system. On the other hand, in advanced equipment and procedures, parameters such as zeta potential of electrolytic system and/or electrophoretic mobility of coagulant or flocculant, their particle size distribution and other physicochemical properties are measured for greater accuracy.

1.6 Advances in Chemical Methods

There are several wastewater treatment techniques to resolve specific issues associated with various pollutants such as physical, chemical, and biological processes. Physical methods are generally listed as part of primary treatment techniques mainly aiming to separate suspended solids. On the other hand, the secondary treatment steps involve both chemical and biological processes to enhance the quality of purified water. Due to the increased complexities with modern wastewaters, most of the time secondary treatment techniques are combined into single-process operation. [21]

Chemical treatments utilize several active additive substances to separate pollutants both dissolved and fine suspended to make industrial wastewater suitable for water reclamation or discharge. Chemical treatment techniques mainly include processes like (a) chemical disinfection, (b) chemical precipitation, (c) neutralization, (d) adsorption, (e) ion exchange, (f) advanced oxidation process (AOP), and (g) nanomaterials. In most of these processes, added chemicals in wastewater forces contaminants that are dissolved to separate more easily after interacting with targeted addition of specific substances. Therefore, added chemical reagents in wastewater precipitate all dissolved substances turning them into an undissolved substance that can be separated out from the clarified liquid part.

(a) *Chemical Disinfection*: Traditionally as well as advanced chemical treatment processes also use specialized chemicals such as hydrogen peroxide, chlorine, sodium hypochlorite (bleach) and sodium chlorite that act as disinfectant agents. Disinfection step is generally applied following primary treatment to sanitize, disinfect, and assist pre-treated wastewater to attain enhanced quality. However, most of the disinfection methods today are applied post-tertiary water treatment techniques. [22]

(b) *Chemical Precipitation*: In this process, the dissolved inorganic pollutants and associated substances are removed by adding an acid or alkali, altering the temperature of influent wastewater through the process of precipitation as a solid mass. As settled precipitate can be removed from the clarified by sedimentation, precipitate flotation, or other solid scraping processes. Although chemical precipitation follows the principle of coagulation and flocculation, the chemical precipitation process is highly recommended to substitute for its simple approach and feasibility. [23]

(c) *Neutralization*: Neutralization process is all about regulating and controlling the pH of the wastewater whether it is alkaline or acidic and to bring it down to near neutral pH 7. In this process, the lack of adequate alkalinity will necessitate the adding of a base to bring down pH or to regulate the pH to the acceptable range. Generally, sodium hydroxide (NaOH) also known as soda ash, sodium carbonate (Na_2CO_3), calcium hydroxide ($Ca(OH)_2$), and Lime (CaO), are the commonly used chemicals to regulate the pH. On the other hand, the lack of sufficient acidity requires the addition of acid regulating chemicals such as carbonic acid (H_2CO_3) and sulfuric acid (H_2SO_4) which will adjust the pH to the adequate acceptable range. [24]

(d) *Adsorption*: Adsorption process exists in both physical as well as chemical processes known as physisorption and chemisorption, respectively. In the physical adsorption process, dissolved pollutants or molecules known as adsorbate are removed by the use of a suitable carrier or via attachment to a solid substrate surface known as adsorbent. To be an efficient adsorbent, the substrate should have an enormously high specific surface area. Generally, there are several natural origin materials that are traditionally used as adsorbents to remove various pollutants from wastewater such as clay colloids, activated carbon, activated alumina, hydroxides and biopolymer-based resins. To initiate the physisorption of pollutants on absorbent, the surface of the adsorbent material should be free from adsorbate. [25] Therefore, the adsorbent surface should be either activated or functionalized before use. For chemisorption application, a wide range of natural as well as synthetic origin organic materials can be used to remove pollutants via adsorption including detergents, drug molecules, heavy meals and other toxic compounds. However, activated carbon has been widely being used for many decades as a physisorption substrate. Most of the activated carbon is prepared following the pyrolytic carbonization process of biomass. Activated carbon is also functionalized to attain partial chemisorption characteristics to attract multiple pollutants from wastewater. Activated carbon possesses several important characteristics such as high surface area with micropores and low-volume pores, feasibility to functionalize the surface by inducing active groups, and most importantly high thermal and chemical stability. Due to their high level of microporosity, a small quantity (1 g) of activated carbon can exert a surface area greater than 500 m^2. Activated carbon can also be produced following various methods and techniques such as steam oxidation and thermal oxidation and adsorbent surface can also be easily recycled and reused for several cycles. The adsorption capacity of activated carbon can be enhanced for various pollutants by altering pH, temperature and stirring conditions of the wastewater during the treatment process. In addition to this, the overall adsorption process can be enhanced following most significant characteristics of an adsorbent can be determined following factors like:

(i) ***Adsorbate concentration***: The adsorption of pollutants on adsorbate is directly proportional to adsorbate concentration.

(ii) ***Particle diameter***: According to one of the basic nano-chemistry principles the adsorption is greatly depends on the particle size of an adsorbate and is inversely proportional to the particle size and higher adsorption is achieved with high surface area adsorbate. Therefore, lower the particle size, higher the surface area.

(iii) ***Molecular weight***: Largely, the adsorption capacity of an adsorbate is inversely proportional to molecular weight of adsorbent. However, the pore size, pore diameter and their configuration of pores affect the diffusion of adsorbent in the adsorbate.

(iv) ***Temperature***: The adsorption process commonly noticed to be directly proportional to temperature.

(v) ***pH***: The adsorption is inversely proportional to pH due to surface charge as the surface zeta varies with pH conditions. Nevertheless, individual adsorbate and pollutant properties greatly influence the adsorption process and rate.

(e) ***Ion exchange***: Ion exchange process is commonly used for ages to attract pollutants from surface water. Ion-exchange resins or polymers have been in practice for drinking water applications. main advantages of ion-exchange materials can undergo a reversible reaction in which a charged ion in wastewater or contaminated solution is swapped or exchanged with an equally potential or charged ion through electrostatic interactions. The most popular application of ion-exchange materials, resins or polymers, find their application in hard water softening applications in which polyvalent cations such as calcium and magnesium are exchanged with sodium. However, ion-exchange materials are increasingly finding a useful venue in wastewater treatment applications. In these advanced practices, wastewater is forced through an ion-exchange bed and specific retention time is provided to facilitate the ion-exchange process. During specific retention time, ionic pollutants from wastewater are attracted by ion-exchange resin or polymer and exchangeable species are replaced within retention time. In advanced, resin preparation techniques, polymeric or polymeric nanocomposite based are manufactured by converting organic compounds (polymers) into porous ion-exchange porous matrix via polymerization process. In most of the cases, typically, sodium (Na^+) is exchanged with cationic pollutants in the wastewater or contaminated solutions. Major disadvantages come with the saturation points of resins in which ion-exchange capacity of ion-exchange resin or polymer diminishes with time and concentration of ionic species in wastewater. In such a situation, the ion-exchange bed or cartridge is replaced when it is saturated with the exchanged sites. Advanced ion-exchange systems provide regeneration possibilities following treatment with concentrated sodium in a regeneration solution. [26]

(f) *Advanced oxidation processes (AOPs):* Advanced oxidation processes is broadly defined as a set of evolving or emerging chemical procedures which helps in removing soluble organic pollutants or contaminants from wastewater or contaminated solutions. In a broad sense, are a set of chemical reagents are designed or used for treatment process to mainly target degradable or easily transformable organic and sometime inorganic pollutants or contaminants from wastewater. For this process, mainly pollutants are removed through oxidative reactions initiated by hydroxyl radicals (OH) from the reagent. Therefore, advanced oxidation process methods or technologies depend on efficient in-situ generations of strong oxidants mainly hydroxyl radicals and sulfate radicals to proceed with oxidative removal of organic pollutants. Recent advances also use ozone and UV irradiation for improved wastewater treatment efficiency. It is now estimated that AOPs are the vital process for environmental remediation. Further, advanced techniques use techniques like Fenton, photo-Fenton, photocatalysis, ozonation, ultraviolet photocatalysis, sonolysis and wet air oxidation. Many of these AOPs techniques have proven successful for the complete degradation of numerous emerging, toxic, and hazardous organic pollutants. [27]

(g) *Nanomaterials*: Advances in nanotechnology provide new avenues that have proven extremely useful for water purification and wastewater treatment application. Nanomaterials possess several unique properties such as large surface areas, tunable surface functionality, selective permeability, affinity for specific pollutants, can be easily tailored to attract specific pollutants, high reactivity, excellent flux rates, excellent recyclability and reusability. In addition to these nanomaterials-based treatment techniques can also be tuned to attain multi-functional properties such as detection, sensing and removal media at the same time. The use of nanomaterials-based biosensors helps in characterizing water quality and level of contamination. Recent advances have shown greater potentials of nanomaterials in trace amount pesticide and heavy metal detection in water. Lately, nanomaterials-based filters have been prepared in cake or fixed bed configuration to easily filter out both conventional as well as emerging pollutants. These loose packed filter beds or filter cakes help in attaining high flux and selective removal of various pollutants from wastewater. [28]

1.7 Advances in Biological Methods

Clean water helps all living beings grow, flourish and survive in their natural ecosystem and environments. Minimum essential amount of clean water is required to water crops to yield a massive harvest when used properly. Nevertheless, fresh water scarcity in major parts of the world necessitates water saving practices, preserving or recycling all used and discarded water. Therefore, water treatment technologists systematically assess and invent newer processes to convert wastewater into safe water. Many of those treatment techniques have been discussed in detail in the previous sections. These established top processes have shown partially successful in handling wastewater following physical and chemical processes. Even though these processes have not been successful in complete recovery and inducing sanity to the recovered water, however both physical and chemical processes enhance the quality of wastewater to maximum level. Thus, with these processes, a larger population would have an experienced hard time in extracting reusable water to maximum extent. While primary and secondary wastewater treatments have proven successful in clarifying wastewater, on the other hand tertiary treatment techniques are adopted as the final step to elevate recovered water quality via various steps like biological treatment (COD & BOD), disinfection, sanitization and membrane filtration processes. These tertiary techniques have been adopted as the final step of treatment after thorough evaluation of health and environmental impact of final recovered water. [29]

Biological wastewater treatment process in tertiary treatment step is designed to degrade mostly organic contaminants dissolved in wastewater using microorganisms. The microorganisms in the treatment process utilize organic substances as nutrients to live and reproduce. In this process all pollutants become nutrients for microorganisms. Interestingly, biological waste contaminated wastewater treatment systems improve the naturally occurring process of microbial degradation and decomposition to break down the organic or biological industrial wastewater contaminants along with other undesirable materials can be removed. Conventionally, there are two main types of biological treatment processes in practice, namely, aerobic treatment and anaerobic treatment. Interestingly, biological wastewater treatment is cost-effective and energy efficient processes for the removal of BOD as microorganisms are used.

While both aerobic and anaerobic treatment methods rely on a process of microbial decomposition and degradation to treat wastewater, the main difference between is that aerobic process requires oxygen, on the other hand, anaerobic systems can work in absence of oxygen. However, in both the cases, microbes are used to perform organic pollutants decomposition process. On one hand aerobic or anaerobic treatment processes consume the dissolved organic pollution and on the other hand both the processes induce oxygen balance in aquatic systems. Compared to physical and chemical treatment methods, biological methods possess several advantages such as:

(a) biological treatment methods have been well tried and tested as well as well understood,

(b) biological treatment processes boost effectiveness in in removing both conventional as well as emerging organic pollutants,

(c) these processes are cost-effective, and

(d) both aerobic and anaerobic treatment processes are environment friendly and safe.

Aerobic wastewater treatment involves decomposition of organic contaminants rich with nutrients like nitrogen and phosphorus being decomposed microbiologically. On the other hand, anaerobic processes which occurs in absence of free or combined oxygen results in sulfate reduction and methanogenesis. Both the processes produce biogas mostly having mixture of methane and carbon dioxide, decomposed or degraded by-product and generate varying amounts of settleable decomposed bio-solids also known as sludge as by-products. In both the biological degradation process, wastewater is subjected to aerated continuously for about a month which allows the complete nitrification. At this stage sludge is activated to be active in the separation process. Therefore, activated sludge is a collective formation that takes place as decomposed bio-solid from domestic compounds composed of bacteria and protozoa. Further, biological treatment has been now integrated to achieve high quality recycled water following;

(a) ***Activated Sludge***: As discussed in this step, the most of inflowing wastewater is treated to near high quality. All dissolved organic pollutants are decomposed via biological degradation process using microorganisms which results in dead cell tissue, water, and nitrogen. The biological reaction that takes place in this step is analogous to the process that occurs at the bottom of ponds, lakes and rivers. Only difference in the wastewater treatment plant process is that the biological treatment process is manipulated to suit the effluent type, whereas underwater degradation takes throughout the year naturally.

(b) ***Secondary Settling:*** In this step, large circular tanks known as secondary clarifiers in which treated wastewater is separated from biological pollutants in aeration tanks. This step yields nearly 90% treated clarified water. On the other hand, concentrated biological and organic pollutants are continuously removed from the bottom of the clarifier tank to the aeration tanks for further activation or concentration, and

(c) ***Filtration***: Once the clarified water is separated it is refined in this stage using 10-micron mesh filter usually made of polyester. All the activated sludge settled on the filter mesh surface is periodically cleaned and separated. Further, after removing settled activated sludge mesh filter is backwashed to make further process sustainable. [30, 31]

Finally at the end, clarified water is disinfected to remove any trace biological contaminants before recommending reuse applications. These disinfection steps include chlorination, ozonation, and ultraviolet (UV). The disinfection process is the last step of the tertiary wastewater treatment process. Disinfection stages also involve chemical treatment by treating the clarified wastewater with a certain disinfectant to eliminate or inactivate the residual pathogens. The main reason or rationale behind disinfection of treated effluent is to protect public health or environment at large by eliminating or deactivating the pathogens like viruses, microbes, and protozoan. This helps in overall elevation of treated water to meet the discharge standards of recycled water. The ideal sanitizer or disinfection agent should have biocidal toxicity, inexpensive, but non-hazardous to handle and be consistent to detect and kill the residual contaminant. Among many chemical disinfection reagents chlorine, ozone, ultraviolet radiation, bromine, and chlorine dioxide are widely used. [32]

Traditionally, chlorine has been used for decades as one of the oldest and efficient disinfection agents. As chlorine has been tried and tested for decades, it has been categorized as one of the safest and reliable. In this, once the chlorine gas is dissolved in influent wastewater, it swiftly hydrolyzes to hypochlorous acid (HOCl) and hydrochloric acid (HCl) following chemical equation: [33]

$$Cl_2 + H_2O = HOCl + H^+ + OCl^- \quad \ldots(1)$$

Even though chlorine has been identified as an effective disinfectant, the overall quality of recycled water depends on chemical composition of wastewater. The major influencing factors include temperature, pH, organic contents, and concentrations in the wastewater as shown in Figure 1.4(a).

Ozone (O_3) is one of the strong oxidants which can oxidize organic material effectively and used typically in wastewater treatment. Ozone can oxidize when dissolved or introduced in contaminated water in which it can produce a broad-spectrum biocide that can destroy all pathogens like virus, bacteria, and cysts. Upon interaction with pathogenic contaminants, ozone produces degraded residues. Typically, ozone in a treatment plant is produced using ozone generator, then it is injected into influent contaminated water. It then immediately starts oxidizing and disintegrating contaminants like virus, bacteria, metals and other organic matter as shown in Figure 1.4(b). Ozone is known for its active role in oxidizing organic material in the membranes of microorganisms such as bacteria, viruses, and parasites. However, as ozone

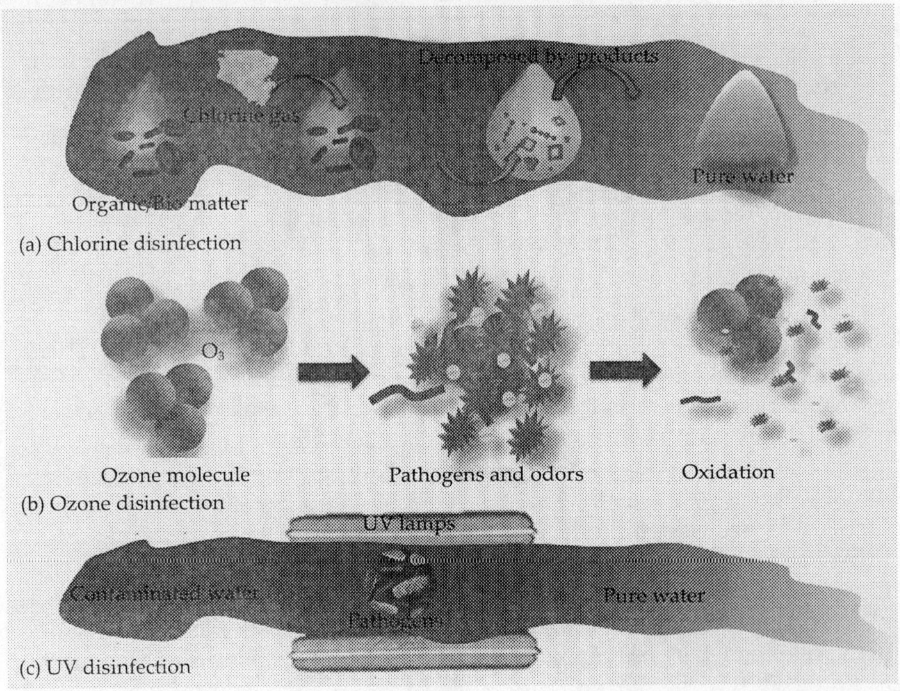

Figure 1.4. Schematics showing different disinfection processes in tertiary wastewater treatment including (a) chlorine disinfection, (b) ozone disinfection, and (c) UV disinfection.

is an unstable molecule, it should be generated or produced at the point of application, in case wastewater treatment in the effluent plant only.

Further, when ozone dissolves or decomposes in contaminated water, the free radicals such as hydroxyl (OH) and hydrogen peroxy (HO_2) are formed that have great oxidizing capacity which plays an important role in the water disinfection process. It is largely understood that the bacteria are destroyed using ozone due to the protoplasmic oxidation resulting in cell wall disintegration. Nevertheless, ozone also effectively removes metals like iron, hydrogen sulfide and manganese contaminations from the water.

Ozone disinfection process provides several advantages such as it is effective over a wide range of pH conditions and upon introduction of ozone into water it rapidly reacts with bacteria, viruses, and protozoans. It alsohas stronger germicidal properties than chlorine. Other major advantages include: (i) ozone disinfection treatment process does not require any other chemical for the treatment process, (ii) ozone can remove a wide range of pollutants of organic, inorganic, microbiological nature, and can drastically reduce odor thereby improving the taste. In advanced setups, ozone has been successfully used for killing microbiological agents and protozoans such as *Cryptosporidium*, *Giardia*, and other water-borne pathogens. However, ozone treatment process also suffers from several disadvantages like high equipment capital and operational costs. Ozonation does not guarantee or induce germicidal or disinfection residual in treated water to prevent or inhibit the regrowth of bacteria. The ozone process may not be effective as a standalone process to remove a wide range of contaminations like hardness which requires pretreatment steps to remove or reduce hardness reduction. In some cases, it necessitates the addition of chemicals like polyphosphate to prevent or avert the formation of carbonate scaling in the treatment plant. Also, compared to chlorine, ozone is less soluble in water, which again requires special mixing techniques to attain optimal ozone dissolution. [34, 35]

Another major concern is the generation of byproducts as a result of degradation remaining in the treated water which are some of the hazardous emerging pollutants with potential carcinogenic impact in nature. These degraded byproducts include brominated residual products like ketones, aldehydes, and carboxylic acids. This necessitates the post-ozone treatment filtration or adsorption using activated carbon steps to eliminate water to make water safe and healthy.

Ultraviolet (UV) assisted disinfection process uses UV radiation which eliminates residual microbial infectants from the treated water. UV radiation is an effective means to eliminate several microorganisms such as viruses, bacteria, protozoa and some emerging and harmful pollutants like pathogens that are difficult to remove from either chlorination or ozone methods. UV disinfection technique has proven effective against protozoa such as Giardia and Cryptosporidium which are resistant to the chlorine disinfection process. However, UV radiation effectively works only when the influent water is un-turbid, clear, and non-colored. The commercial variety of UV lamps have been designed to suit the end use applications. However, post-treatment UV disinfection systems use low or medium powered (wavelength of 354 nm) UV lamps.

1.8 Advances in Thermal Methods

In the industrial sector, freshwater is consumed for various purposes such as process running, rinsing, scrubbing and cooling applications. This indicates that there are large amounts of industrial wastewater with various contaminations of different compositions at the end of product development. In many of these cases, this wastewater needs to be treated in an in-house effluent plant before it can be discharged or disposed of into the environment or landfilling, respectively. However, modern approaches seriously consider economical and environment- friendly wastewater management and treatment techniques that should be integrated to advanced approaches to improve the overall process. At the end of the tertiary wastewater treatment step, generally large and highly concentrated sludge is produced in most of the effluent treatment plans. However, this highly concentrated solid mass needs to be dried appropriately before its disposal using techniques like thermal treatment. Thermal treatment is a process by which heat is applied to the waste to sanitize it. The primary function of thermal treatment is to convert the waste to a stable and usable end product and reduce the amount that requires final disposal in landfills. [36, 37]

Interestingly, many of the treatment techniques discussed above have either been used as standalone or in combination methods. The choice of method mainly depends on the purpose, type of wastewater, and contamination level. Modern industrial activities and improved wastewater treatment approaches have led to the production of highly concentrated sludge from wastewater. However, serious concerns are identified associated with sludge management and disposal. For this, there are several attempts at individual level or larger technological innovations. This is coming through the innovative strategies or optimization of the traditional sludge management techniques. To make sludge management easier, the concentrated sludge is subjected to drying before disposed of to landfills using various thermal drying techniques such as energetic valorization processes, direct combustion in a burning or incineration plant, gasification, and pyrolysis. Also, there are several recent innovative processes to impart the greener and sustainable aspects to the dried sludge management. This also considers the use of minimum energy inputs. Further, there have been several attempts to use the dried sludge for value added applications. On the other hand, this is expected to reduce the load on fresh natural resources in many futuristic applications. From this point of view, it is significant to consider that, in disagreement to the continuously increasing difficulty to produce energy from conventional fossil sources, the possibility to obtain energy from concentrated wastewater sludge such as sewage sludge. This signifies an interesting means for both the reasons that the new source is certainly renewable and consistent with new sustainability directives giving economic aspects an important consideration. On these considerations, future thermal waste treatment techniques will adopt specific innovative thermal treatment approaches such as thermal treatment with direct combustion in incineration plants and gasification.

On the other hand, the thermal separation of contaminant is mainly suited if the wastewater is dangerous to handle or cannot be clearly defined. Ingredients are complex to treat using other techniques, directly dehydrate the constituent pollutants, extreme pH value or constituent solute size and nature changes frequently and to obtain highly concentrated or dried end mass. Among many approaches to achieve these targets, evaporation and distillation plants with specific process design and other numerous specifications must be taken into consideration. In one such approach, wastewater with specific contaminants and their concentrations, energy requirements and overall treatment parameters specific to the treatment plant are taken into consideration. In all cases, generation of optimal data, analysis of process engineering parameters and operational data influence the design of treatment plant, capacity, and type of construction. At the end, energy consumption essentially determines and regulates the operating costs of evaporation and distillation wastewater treatment plants. Therefore, several attempts are being made to develop ingenious thermo-technical treatment plants with minimum energy requirements and lowest carbon footprints.

In many industries, wastewater generation and management are major environmental and economic concerns. Wastewater treatment process involves costing of waste segregation, waste collecting, transporting, treating, and disposing of effluent plants is a major contributor to manufacturing costs. Advanced approaches in thermal treatment systems offer a way to control and regulate the evaporation of the wastewater, retain concentrated solute, collect solid residues, segregate non-hazardous particulate from harmful content, capture the volatile organic compounds (VOCs) and control the exit of hazardous air pollutants (HAPs) in a single integrated equipment. However, specific airstreams containing trace pollutants with relatively high air-water partition coefficients (Henry's law coefficients) can be stripped out with economically feasible air strippers. Nevertheless, a typical concentrated wastewater stream comprising several constituents depends on the origin of wastewater with relatively low vapor pressure or a low aqueous solubility. [38]

On the other hand, the process of thermal oxidation is a relatively simple technique in which the untreated process stream temperature is elevated to a stage where the chemical bonds that hold the VOC molecules together are broken. Meanwhile, the VOCs in the process are converted to water, carbon dioxide (CO_2), and thermal energy due to the high temperature of the combustion process which can be utilized for further applications. Further, thermal oxidation processes which operate at high operating temperatures immediately evaporate the water from wastewater sources during treatment leaving the raw and untreated organic contaminants exposed to the combustion process. On the operational and efficacy point of view, appropriate system design and identification of wastewater sources can make thermal treatment techniques an environment-friendly process. Interestingly, the thermal oxidation routes have smaller footprint compared to physical or chemical treatment technologies, which frequently require a set of specific equipment packages. If an effective alternative post-treatment step is found, thermal treatment technology is a zero-discharge or zero-disposal process with a complete system with complete recovery of water, steam, solid and value-added products.

On the other hand, advances in thermal treatment techniques are incomplete without the effective sludge management infrastructure and protocols in place. Following section lists the important steps involved for sludge management as part of the thermal wastewater treatment process.

(a) **Immission of landfill:** Thermal treatment process encourages adoption as the process offers environment-friendly and minimum negative impact on overall waste residues at the end of the treatment. This process limits environmental impact of sludge treatment and stabilization before the residual immission.

(b) **Agriculture use:** Post-thermal treatment results in carbonaceous residues rich in organic substances which can be used as value-added products for applications in the agriculture sector. However, the main disadvantage is associated with possible presence of hazardous and toxic traces such as heavy metals and micro-pollutants accumulated in final residues. Further, it may require detailed study or the optimization of conditions and the availability of agriculture sites for effective utilization of residues. Nonetheless, thermal waste treatment technique is a unique and standalone technique which is gaining widespread applications in waste management both in medium to large-scale applications. Detailed literature data and critical assessment reveals thermal treatment technology is promising for the wastewater sludge treatment. [39]

With advancement of sustainable technologies in the form of multi-effect evaporators, thermal vapor recompression, and mechanical vapor recompression. Further, these treatment processes offers useful features like possibilities of tailor-made plant designs, combining multiple processes with this step for maximizing solid-waste concentration, this process uses minimum chemical substances, results in complete separation of solid matters from liquid component, high potential of recovering high purity water from wastewaters, low quantities of end-residues, low disposal costs, offers recovery of value added product from condensed solid waste, high operational reliability and long shelf-life of equipment.

1.9 Recent Developments in Membrane-based Separation Processes

Membrane-based processes have revolutionized the separation and purification sector in separating liquid/liquid, liquid/solid and gas/gas components of industrial importance. Even though membrane-based separation processes are in use for decades, modern approaches in membrane preparation, module design and new emerging applications constantly encourage membrane technologists to innovate sustainable routes in this field. Early membrane developmental activities were inspired by biological systems as semi-permeable membranes or tissues in living beings have proven efficient mass transport media. In recent times, innovations in the field of nanoscience and nanotechnology have provided new opportunities to membrane technologists to adopt new membrane materials and technological approaches. Membrane science and technology has gained significant importance in the last decade as one of the most important, useful, and rapidly modernizing branches in the materials science and engineering sector. In addition to this, membrane separation processes are also expanding their reach to find new and wider applications for which rigorous research, development, innovation in engineering and manufacturing of support equipment is extensively happening across the board. These developments are targeting emerging applications including addressing issues of persistent pollutants like active pharmaceutical ingredients (APIs), macromolecular separation, dairy products, desalination, petroleum industry, groundwater treatment, extremely contaminated wastewater treatment applications, applications controlled-release media of pharmaceutical formulations, dialysis, artificial kidney, gas separation, etc. [40] At the materials front, popularly polymer-based membranes have been widely used for a series of membrane separation applications. However, polymers have shown several limitations in terms of their shelf-life and deterioration of membrane properties when they are used in the long term. Polymers being organic in nature are susceptible to interaction with a series of pollutants which may in long term deposit thereby blocking the membrane pores. Nonetheless, membrane processes have now reached the developed stage and successfully employed for seawater desalination, process water concentration and selected industrial wastewater treatment. Conventionally, synthetic polymer-based membranes were prepared following various techniques which allow precursor polymer material to attain desired characteristics such as porosity, morphology, and their distribution.

Conventionally, before casting dope polymer solution on a glass plate or polyester substrate, desired membrane characteristics are pre-decided for their pore size range and membrane thickness following non-solvent induced phase inversion (NIPS) method. Among many considerations, the nature of polymer precursor material and final morphology membranes are the main features. However, advanced membrane preparation techniques require the extended parameter like specific feed that the membrane is expected to face, applied operational pressure ranges, cleaning properties and surface reproducibility criteria. However, the most frequently used method for fabricating polymer membranes is non-solvent assisted phase inversion technique. In this method, phase inversion polymer solution is achieved via the miscibility and demixing of solvent and non-solvents in a coagulation bath. Nonetheless, there has been very little change in solution-casting methods since their discovery by Loeb and Sourirajan in the 1960s. However, recent innovations have adopted new routes of controlled porous membrane. Among these advanced techniques, thermally induced phase inversion (TIPS) method has been rediscovered to solve modern-day membrane technology related problems. [41]

Even though the thermally induced phase inversion technique was first developed in the mid-1980s, however, large scale potentials of TIPS membranes have been realized very recently. Earlier TIPS-based membranes received little attraction as NIPS-based membranes were considered convenient and cost effective. NIPS-based membranes also provide much needed flexibility and can be prepared on a wide range of porous polymeric support materials. Howeve,r recent advances in TIPS research is regaining attraction due to innovation in membrane contactors that offer a number of distinct advantages such as process simplicity, a low propensity to form defects, high reproducibility, and control over high porosity. In addition to this, TIPS also offers to create tailor-made pore architecture of interesting microstructures and narrow pore size distribution patterns. These features make TIPS a potential technique to replace conventional NIPS series membranes. In addition to this, technologists now design advanced TIPS membrane fabrication equipment that can produce reliable microporous membranes for large-scale applications. Furthermore, the TIPS process also provides an opportunity to manipulate the polymer chain reassembling during thermal annealing and evaporation which helps in controlling pore size distribution. [42]

On the other hand, growing demand for advanced anti-fouling and self-cleaning membranes with controlled structure for a variety of applications such as seawater desalination and industrial wastewater treatment applications necessitate a novel membrane preparation approach. Membrane scientists majorly look for features in preparation that can be used to manipulate the pore structures at the molecular level. Further, membrane surface chemistry also plays an important role. Pore manipulation features in combination with surface chemistry can achieve optimal physicochemical properties in novel membranes to enhance the properties for specific applications. This is generally achieved in the so-called next generation block copolymer membrane through a process called self-assembly technique. Recent studies show promising potentials of block copolymers that are now opening new venues and possibilities to overcome the limitations and demerits of traditional membranes. The main advantage with block copolymers precursor in membrane preparation is their control over morphology that can be easily tuned to adjust pore size distribution via degradable polymer block. In addition to this self-assembly property of diblock can also be extended to triblock, tetrablock, and pentablock copolymers of which can be further used to control pore texture and tortuosity, and periodic geometry. These features are hard to achieve in both NIPS and TIPS-based techniques. Further, with suitable selection of degradable block size in block copolymer, it is easy to control pore size in nanoscale. [43]

On the other hand, functional polymer-based precursors have attracted enormous attention from membrane scientists and technologists. Functional polymer materials offer several functions when used as contaminant separation media via degrade, adsorb, and/or deactivate pollutants and rejection or selective separation of pollutants or solutes via sieving mechanism. Membranes are prepared using functional polymers typically engineered and optimized to suit the specific applications. These functional polymer-based membranes are generally used for emerging pollutant separation or extraction such as organic contaminants, heavy metals, pharmaceuticals as well as biological contaminants. Functional polymer-based membrane preparation involves polymer processing to functionalization followed by membrane casting to achieve stable membranes. The polymer functionalization also focuses on achieving hydrophilicity, antistatic properties, pore or particle size in end product, compatibility and efficiency. Functionalization of polymers is also achieved either inducing nanomaterials into the polymer matrix or direct polymer backbone functionalization of desired end property.

Polyelectrolytes complexes (PECs) are another class of membrane materials that attracted enormous attention from membrane scientists due to their unique charge and surface properties. The PECs can be easily adopted for contaminant separation, purification or in general water treatment applications. Earlier, PECs prepared using polycationic and polyanionic polymer combinations were largely used for pervaporation and nanofiltration applications. However, polyelectrolyte complexes have a strong affinity to adsorb pollutants on the surfaces in an aqueous effluent which requires careful attention during their property tuning. Precursors used for the preparation of polyelectrolytes complexes are generally economical and can be prepared easily with controllable thickness. Furthermore, PECs are classified based on their electrolyte behavior (weak or strong), molecular structure, charge density and molecular weight. Components in PECs pose a unique ability to repel each other due to electrostatic repulsion forces to separate via different mechanisms. Interestingly, when PECs are used in water treatment applications, they can provide extremely thin but efficient barriers for effective separation of pollutants from contaminated water. Further, recent advances also reveal the use of PECs as a selective layer on different support substrates. This thin-film composite form of PEC membranes has also shown promising improvements in separation and purification performances. Another major advantage with PEC membranes is that they offer tunable surface charges that can be effectively used to reject ionic pollutants' repulsion mechanisms. Therefore, PEC-based membranes are now frequently prepared using a combination with a series of biopolymers that are biocompatible and non-toxic in nature. [44, 45]

Aerogels are another new class of ultra-high porous materials prepared via sol-gel technique that offers specific pore and surface properties. Initially, aerogels have been developed using inorganic precursors like silica, however recently polymer-based highly porous aerogels are gaining significance. This is because polymers offer greater potential to functionalize and induce target specific functionality. This is simultaneously induced by retaining ultra-high porosity in aerogels. Aerogels are prepared following sol-gel solid state material preparation technique. In this technique, colloidal polymer sol is further converted into a gel in pre-defined shape and composition. During this process, the liquid component is frozen and sublimated using a lyophilization process to result in an ultra-porous 3D network. During this process, the semi-solid gel 3D network liquid part is completely replaced with a gaseous component to yield a stable solid network structure. Therefore, aerogel is a three-dimensional (3D) solid structure of air-filled pores in the form of crosslinked structures. These 3D network structures of aerogels can be prepared using various precursors including both synthetic and natural polymers. [46, 47]

Further, polymer-based aerogels in pristine form as well as composite forms have been widely used for various types of water treatment. As discussed previously, among many important characteristics in aerogel, two distinct properties, namely, unique structural features and functional characteristics play an important role in retaining a series of pollutants from contaminated water. Another important feature of polymer-based aerogels comes from their tunable super-hydrophilicity and/or hydrophobicity, high porosity, low density, high specific surface area and flexible mechanical properties show great potential in retaining both conventional as well as emerging pollutants from contaminated water.

1.10 Sustainability of Present Methods

Future technologies demand smart, futuristic, and sustainable materials that can be manipulated to achieve specific response during their application. These materials can also be modified to respond to external stimuli such as pH, temperature mechanical stress, and chemical compositions among others. There are several futuristic or smart materials including functional polymers, aerogels, nanomaterials, metamaterials, carbonaceous materials, natural fibers, biopolymers and so on. Some of these materials and substances may be years away from being of use extensively, however, it is definite to happen in the field of water treatment technologies.

On the other hand, increased awareness for greener and sustainable materials and processes paved the way to adopt bio-origin functional materials and polymers. In this direction, researchers have taken serious note on finding the right recipe of natural materials for sustainable water treatment applications. In addition to this, naturally abundant clay materials have also been greatly explored for water treatment applications. These natural clay materials can be abundantly found in volcanic explosions or eruptions and rock weathering or erosion. Further, another class of natural material can be extracted from anthropogenic sources such as unprocessed discarded wastes, terrestrial biomass, and seaweed biomass that have shown promising potential in water purification applications. Nonetheless, researchers have already proven potentials of natural-origin material for water treatment applications. These materials have been used in the form of separation membranes, electrodialytic membrane, nanofibrous membrane, liquid membrane, etc. Nonetheless, future water treatment technologies are sure to adopt innovative approaches and advancements such as integration of solar technology, high-tech filtration water ultrasound technologies, nanotechnology and so on. These futuristic innovations are briefly listed in Figure 1.5 and discussed as follows.

(a) **Advanced filters:** On the other hand, recent break throughs have been achieved in realizing advanced water technology for filtering salt water from the sea to produce high quality potable water. In this direction, graphene oxide-based membranes have been extensively studied in which researchers have proven to control precise pore size suitable for target specific applications. This technological breakthrough offers to filter out the sea salts from ubiquitous seawater to convert seawater to drinking water.

(b) **Ultrasound technique:** In another attempt, researchers have successfully tested potentials of ultrasound waves for the purification of seawater with precise desalination of saltwater. This ultrasound system works by using ultrasound waves to 'explode' the salt water into particles less than ten microns. These microns selectively evaporate water molecules and further condense thereby resulting in pure water collection.

(c) **High-tech filtration:** Researchers have also explored new sources of water and intend to extract high quality water from untapped sources using innovative ideas. In one such innovative technique, researchers have developed systems to filter water from untapped water sources like cow manure. This extracted water may not be useful for domestic water supply applications however can be utilized as fertilizer. Researchers have in detail studied potentials of cow manure which is predominantly water and it is estimated that one thousand cows can provide nearly ten

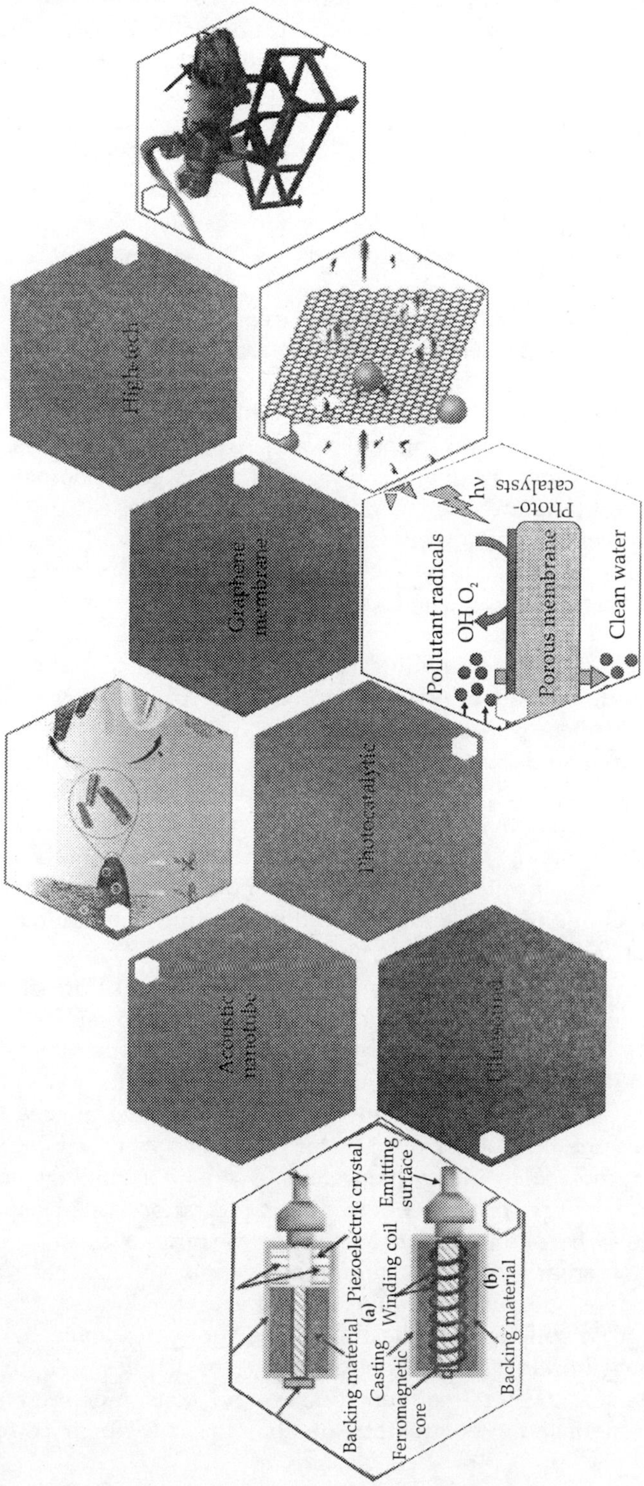

Figure 1.5. Spectrum of futuristic water treatment technologies.

million gallons extractable water, annually. In this high-tech purification technique, the cow manure is subjected to anaerobic digestion to produce energy from at one point and on other hand ultrafiltration (UF) process is adopted to extract clarified water as byproduct. Nevertheless, water purification technologies demand cleaner energy sources to operate. Among many greener approaches, solar energy technology has been widely explored for water purification and treatment applications. [48]

(d) **Nanotechnology:** Nanotechnology water treatment techniques involve various approaches and protocols to adopt nanomaterials and nanotechnology-based devices for water treatment applications. Nanomaterials have already been tested successfully at laboratory scale at both atomic and molecular scale. Research results so far reported in the literature reveals that nanotechnology-based water purification and treatment processes are modern, highly effective and cost-effective in comparison to conventional water treatment methods. In the last decade, conventional technologies adopted nanomaterials as one of the components in water treatment processes in the form of copper, silver, nanoadsorbents and zero-valent iron (ZVI) nanoparticles, photocatalysis using nanostructured materials, nanomaterials induced membranes and so on. As in individual components, nanomaterial with large surface-to-volume ratio enhances the adsorption of conventional chemical and biological pollutants even at trace concentrations. On the other hand, nanoadsorbents feature specific physical, chemical, and physicochemical properties for the selective removal of heavy metal and organic pollutants from water. For this purpose, CNTs-based nanomaterials in both pristine as well as composite forms are one of the eminent nanomaterials classes for water treatment applications. On the other hand, several working technologies have been developed using nanocomposite membranes for both complete removal and selective elimination of emerging pollutants from contaminated water sources. [49, 50]

(e) **Acoustic nanotube technology:** In this technology like pressure driven membrane process employs acoustics in place of pressure to direct or push water through small-diameter carbon nanotubes thereby separating high quality water from contaminated sources. This technology utilizes a unique acoustically driven molecular barrier integrated with CNTs that allow the passage of pure water molecules while retaining larger pollutant molecules and contaminants. Acoustic technology consumes minimum energy compared to traditional filtration systems and drives water away by eliminating the need for flushing or cleaning the filter system. Currently, acoustic nanotube technology has been used in preliminary stages for the treatment of municipal or domestic wastewater treatment, distilleries, medical facilities, both academic as well as industrial laboratories, desalination plants, industrial water purification facilities, wastewater treatment plants, and consumer segment. This innovation has shown greater potential to scale-up in combination with other existing technologies to be cost-effective solutions for the future. [51]

(f) **Photocatalytic water purification:** Even though traditionally the UV-assisted photocatalytic water treatment process has been in use for water treatment for decades, visible light assisted photocatalytic water treatment has gained prominence in recent times due to its efficiency in treating contaminated water. The method utilizes efficient photocatalyst and ultraviolet (UV) rays as well as visible light wavelength to catalytically degrade and remove toxic substances from contaminated water. Traditionally, titanium dioxide-based photocatalysts have been used at commercial level as successful adsorbents. However, in recent times natural materials and synthetic clay-type catalysts such as zeolite have been developed with varying compositions to ensure effective separation of contaminants. Further, future approaches have been focusing on recovery and reuse of photocatalysts to make the overall process sustainable. Future nanomaterials-based photocatalysts are also being developed to break down a range of hazardous organic substances, materials, synthetic steroids like estrogens, highly toxic pesticides, crude oil, smaller organic molecules like dyes, and microbes. Also, nanomaterials-based photocatalysts have also been focusing on making them effective against biological contaminants like viruses, chlorine-resistant pathogens, and inorganic compounds such as nitrous oxides. Therefore, photocatalytic water treatment systems have shown great potential and are still estimated to grow further due to the innovations in nanotechnology solutions. Improved properties of nanomaterials are also expected to further enhance photocatalytic properties of novel materials making them best candidates for sustainable water and wastewater treatment media. [52]

(g) **Aquaporin technology:** Aquaporin is a membrane protein which is generally found in all living beings from bacteria to humans. Aquaporin facilitates transmembrane diffusion of water from one tissue to other and allows the selective passage of small solutes via bidirectional mechanisms that are widely distributed in different human tissues. Synthetic aquaporin-based membranes are gaining increased attention due to their bio-mimetic water treatment potential. These membranes are designed taking inspiration from nature as an effective separation barrier. It is well understood that aquaporin enables robust and highly preferential water transfer across the membrane. Aquaporin also allows the protein assemblies to regularize its volume and internal osmotic pressure according to the hydrostatic and osmotic pressure gradient. Uniqueness of aquaporin comes from its distinct channel architecture that allows the selective passage of water molecules simultaneously retaining other contaminants or larger substances. Recent advances also show that natural bio-mimetic membranes have also been adopted as base materials for the development of artificial biomimetic membranes. Many of these artificial aquaporin membranes have been so far used for both reverse osmosis (RO) and forward osmosis (FO) applications. Aquaporin membranes have been projected to find new venues in coming days for sustainable water treatment applications. [53]

(h) **Automatic variable filtration (AVF) method:** Automated variable filtration is a simple process in which upward flow of influent wastewater is infused and cleaned by downward flow through filter media. This method employs continuous descending bed filters containing various arrays and compositions as cleaning media. The AVF technology exists in two-stage configuration that integrates two sets of media filters that can act either in series or parallel mode. The AVF method produces high quality water quality equivalent to that of microfiltration (MF) technology at a fraction of the cost making it one of potential future technologies. Another major advantage with this technology is it features immovable parts and requires less power to operate making one low operational cost and maintenance process in water treatment applications. The AVF technology has now proven efficient against municipal drinking water and domestic wastewater treatment, industrial wastewater treatment and recycling applications.

Therefore, there have been increasing efforts to adopt new water treatment methods and sustainable materials to achieve efficient water treatment, reuse to support a circular economy. As discussed above, efforts are being made to achieve target waste reduction and high quality water recovery from waste sources. This also shows that researchers and policy makers are worried and investing a lot of time and energy to address increasing water pollution issues which are also expected to influence the climate change pattern. Further, at local level overexploitation of resources specially contaminated water causing serious worries, however innovative water treatment technologies ensure safer future in treatment of wastewater and recycling to the level in which the recycle economy booms in near future. [54]

On the other hand, water treatment technologies and future environmental technologies are greatly influencing the traditional investments, planning, design, and operation to adopt sustainable water treatment technologies for the future. These approaches also consider monitoring and management of water from the source, transportation channel, usage pattern and discharge points. In addition to this, wastewater treatment, sludge concentration approaches are being looked at with different views to extract highly value-added product and resource recovery. These efforts aim to achieve a recycle economy, create local businesses opportunities through modern wastewater treatment models, methods, and innovations.

1.11. Conclusion

For decades, conventional water treatment technologies have served human-kind and society. Historically, conventional water treatment techniques have been successful in addressing issues associated with hardness, groundwater contamination, biological contamination as well as industrial pollutants to some extent. However, modern lifestyle and increasing industrialization has led to the production of a series of new pollutants and problems. These modern problems are severely affecting the ecosystem and environment at large. Conventional technologies have shown several limitations in addressing modern problems such as complex wastewaters, groundwater contamination, treatment of emerging pollutants and so on. However, several conventional water treatment technologies have greater potential in treating advanced pollutants with suitable modifications and modernization. This chapter has given detailed account of conventional water treatment used in the past, present and potentially used in the future with suitable modifications.

References

1. He, C., Liu, Z., Wu, J. et al. Future global urban water scarcity and potential solutions. Nat Commun 12, 4667 (2021). https://doi.org/10.1038/s41467-021-25026-3.
2. https://www.globenewswire.com/en/news-release /2022/06/07/2457330 /0/en/Water-Treatment-Chemicals-Market-to-reach-USD-78-4-Billion-by-2030-at-a Compound-Annual-Growth-Rate-CAGR-of-10-8-Reveals-MDC-Research-Study.html
3. Keisuke Ikehata, Yuanyuan Zhao, Harshad V. Kulkarni,Yuan Li, Shane A. Snyder, Kenneth P. Ishida, and Michael A. Anderson Water Recovery from Advanced Water Purification Facility Reverse Osmosis Concentrate by Photobiological Treatment Followed by Secondary Reverse Osmosis, Environ. Sci. Technol. 2018, 52, 15, 8588–8595.
4. https://education.nationalgeographic.org/resource/point-source-and-nonpoint-sources-pollution.
5. Chislock, M.F., Doster, E., Zitomer, R. A. & Wilson, A.E. (2013) Eutro phication: Causes, Consequences, and Controls in Aquatic Ecosystems. Nature Education Knowledge 4(4): 10.
6. https://www.business-standard.com/article/international/sea-levels-to-rise-10-12-inches-by-2050-rate-alarming-report-122021700192_1.html
7. Emiko F, Will M. Economic Growth, Convergence, and World Food Demand and Supply, Policy Research Working Paper 8257, World Bank; 2017.
8. https://sdgs.un.org/goals
9. Nataraj, S.K. (2022). Emerging Pollutant Treatment in Wastewater (1st ed.). CRC Press. https://doi.org/10.1201/9781003214786, 5 Howick Place, London, SW1P 1WG, UK.
10. Malcolm J. Brandt, K. Michael Johnson, Andrew J. Elphinston, Don D. Ratnayaka, Twort's Water Supply, 7th Edition, ISBN 978-0-08-100025-0, Butterworth-Heinemann, 2017, ISBN 978-0-08-100025-0.
11. R. Saravanane, Vivek V. Ranade, Vinay M. Bhandari, A. Seshagiri Rao,Chapter 7 - Urban Wastewater Treatment for Recycling and Reuse in Industrial Applications: Indian Scenario, Editor(s): Vivek V. Ranade,

Vinay M. Bhandari, Industrial Wastewater Treatment, Recycling and Reuse, Butterworth-Heinemann, 2014, Pages 283-322, ISBN 9780080999685, https://doi.org/10.1016/B978-0-08-099968-5.00007-6.

12. Elorm Obotey Ezugbe* and Sudesh Rathilal, Membrane Technologies in Wastewater Treatment: A Review, Membranes (Basel). 2020 May; 10(5): 89. Published online 2020 Apr 30. doi: 10.3390/membranes10050089.
13. Abdul Wahab Mohammad and Wei Lun Ang, Integrated and Hybrid Process Technology for Water and Wastewater Treatment, ISBN 978-0-12-823031-2, Elsevier DOI https://doi.org/10.1016/C2019-0-04696-1.
14. Crites, R. and G. Tchobanoglous, 1998. Small and Decentralized Wastewater Management Systems. The McGraw-Hill Companies. Boston, Massachusetts.
15. Sandrine Boivin, Shuji Tanabe, Takahiro Fujioka. Online evaluation of bacterial cells in sand filter effluents during full-scale treatment of drinking water. Science of The Total Environment 2022, 814, 152508. https://doi.org/10.1016/j.scitotenv.2021.152508.
16. Chapter 5 - Removal of suspended solids, Editor(s): Nelson Leonard Nemerow, Industrial Waste Treatment, Butterworth-Heinemann, 2007, Pages 53-77, ISBN 9780123724939, https://doi.org/10.1016/B978-012372493-9/50040-5.
17. Kesari, K.K., Soni, R., Jamal, Q.M.S. et al. Wastewater Treatment and Reuse: a Review of its Applications and Health Implications. Water Air Soil Pollut 232, 208 (2021). https://doi.org/10.1007/s11270-021-05154-8.
18. Chee Yang The, Pretty Mori Budiman, Katrina Pui Yee Shak, , and Ta Yeong Wu, Recent Advancement of Coagulation–Flocculation and Its Application in Wastewater Treatment, Ind. Eng. Chem. Res. 2016, 55, 16, 4363–4389.
19. Lian-Wei Zhang, Jia-Rui Hua, Wen-Jing Zhu, Lin Liu, Xiao-Lin Du, Ran-Ju Meng, and Ju-Ming Yao, ACS Sustainable Chemistry & Engineering 2018 6 (2), 1592-1601, DOI: 10.1021/acssuschemeng.7b02343.
20. Johannes Völker, Michael Stapf, Ulf Miehe, and Martin Wagner, Systematic Review of Toxicity Removal by Advanced Wastewater Treatment Technologies via Ozonation and Activated Carbon Environmental Science & Technology 2019 53 (13), 7215-7233, DOI: 10.1021/acs.est.9b00570.
21. Alicia An, Vinay Tyagi, Manish Kumar, Zeynep Cetecioglu, Clean Energy and Resource Recovery Wastewater Treatment Plants as Biorefineries, Volume 2, 2022, ISBN 978-0-323-90178-9, DOI https://doi.org/10.1016/C2020-0-03253-6.
22. Jalali Milani, S., Nabi Bidhendi, G. A Review on the Potential of Common Disinfection Processes for the Removal of Virus from Wastewater. Int J Environ Res 16, 9 (2022). https://doi.org/10.1007/s41742-021-00387-1.
23. Wang, L.K., Vaccari, D.A., Li, Y., Shammas, N.K. (2005). Chemical Precipitation. In: Wang, L.K., Hung, YT., Shammas, N.K. (eds)

Physicochemical Treatment Processes. Handbook of Environmental Engineering, vol 3. Humana Press. https://doi.org/10.1385/1-59259-820-x:141.

24. Chapter 3 - Neutralization, Editor(s): Nelson Leonard Nemerow, Industrial Waste Treatment, Butterworth-Heinemann, 2007, Pages 35-44, ISBN 9780123724939, https://doi.org/10.1016/B978-012372493-9/50038-7.
25. Ali, I., Gupta, V. Advances in water treatment by adsorption technology. Nat Protoc 1, 2661–2667 (2006). https://doi.org/10.1038/nprot.2006.370.
26. J Gregory, R.V Dhond, Wastewater treatment by ion exchange, Water Research, Volume 6, Issue 6, 1972, Pages 681-694, ISSN 0043-1354, https://doi.org/10.1016/0043-1354(72)90183-2.
27. Deng, Y., Zhao, R. Advanced Oxidation Processes (AOPs) in Wastewater Treatment. Curr Pollution Rep 1, 167–176 (2015). https://doi.org/10.1007/s40726-015-0015-z.
28. Kumar S, Ahlawat W, Bhanjana G, Heydarifard S, Nazhad MM, Dilbaghi N. Nanotechnology-based water treatment strategies. J Nanosci Nanotechnol. 2014 Feb;14(2):1838-58. doi: 10.1166/jnn.2014.9050. PMID: 24749460.
29. Rania Ahmed Hamza, Oliver Terna Iorhemen, Joo Hwa Tay, Advances in biological systems for the treatment of high-strength wastewater, Journal of Water Process Engineering, Volume 10, 2016, Pages 128-142, ISSN 2214-7144, https://doi.org/10.1016/j.jwpe.2016.02.008.
30. Johannes Völker, Sandro Castronovo, Arne Wick, Thomas A. Ternes, Adriano Joss, Jörg Oehlmann, Martin Wagner, Advancing Biological Wastewater Treatment: Extended Anaerobic Conditions Enhance the Removal of Endocrine and Dioxin-like Activities, Environ. Sci. Technol. 2016, 50, 19, 10606–10615.
31. Neyens E, Baeyens J, Dewil R, De heyder B. Advanced sludge treatment affects extracellular polymeric substances to improve activated sludge dewatering. J Hazard Mater. 2004 Jan 30;106(2-):83-92. doi: 10.1016/j.jhazmat.2003.11.014. PMID: 15177096.
32. John D. Keenan, David A. Hegemann, Chlorination and ozonation in water and wastewater treatment, Chemosphere, Volume 7, Issue 1, 1978, Pages 9-28, ISSN 0045-6535, https://doi.org/10.1016/0045-6535(78)90027-9.
33. Christina K Remucal and Devon Manley Emerging investigators series: The efficacy of chlorine photolysis as an advanced oxidation process for drinking water treatment, Environ. Sci.: Water Res. Technol., 2016,2, 565-579. DOI:10.1039/C6EW00029K.
34. Chhaya V. Rekhate, J.K. Srivastava, Recent advances in ozone-based advanced oxidation processes for treatment of wastewater- A review, Chemical Engineering Journal Advances, Volume 3, 2020, 100031, ISSN 2666-8211, https://doi.org/10.1016/j.ceja.2020.100031.
35. X. Huang, Y. Xu, C. Shan, X. Li, W. Zhang, B. Pan, Coupled Cu (II)-EDTA degradation and Cu (II) removal from acidic wastewater by ozonation: performance, products and pathways Chem. Eng. J., 299 (2016), pp. 23-29.

36. Śmiech, K.M., Kovács, T., Wildschut, R.F. et al. Thermal disinfection of hospital wastewater in a pilot-scale continuous-flow system. Appl Water Sci 10, 101 (2020). https://doi.org/10.1007/s13201-020-01181-8.
37. Mika Horttanainen, Juha Kaikko, Riikka Bergman, Janne Nerg, Analysis of Power Generating Sludge Combustion Plant and Comparison against Other Sludge Treatment Technologies, Applied Thermal Engineering 30(2-3): 110-118.
38. David E, Niculescu VC. Volatile Organic Compounds (VOCs as Environmental Pollutants: Occurrence and Mitigation Using Nanomaterials. Int J Environ Res Public Health. 2021 Dec 13; 18(24): 13147. doi: 10.3390/ijerph182413147. PMID: 34948756; PMCID: PMC8700805.
39. Jin-Won Park, Ho-Chul Shin, Surface emission of landfill gas from solid waste landfill, Atmospheric Environment, Volume 35, Issue 20, 2001, Pages 3445-3451, ISSN 1352-2310, https://doi.org/10.1016/S1352-2310(01)00118-2.
40. Angelo Basile, Alfredo Cassano and Navin K. Rastogi, Advances in Membrane Technologies for Water Treatment Materials, Processes and Applications, ISBN 978-1-78242-121-4. DOIhttps://doi.org/10.1016/C2013-0-16469-0.
41. Tasselli, F. (2014). Membrane Preparation Techniques. In: Drioli, E., Giorno, L. (eds) Encyclopedia of Membranes. Springer, Berlin, Heidelberg. https://doi.org/10.1007/978-3-642-40872-4_1825-1.
42. Ashok Shrishail Maraddi and S.K. Nataraj, Polymer-Based Membranes, Advanced Functional Membranes: Materials and Applications edited by Inamuddin, Tariq Altalhi, Mohd Imran Ahamed, Mohammad Luqman, Materials Research Forum LLC, 15-Mar-2022.
43. Hongbo Feng, Xinyi Lu, Weiyu Wang, Nam-Goo Kang and Jimmy W. Mays, Block Copolymers: Synthesis, Self-Assembly, and Applications Polymers 2017, 9, 494; doi:10.3390/polym9100494.
44. Senuma M, Kuwabara S, Kaeriyama K, Hase F, Shimura Y. Polymer complex from copolymers of acrylonitrile and ionic vinyl benzyl compounds. J Appl Polym Sci. 1986; 31: 1687–97.
45. Jabin, S., Kapoor, J.K. (2020). Role of Polyelectrolytes in the Treatment of Water and Wastewater. In: Inamuddin, Asiri, A. (eds) Sustainable Green Chemical Processes and their Allied Applications. Nanotechnology in the Life Sciences. Springer, Cham. https://doi.org/10.1007/978-3-030-42284-4_10.
46. Jai Prakash Chaudhary, Nilesh Vadodariya, Sanna Kotrappanavar Nataraj, and Ramavatar Meena, Chitosan-Based Aerogel Membrane for Robust Oil-in-Water Emulsion Separation, ACS Appl. Mater. Interfaces 2015, 7, 44, 24957–24962.
47. Mruthunjayappa, M.H., Mondal, D., Nataraj, S.K. (2022). Polymer Aerogels for Energy Storage and Water Purification Applications. In:

Subramani, N.K., Nataraj, S.K., Patel, C., Shivanna, S. (eds) Polymer-Based Advanced Functional Materials for Energy and Environmental Applications. Energy, Environment, and Sustainability. Springer, Singapore. https://doi.org/10.1007/978-981-16-8755-6_3.

48. Chuyang Y. Tang, Zhe Yang, Hao Guo, Jason J. Wen, Long D. Nghiem, and Emile Cornelissen, Potable Water Reuse through Advanced Membrane Technology, Environmental Science & Technology 2018 52 (18), 10215-10223. DOI: 10.1021/acs.est.8b00562.

49. Yulong Ying, Wen Ying, Qiaochu Li, Donghui Meng, Guohua Ren, Rongxin Yan, Xinsheng Peng, Recent advances of nanomaterial-based membrane for water purification, Applied Materials Today, Volume 7, 2017, Pages 144-158, ISSN 2352-9407, https://doi.org/10.1016/j.apmt.2017.02.010.

50. Smitha V. Kamath, Manohara Halanur Mruthunjayappa, Dibyendu Mondal and Nataraj Sanna Kotrappanavar, Nanocomposite-based high-performance adsorptive water filters: recent advances, limitations, nanotoxicity and environmental implications, Environ. Sci.: Nano, 2022,9, 2320-2341.

51. https://www.nasa.gov/centers/johnson/techtransfer/technology/MSC-24180-1_Water-Filtering-Device.html

52. Ren G, Han H, Wang Y, Liu S, Zhao J, Meng X, Li Z. Recent Advances of Photocatalytic Application in Water Treatment: A Review. Nanomaterials (Basel). 2021 Jul 12;11(7):1804. doi: 10.3390/nano11071804. PMID: 34361190; PMCID: PMC8308214.

53. Wenyuan Xie,a Fang He,a Bingfang Wang, a Tai-Shung Chung,a Kandiah Jeyaseelan, b Arunmozhiarasi Armugamb and Yen Wah Tong, An aquaporin-based vesicle-embedded polymeric membrane for low energy water filtration, J. Mater. Chem. A, 2013,1, 7592-7600.

54. Gopalkrishna V Gaonkar, Pavan P S A, Review on Wastewater Treatment Techniques JETIREW06015 Journal of Emerging Technologies and Innovative Research (JETIR) JETIR March 2019, Volume 6, Issue 3, 102-109.

Chapter 2

Advances in Water Recycling and Resource Reclamation

— Abstract —

This chapter explores the recent advances in the separation and purification technology used for water treatment and resource reclamation. This section includes both traditional separation technology evaluation and advances in the novel protocols and approaches in recycling water from wastewater sources. This chapter also highlights the adaptation of environment-friendly water treatment technologies with novel design and approaches to improve the effectiveness and sustainability of water recycling processes. This involves the development of smart materials, low capital and maintenance process designs and innovative thermodynamic models to optimize and maximize water reclamation. This section discusses advances in water recycling aspects covering conventional methods used for water recycling and recovery, advances in waste management and disposal, in value addition during water treatment, in waste segregation, in process water recovery, and reuse.

2.1 Introduction

World is today facing severe water scarcity and has been predicted to deteriorate in near-term decades. Interestingly, the water scarcity pattern has been continuously monitored to understand the impending crisis. The water scarcity has originated due to the lack of adequate accessible water resources to meet the demands within a region or provenance. In some regions, water scarcity arises due to drastic change in the weather conditions amplified by climate change. Increasing climate change effects are resulting in unforeseen droughts or floods. Further, increased water pollution, industrial activities, and excessive use of water and lack of suitable water treatment facilities add to the severity. Even in countries or regions with adequate freshwater resources, water scarcity is not uncommon. This may be due to the several individual and/or cumulative factors such as lack of strategic water management systems, collapsed infrastructure, increasing pollution, lack of water storage and distribution systems, adulteration, cross-boundary conflicts, and lack of human and financial resources invested in managing water resources.

Majority of the developing world is suffering from lack of adequate clean and safe drinking water. According to some estimations, water scarcity is also spreading even to the developed world covering every continent. Consequently, nearly 1.2 billion people, or effectively one-fifth of the world's population is exposed to moderate to high physical scarcity, and this situation is deteriorating by every passing day affecting over 500 million added population. On the other hand, nearly one quarter of the world's population (1.6 billion) spread across continents arev facing economic water stress. Therefore, water scarcity is not only concerned or restricted to a single living being or a specific geological location, region, or duration, rather it is affecting a larger population within a specific geographic region, biodiversity and countries pertaining to broader timescales extending to years. Hence, increasing water scarcity is one of many serious crises faced by many societies, countries, and the world at large in the 21st century. With increasing world population and industrialization water stress is expected to intensify further. [1, 2]

Nevertheless, water shortage is both a natural and a man-made phenomenon. Even though there is enough freshwater available on the earth for the entire population to live on, freshwater resources are unevenly distributed. On the other hand, overexploitation of freshwater resources has left large volumes of it either wasted, polluted, or left untreated. This limits access to clean and safe water for drinking and for practicing rudimentary hygiene at home, workplace and in communities. Further, deficiency of municipal water supplies can clog sewage systems and disrupt sanitary infrastructure resulting in surge of many water-borne diseases like cholera, fever, and typhoid. Moreover, water scarcity will certainly cause an intense competition between agriculture and industries to access water resources. On the socio-economic front, water scarcity induces social inequality and access to water resources becomes expensive. On the other hand, water scarcity at large affects agricultural productivity leading to food shortages resulting in levitation of commodity prices and inflation. This also at large hinders the trade between developing economies. Overall effect of food shortage, inflation, declining trade in long run may cause civil unrest. Further, the severe situation destabilizes and significantly distresses the domestic food production and processing industrial economy. Collective impact resulting in increased political instability borne by the escalation of climate change, migration, and trans-boundary influx of refugees seeking asylum. Therefore, efficient water management and adopting a water recycling economy can offer significant benefits to modern societies. [3]

Water recycling or the reclamation and reuse is the process of identifying, monitoring, collecting, and treating of wasted water or wastewater, mainly from municipalities, industries, food processing units, and agriculture discards. The reclaimed or recycled water can be used for domestic purposes, industrial activities, irrigation, or industrial activities as well as gardening purposes subjected to appropriate treatment. On the other hand, recycling of water can offer considerable paybacks to micro and macro socio-economic groups. Ecological benefits include reduction of soil contamination, retention of nutrients and decrease in influx of contaminants into ponds, large reservoirs, rivers, and finally into oceans. Further, recycling efforts lead to providing high quality reusable water for domestic usages and replacing freshwater with recycled water for irrigation of commercial crops and amenity. On the economic front, sludge or concentrated leftovers can be utilized as fertilizers or precursors for energy materials. This also adds additional environmental and health benefits imparting sanity to the ecosystem and environment at large. Nevertheless, awareness in water reclamation and reuse among the general public is increasing every passing day. [4]

More significantly, recycling and reclaiming of water from waste sources protects the fresh water resources which are dwindling every passing day. Water recycling also helps in reducing the stress on freshwater drawing from various sources, reducing human or industrial activities near freshwater sources, and containment of pollution. Civic response and attitude toward the reusable water reclamation for both domestic as well as commercial application is increasingly growing. But, application of recycled water for irrigation of food crops and agricultural products still finds multi-step scrutiny. Throughout the world it has been noticed that there have been several inhibitions and conflicting studies exist on both direct and indirect use of recycled water for potable purposes. Fortunately, increasing efforts by researchers, public outreach activities, efforts from socio-economic groups and increased educational programs in the sector of water treatment technologies have enabled public awareness about reuse of recycled water. On the other hand, studies confirm that the effects of extreme weather induced global warming on imminent water supplies is not likely to be uniform everywhere, however, these uncertain factors will be drastically limiting in certain parts of the dry or wasteland and arid regions of the world. These factors force the urgency for efficient water management protocols and water use competence measures like water recycling and reuse in affected regions. [5, 6]

In such a situation, water recycling, reclamation and reuse become ever more important in the years to come for accomplishing a sustainable world. In addition to this water reclamation from wastewater treatment route can fetch economic benefits through recycle economy in various sectors. For this, advanced water treatment technologies have been developed and implemented in various sectors. On the other hand, several guidelines and protocols have been developed to make it possible to use recycled water with all required norms for all industrial, domestic, and personal uses. However, even with several attempts to educate the population, perceptions related to recovered water still inhibit the direct use of reclaimed water. Such concerns also extend to long-standing reserves and inhibit the bodily functions of regulators and community apprehensions around health risks. This negative connotation associated with wastewater comes from the fear of industrial hazard, sewage contamination and unknown unease about emerging pollutants discharged daily. To address these concerns school, college and community level educational programs need to be devised and implemented. [7, 8]

Nonetheless, water reclamation and reuse offer a unique and feasible opportunity to augment conventional water supplies. Modern pollutant removal and overall treatment of contaminated water requires a multifaceted approach with multi-disciplined expertise involved in managing important state of water resources. Modern contaminated water is complex and tedious to analyse for contaminant levels. However, proper monitoring, tracking protocol and management tools can help in closing the loop between water supply chain and wastewater discharging. Further, the effective water recycling strategies require integration of wastewater collection, conditioning, pretreatment, and recycling routes. In addition to this, successful development of a reliable treatment strategy depends on close examination and adopting right treatment elements and materials stocks. This helps in developing a method or technology via a tailored assembling, facilities planning, selecting appropriate wastewater treatment plant location for an effective integrated infrastructure. With this, reliable, economical, easy to maintain and environmentally beneficial water treatment and recycling facility can be a reality with larger public acceptance. [9, 10]

2.2 Advances in Water Recycling

Water recycling from wastewater, treatment and reuse is central to future scientific research programs. This is being proposed involving a futuristic approach of creating sustainable technologies at affordable cost. The approaches with low-cost are forecasted to reach wider population in recycling effort of wastewater. Therefore, water treatment technologies are being devised keeping maximum water waving and reclamation in mind such as reduction in water consumption, treatment of wastewater and recycling of wastewater. Conventional at present, several water treatment operations and processes are in place as individual and combined techniques to provide recycling the water from wasted sources. These processes are broadly categorized as primary treatment, secondary treatment, and tertiary treatment. However, recent advances in separation and purification technologies have adopted a series of new materials and methods to reclaim reusable quality water from wasted sources which can be categorized as advanced treatment techniques. Therefore, innovation in sustainable water usage, management and recycling means is the way that would focus on including every stakeholder including the existing resource stock, supply, ecological conditions, social demand, and economic conditions without compromising the capability to meet those demands in the future. Sustainable treatment and water reclamation specification involves the "a typical treatment method or combination of various types of treatments that can reclaim the reusable quality raw material or water in order to preserve the natural resources from all discarded waste or wastewater. [11, 12]

These treatment techniques categorized as shown in Figure 2.1 aim to remove every possible contaminant first including COD, BOD, organic contamination, phosphorous, nitrogen, pharmaceutical, suspended solids, other emerging pollutants and biological contaminants like bacteria, virus, hormones, etc., among all, choice of treatment method or methods depends on the complexity of the wastewater, composition of wastewater, source of contamination, source of pollutant and economic considerations of the treatment site and feasibility.

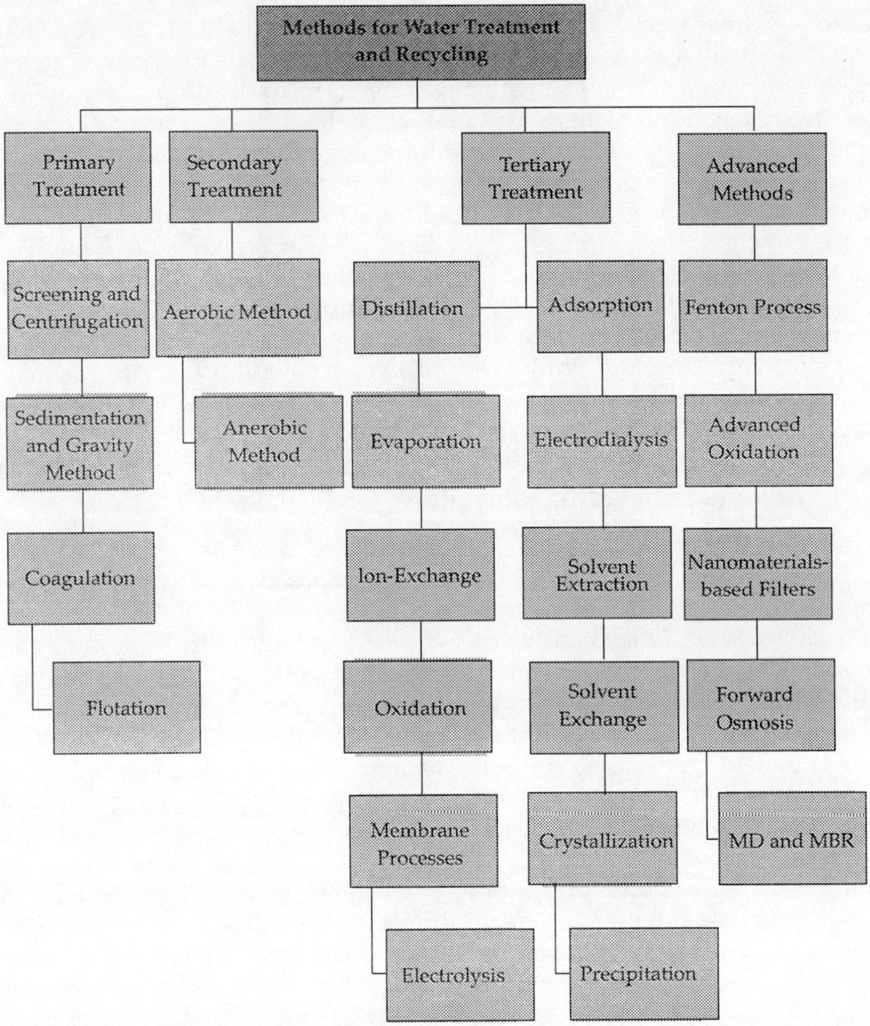

Figure 2.1. Spectrum of methods used for water treatment and recycling.

Nonetheless, the wastewater treatment methods and water reclamation process are broadly classified as primary treatment, secondary treatment, tertiary treatment, and advanced treatment methods. The primary water treatment includes preliminary purification processes comprising methods mainly having physical, physicochemical, and chemical nature. The secondary water treatment methods mainly deal with the biological treatment and broadly used as pretreatment steps for further advanced treatment of wastewater. The tertiary water treatment processes are then adopted to further treat the pretreated wastewater at the stage of either primary or secondary treatment level. In tertiary water treatment routes convert pre-treated wastewater from primary and secondary treated wastewater into good quality water. The tertiary treated water can be directly used for municipal or industrial purposes. Interestingly, the tertiary treatment process is known to provide high quality of treated water with over 99% purity removing all pollutants to meet water reusability criteria set by several regulatory agencies. The tertiary methods are generally selected and applied to pretreated wastewaters either as individual or combination of methods for producing high quality clean and safe reusable water. [13]

On the other hand, evolution in the field of commodity chemicals and products in the domestic market add highly toxic and mysterious emerging pollutants to the environment by each passing day. Currently, most of the existing water treatment techniques or methods are known to be ineffective against emerging pollutants. Despite the presence of several effective methods and development of various materials to suit the evolving challenges, currently existing technologies like primary, secondary, and tertiary water treatments lack effectiveness against emerging pollutants. These methods are also known to fail to effectively recycle water contaminated with emerging and complex pollutants based on economic considerations. For this, advanced materials and methods are adopted to suit the pollutant nature and level of contamination. This approach also considers the effective ways and rapid water treatment routes to make the water reclamation process commercially viable. Through these approaches, real-time monitoring and applications can be handled with effectiveness. Therefore, systematic advanced approaches of water reclamation, treatment and recycling technologies should include understanding of the novel approaches with detailed materials specifications, design, and operational cost. Considerations should also be given for the inclusion of long run the maintenance charges and management of removed solid waste or pollutants. Further, a comparison of these water treatment and reclamation technologies with respect to their practicality, appropriateness and economics is discussed in the following sections. [14]

2.3 Conventional Methods used for Water Recycling and Recovery

Traditionally, water treatment technologies have been classified mainly into three categories, primary treatment, secondary treatment, and tertiary treatment. However, due to the presence of complex pollutants and wastewater nature, most of the conventional methods fail to fulfill the high standard of recycled water. Therefore, recent approaches in developing effective materials and methods to tackle complex wastewater sources into fourth category treatment techniques. (as classified in Figure 2.1). However, the choice of technique to be adopted for the treatment of wastewater needs specific contaminant nature and level of contamination. Nevertheless, conventional methods are still preferred as pre-treatment purposes in which considerable levels of contaminants are removed in primary or secondary treatment steps to aid the further reclamation of high-quality water.

2.3.1. Primary water treatment technologies

Primary water treatment techniques mainly involve physical separation of pollutants from wastewaters. In this category, wastewater is treated at the primary treatment level using screening, mesh filtration, screening, centrifugation, sedimentation, gravity separation, coagulation, and floatation methods. Generally, primary treatment techniques are used for pretreating highly contaminated complex wastewater. However, primary treatment techniques do not provide sufficient quality *t* water. Following are the brief descriptions of primary water treatment and techniques conventionally used. [15]

(a) **Screening:** Physical screening of wastewater is widely used in wastewater treatment plants as the first step to retain all suspended solids. Screening step is also a first unit operation generally used in almost all wastewater treatment plants. In this method, a simple metal screen tailored device is used to retain all suspended solids found in the inflowing wastewater to the treatment plant. The key purpose of screening the wastewater at the first stage is to remove all suspended solid materials which could cause severe blockage or clogging of further treatment steps. This can also cause damage to the equipment. Further, the screen step is ideally used for the removal of larger solids at first including pieces of plastics, cloth fibers, paper discards, wood chops, cork, hair, fiber, kitchen waste, fecal solids and so on from wastewater. In this equipment the screens of various channel sizes are used for the purpose of pre-filtration and the choice of screen size and dimension are selected based on the nature of wastewater and contamination level. During the pre-filtration step, screening devices trap solid trash, coarse and macro-debris present in the wastewater using smaller apertures such as grates or screens as shown in Figure 2.2(a). Screening is best credited for treating fecal matter and suspended organic matters and further passed onto the next step. [16]

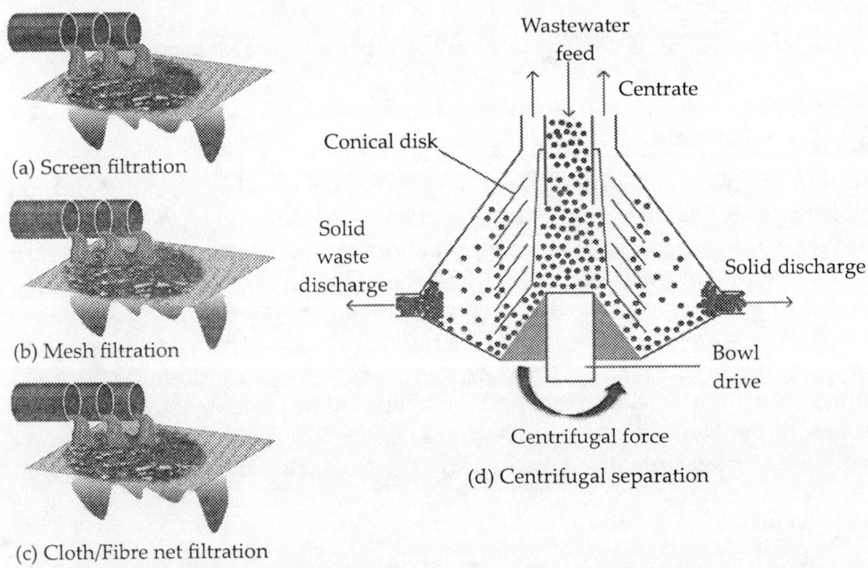

Figure 2.2. Primary treatment technique used for physically removing solids and clarification of water.

Mesh filtration: In mesh filtration pretreatment step wastewater is passed through a filtration media having fine mesh or cloth or net pores as shown in Figure 2.2(b&c). Generally, mesh filters are made up of pore size ranging from 0.1 to 0.5 μm for the purpose of fine suspended particles or solid retention including macro-suspended solids, oils, bacteria, biological contaminants, greases, and so on. In this category, macroporous membrane filters, cartridges and porous plastic meshes are generally used. In this method, suspended solids of size <100 $mg.L^{-1}$ and removal oil contaminants with 25 $mg.L^{-1}$ can be achieved up to 99% efficiency. The mesh filtration pr-treated wastewater is further directly passed on to process steps like adsorption, ion-exchange or membrane-based separation processes for obtaining high quality reusable water recovery. [17]

(b) **Centrifugal separation**: In this technique, centrifugal force created by high-speed axial rotations is used for separating separate suspended solids from liquid wastes. Here, centrifugal force moves denser material outwardly on the centrifugal orbit toward the wall of the centrifuge vessel. While solids are thrown towards the orbit with the centrifugal force, simultaneously the clarified liquid overflows from the centrifuge vessel which is separated by a skimmer. Centrifugal separation pretreatment step is used to remove suspended non-colloidal solid matter having size up to 1 μm. Specially designed sludge dewatering centrifuge systems in wastewater treatment plants use a fast rotation of a 'cylindrical bowl or vessel' to segregate wastewater constituent liquid from solids as shown in Figure 2.2(d). The wastewater treatment centrifuge generally shows greater dewatering efficiency in removing high percentage of clarified water than any other pretreatment technique leaving behind highly concentrated sludge known as 'cake'. One of the main advantages with centrifugal separation techniques is separation units consume minimum space and appropriate design can fit medium to large scale treatment capacities. Also, the process of solid/liquid segregation can be achieved with different applied conditions to centrifugal devices and rotating speeds. This helps in adopting this pretreatment technique to suitable feed conditions. It is well proven that the extent of suspended solids separation is directly related to their densities and the speed of the centrifugal apparatus is also responsible for the removal efficiency of suspended solids. This technique is currently used successfully for the separation of oils and greases from contaminated suspended solids. [18]

(c) **Sedimentation and gravity separation:** In sedimentation and gravity separation processes, the suspended solids, silts and grits are segregated from liquid phase by letting water to be unobstructed or semi-disturbed for various time intervals in different types of separation tank. Here, the suspended solids are allowed to settle under the influence of gravity. The settling time of solid material depends upon the size and density of the solids matter or the velocity of the wastewater influent in motion. In special cases, additional aggregators like alum are added to supplement the sedimentation or settlement process. It is generally noticed that the gravity separation process can reduce the suspended solids by 60%. Both gravity separation and sedimentation processes are used as pre-treatment to clarify the wastewater in conventional wastewater treatment plants for the treatment of paper industry, distillery industry and refinery industries effluents. [19, 20]

(d) Coagulation: After settling all macro and micron size suspended solids, the colloidal suspension does not easily settle down under the under the gravity and, hence, non-settleable or colloidal suspensions are forced to sediment under the influence of charge imbalance adding certain chemicals which is broadly known 'coagulation process'. in his process to create surface charge imbalance or to break the charge neutrality, coagulants like alum, iron-based materials, starch, activated silica and aluminum salts are used. Further, in special condition ionic polymer bearing cationic, anionic, zwitterionic and non-ionic functionalities are effectively used. However, inorganic coagulants are known to be economical compared to polymeric coagulants. In this process, pollutant charge, concentration, pH condition of wastewater, temperature, and contact time play an important role in controlling the coagulation process. Interestingly, the hybrid approach comprising biological treatment elements, complex organic nutrients, microbes, and oily contaminated pollutants are removed with the help of suitable coagulants. In many cases, the coagulation step is adopted as an integral component of wastewater treatment plants. Here, increasing the charge density of the counter ions or coagulants can lead to the stronger the flocs formation which leads to the breaking of the colloidal system. The formation of flocs or aggregates in this physicochemical method can be attributed to four mechanisms, namely, charge neutralization, sweeping coagulation, bridging, and patch flocculation or electrostatic patch as illustrated in Figure 2.3. This step is helpful in reducing suspended solid load in domestic wastewater, industrial wastewater, and domestic wastewater with simultaneous recycling of high-quality reusable water and pollutant segregation. [21, 22]

Among many approaches in coagulation processes, the charge neutralization method refers to athe state in which the net electrical charge on pollutants like colloidal particles, fibers and polyelectrolytes in wastewater sources cancel net charges to retain an equal number of opposite charges via the adsorption process as shown in Figure 2.3 (a). Typically, excessive opposite charges for the purpose of balancing or neutralization are supplied by a multivalent chemical. Among them are the polymeric flocculant and inorganic coagulants which adsorb on to the surface of the oppositely charged pollutant particle leading to neutralization in overall surface groups and particle aggregation. Likewise, cationic coagulants will be able to neutralize the anionic surfaces of pollutants in wastewater like kaolin clay, resulting in flocculation of the pollutant matter. In charge neutralization route, it

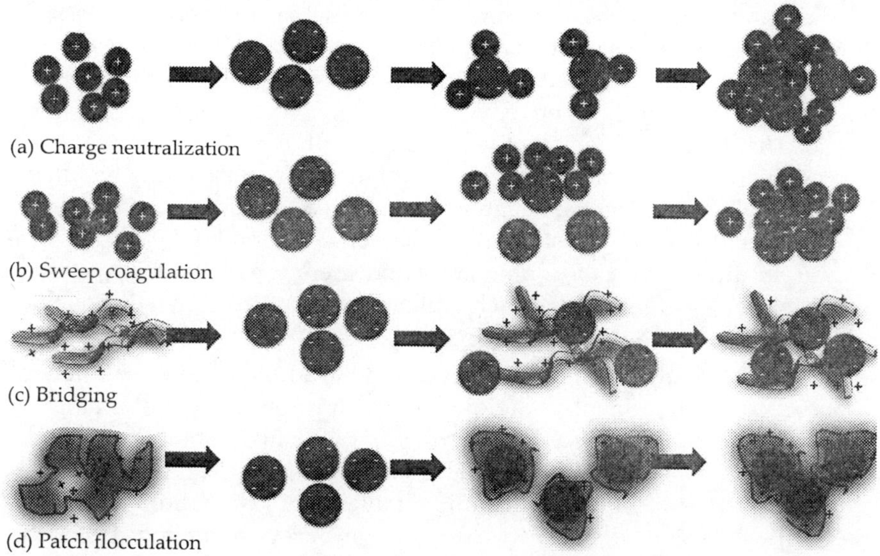

Figure 2.3. Coagulation mechanism in separating pollutants from wastewater sources involving (a) charge neutralization, (b) sweep coagulation, (c) bridging, and (d) patch flocculation.

is expected that cationic pollutants can be neutralized with the help of anionic coagulant causing flocculation of the pollutant particle. However, the rate of flocculation is influenced as a function of the charge density which directly relates to the number of cationic charges induced for the charge neutralization process. [23]

Inversely, excessive addition of coagulants results in rapid charge reversal and re-stabilization of the pollutants in the effluent. When a metal salt-based coagulant is introduced to wastewater at an adequately high concentration to cause the precipitation of amorphous metal hydroxide ($M(OH)_3$) colloid pollutants to force them to be entangled in the precipitates, the process is called 'sweep coagulation' as shown in Figure 2.3 (b). Further, when high concentration of counterions is induced into that already stabilized colloidal suspension, the added counter-ions will infiltrate into the stern layer. The stern layer is the dense layer of counter-ions fixed immobile to the surface of the primary pollutant, following this there exists a more diffused layer. However, in influent effluent only the bound layer travels with the pollutant particle. However, there is a shear plane between the bound layer and the diffuse layer and the potential difference between the shear plane and the bulk solution is known as the 'zeta potential'. Such phenomenon results in compression and repulsion between colloidal pollutants enabling aggregation which then form floc, further leading to flocculation. Therefore, the higher the charge density of the coagulant, the stronger the flocs formed. Further, in absence of rapid re-stabilization in presence of excessive coagulants refer to a bridging mechanism as shown in Figure 2.3 (c). In such a situation, it is possible that larger pollutants facilitate the adequate surface for the adsorption of several pollutant matters connecting them together. Therefore, the coagulation process is effective or relevant when oppositely charged flocculants are added to the colloidal effluent. Positively charged coagulants or polymers strongly adsorb on the surface of the pollutant, eventually canceling the negative charge of the pollutant at optimal pH. As a result, the electrostatic repulsion between pollutants disappears due to diminishing of double-layer and therefore, colloidal pollutants get closer and form flocs which will eventually settle down due as flocs get denser as shown in Figure 2.3(d). [24]

(e) **Flotation:** Flotation is another common and essential component used as pretreatment technique in wastewater treatment plants of a conventional primary treatment plant. In this technique, the oils, greases, suspended solids, biological solids, etc., are removed by passing air through the floatation tank to which pollutant adheres forming larger globule which helps in easy removing from the tank. When air or gas is supplied to a flotation tank containing influent wastewater the suspended solids get adhered to inflowing air resulting in agglomerates, which at the end of the process accumulate at the water's surface and which can eventually be skimmed off from the system. To facilitate the faster removal of pollutants, chemicals such as alum, activated silica, etc., which help in aggregation or formation of flocs are added to the floatation tank. In this flotation tank compressed air is made to pass through the contaminated water to help in floating the pollutants. Among several approaches, pressurized dissolved air flotation technique is widely used for wastewater treatment in which suspended solids are easily separated from clarified water as shown in Figure 2.4. Further, compressed air bubbles forced into the floatation

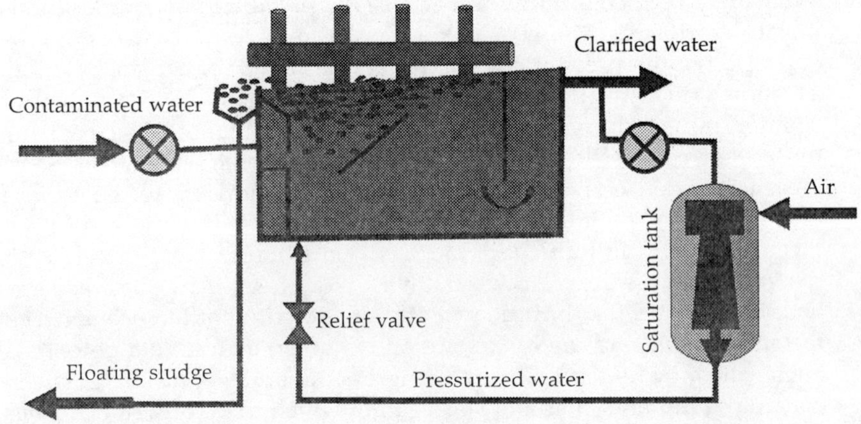

Figure 2.4. Floatation pretreatment method used in wastewater treatment plants.

tank in high velocity gather solid pollutant particles by adhering to the air surface to make them float to the surface of the where they are removed either manually or mechanically. Simultaneously, to make the overall process sustainable, the air bubbles flow from the pressurized air tank is controlled to enrich water by lowering the compressed air tank pressure downstream using a relief valve. In advanced floatation tanks, electro-flotation or electro-flocculation methods are adopted to effectively remove the suspended solids with simultaneous reusable water recycling. Through advanced floatation equipment designs close to 75% of suspended solids contaminants are removed. On the other hand, up to 99% of oily, grease or fat-based contaminants are efficiently removed. This method is popularly used at industrial scale for the treatment of oil refinery wastewater, paper industry, dairy industry, and distillery industrial wastewater. [25-27]

2.3.2. Secondary water treatment technologies

Secondary water treatment method is detecting and removal of biodegradable organic matter both in solution or suspended from contaminated water and wastewaters like domestic effluents. This approach aims to achieve a high degree of effluent treatment quality in biologically contaminated water treatment plants. Therefore, secondary water treatment mainly comprises the water treatment via biological routes to remove the soluble and insoluble or suspended biological pollutants. The secondary water treatment step works well as subsequent steps to primary treatment techniques and increases the quality of recovered water by removing biological contaminants to greater extent. The secondary water treatment is also known as biological treatment since it is intended to significantly remove or degrade the biological pollutants through aerobic or anaerobic biological processes. Through this step all dissolved organic matter is treated and removed using biological agents. For this purpose, water is circulated in a biological treatment reactor that sustains a high concentration of biological agents. Among many biological agents bacterial, fungal strains and other microbes convert both suspended and dissolved organic matter into clarified water, carbon dioxide and ammonia gas. Through this process some organic matter is transformed into degraded products like glucose, nitrate, alcohol, etc. Furthermore, activities of degraded organic matter to make it less-toxic and with suitable post-treatment contaminated water can be completely recovered from toxic inorganics and organic matters. The secondary treatment steps involve two sub-treatment biological treatment steps, namely, the aerobic and anaerobic digestion of contaminated water or wastewater. [28]

1. **Aerobic processes:** In this approach, air or oxygen is spontaneously supplied for breaking of dissolved organic matter in contaminated water or wastewater using aerobic and facultative microorganisms or bacteria. It is known as aerobic decomposition. The rate of aerobic digestion or biodegradation process mainly depends on the accessibility of oxygen, temperature of the feed, retention time and the biological activities

of the microorganisms or bacteria. A brief representation of aerobic decomposition of dissolved organic matter in the water treatment system is given by the following equation. [29, 30]

Dissolved organic matter (DOM) + O_2 + Bacteria → CO_2 + H_2**O + Bacteria + Byproducts**

In addition, the extent of the biological oxidation in dissolved organic matter pollutants can be increased by the addition of some bacterial growth chemicals or nutrients. This technique is generally adopted to effectively remove volatile, dissolved organics, suspended organics matter, nitrates, phosphates, biological oxygen demand (BOD) and chemical oxygen demand (COD), etc. ,in this secondary treatment step concentration of biodegradable organic matter can be decreased up to 90%. The drawback of aerobic biodegradation method is the formation of a large quantity of biosolid waste or bio-sludge that demands subsequent treatment or disposal step which may add to the treatment and waste management cost. The aerobic biodegradation is generally carried out using trickling filters or activated sludge processes or oxidation ponds.

2. **Anaerobic process:** The secondary water treatment technique involving the degradation of both dissolved and suspended organic matter using microorganisms or microbes in the absence of oxygen is known as anaerobic biodegradation. Whereas aerobic degradation uses oxidative reactions to degrade organic pollutants, on the other hand, anaerobic bacteria take the route of reductive types of reactions. Also, the anaerobes degrade organic compounds by reductive conversions. The anaerobic microbes and facultative bacteria transform both the suspended and dissolved complex bio/organic matter into simpler organic compounds containing carbon, nitrogen, carbon, and sulfur. Through this process important gasses like ammonia, methane, nitrogen, and hydrogen sulfide are evolved during different intervals of the process. Therefore, anaerobic method is used to reduce the biological load of bio-organic matters contaminated water or domestic wastewater. The anaerobic degradation process is generally represented by the following equation.

   Organic matter + Bacteria → CO_2 + CH_4 + Bacteria + Byproducts

   Here, sometimes the microorganisms use a chemical in the absence of oxygen as an electron acceptor to initiate the biodegradation process known as 'putrefaction'. [31]

2.3.3. Tertiary water treatment technologies

Following the pretreatment of contaminated water from primary and secondary treatment, most of the wastewater treatment plants adopt tertiary treatment techniques to improve the quality of clarified wastewater before it is given the status of recycled water, or reusable quality water or dischargeable quality of water to the environment. Industrial wastewaters are characterized by high concentrations of pollutants like inorganic compounds, organic matters, and biological moieties. Upon direct discharge or disposal of these highly contaminated wastewater into natural waterbodies like lakes, ponds, and rivers cause severe damage to the environment. Primary and secondary treatments most of the time fail to address the inorganic contamination to the acceptable level. Therefore, tertiary water treatment technologies are adopted to increase the quality of pretreated water in preceding techniques to achieve reusable quality standards. However, the tertiary treatment needs to be used strategically to suit the contaminant nature and quantity to make it an effective step. The tertiary treatment techniques are subdivided into distillation, evaporation, ion-exchange, oxidation, membrane-based separation processes, electrolysis, electrodialysis, solvent extraction, solvent exchange, crystallization, and precipitation, etc.

1. **Distillation:** Distillation is a process that relies on thermal evaporation of pretreated effluent to purify and extract high quality reusable water. In this method, polluted water is heated to form steam which is then condensed to get high-quality water. When pre-treated effluent is heat treated up to 100°C, above boiling point of water at which water is vaporized, inorganic compounds, large non-volatile and semi-volatile organic compounds do not evaporate along with the water, instead are left behind. The steam then cooled and condenses to form purified water. The vapors thus generated are cooled into liquid water. Therefore, at the end of the distillation process, effluent is made free from volatile, semi-volatile and inorganic impurities resulting in high quality water, ~99% free from pollutants. In recent advances, various novel approaches have been adopted to design efficient distillation or boilers to achieve the quality of water products. Also, the size of the distillation units depends on the size and extent of water quality and quantity. The product water extracted from the distillation process may be suitable for the preparation of drinkable water from the seawater, brackish water and can also be directly used for laboratories and medical applications. [32-34]
2. **Evaporation:** Evaporation is another secondary treatment technique being considered as an alternative process for distillation treatment in treatment plants. Like distillation, evaporation technique is helpful in concentrating salts, heavy metals and various hazardous materials from the wastewater. However, unlike the distillation processes, evaporation is driven by heat transferred from condensing steam to a contaminated feed at a lower temperature across a metallic heat transfer surface in an evaporation treatment system. Further, the riveted heat from steam causes vaporization of the water which increases the concentration of solute or pollutants in feed. As produced water may be then directly condensed to collect as recycled water for reuse purpose.
3. **Ion-exchange:** Ion-exchange process in secondary treatment technique plants is a wastewater treatment where multiple ionic contaminants are selectively removed from contaminated water via the exchange with another exchangeable ionic substance known as ion-exchange resin or polymers. Ion-exchange also defines a specific chemical process

in which ionic pollutants dissolved in water and wastewater such as sulfate, nitrate, fluoride, sulfate, heavy metals, and arsenic, etc., are exchanged or swapped for other ions with a similar charge in ion-exchange substrate. An ion-exchange polymer or resin is a substrate that acts as a medium for ion-exchange having tailor-made exchangeable functional groups. Broadly, ion-exchange resins or polymer are classified into (a) cation-exchangers and (b) anion-exchanger, cation exchangers facilitate exchange of cationic pollutants (positively charged ions), on the other hand, anion-exchangers enable the anionic (negatively charged ions) pollutants. Typical ion exchangers are made up of functional porous natural as well as synthetic materials or polymeric gels such as montmorillonite, zeolites, clay, kaolinite, bentonite, vermiculite, sodalite, zeolites, chabazite, analcite, clinoptilolite and soil humus having robust ion exchange properties. Among all, natural zeolites and clay materials were the first set of materials extensively used for the ion-exchange processes. In domestic water treatment applications, ion-exchange units or cartridges are used for the removal of hardness in water mainly caused by magnesium, calcium, manganese, and iron salts known as water softening. [35]

In wastewater treatment plants, ion-exchange resins or polymers are used for the removal of dissolved ionic pollutants from contaminated water. In the ion-exchange process, both anionic or cationic pollutants are swapped with the help of ion-exchange materials. These ion-exchange resins or polymers can be used as standalone materials in batch experiments or as packed cartridges in treatment plants to eliminate ionic contaminants from effluents. Compared to other primary and other secondary treatment techniques, ion-exchange method is particularly pertinent in removing a specific contaminant at low concentration like boron from well or groundwater, heavy metals from both surface and industrial wastewater.

Interestingly, unlike other methods, ion-exchange materials pose a finite exchange capacity, in which each individual exchange site will saturate in prolonged use. However, in many cases these functional groups can be regenerated and reused for the same purpose. Nonetheless, when functional groups become saturated with exchanged groups and ion-exchange process no longer feasible, the resin or polymers are subjected to recharging or regeneration to restore it to its original condition. The ion-exchange materials can be recycled and regenerated using chemicals like sodium chloride, hydrochloric acid, sulfuric acid, and sodium hydroxide. [36]

4. **Oxidation:** In the chemical oxidation process, organic pollutants are oxidized and degraded into water and carbon dioxide. In general, oxidation processes degrade organic products like carboxylic acids, alcohols, ketones, and aldehydes into easily biodegradable end products. Chemical oxidation process is initiated with the help of catalysts like hydrogen peroxide (H_2O_2), Fenton's reagent in which

mixture of H_2O_2 and Fe are used as conventional catalysts, potassium permanganate, ozone, chlorine, and chlorine dioxides. Although photo-assisted Fenton reactions are now considered advanced treatment methods, catalytic degradation of pollutants has a trace in secondary treatment techniques. Nonetheless, the rate of chemical oxidation in effluents depends on the nature of oxidants, level of contamination and nature of the pollutants present. In addition to this, rate of chemical oxidation depends on effluent temperature, reactor design, pH, and presence of catalyst and nature, etc., oxidation process can also address the effluents containing pollutants like dyes, organics, ammonia, phenols, and other hydrocarbons economically. [36, 37]

5. **Membrane-based separation processes:** Membrane-based separation processes are categorized as one of physical water treatment techniques in which semi-permeable membrane is used as separating media as illustrated in Figure 2.5. In this process, semi-permeable membrane separates feed or contaminated water components from purified water. Membrane is a semi-permeable barrier which allows the passage of liquid phase rejecting solute phase concentrated in the feed compartment. In pressure-driven membrane processes applied pressure acts as a driving

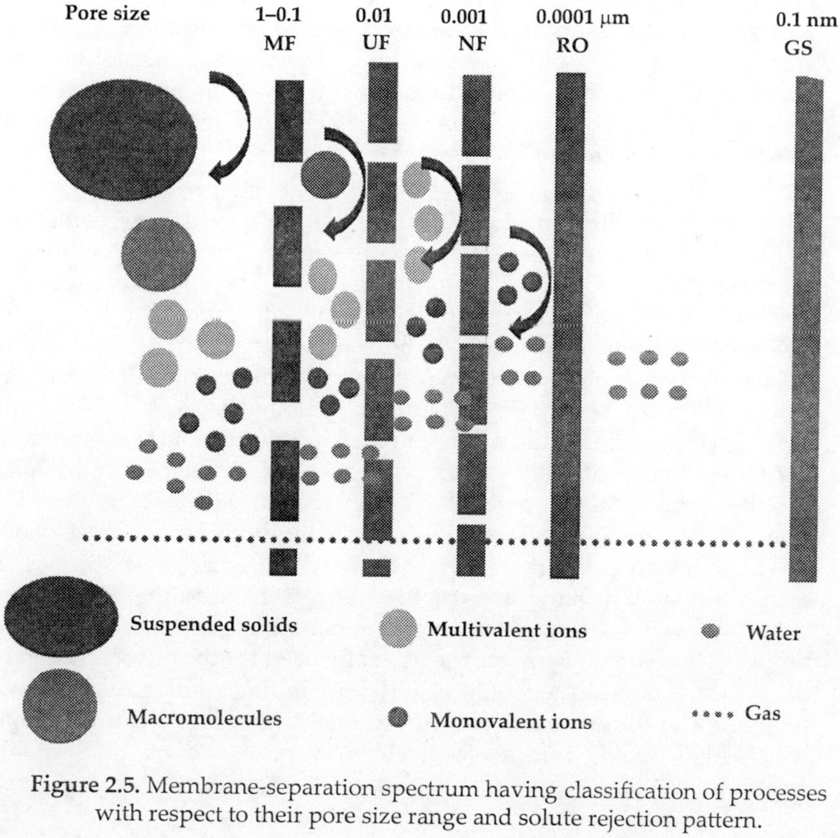

Figure 2.5. Membrane-separation spectrum having classification of processes with respect to their pore size range and solute rejection pattern.

force to separate pollutants from pure water. In addition to this, pressure-driven membrane separation processes are further categorized based on membrane pore size and their molecular weight cut off (MWCO), namely, microfiltration (MF), ultrafiltration (UF), nanofiltration (NF), and reverse osmosis (RO). [38, 39]

6. **Microfiltration:** In membrane separation spectrum, microfiltration (MF) process lies on the upper end of the band. In this low pressure-driven membrane technique, pore size of the membranes poses the largest pore size among all. The microfiltration membrane pores lie between 0.1-10 μm and this process is capable of separating solute more than pore size including suspended solids, colloidal suspension, and other pollutants above 0.1-10 μm size. Due to the high pore size, often low pressure (0.5 to 1.5 bar) is used to drive the purified water from the effluent. The microfilter membranes are often made of cartridges of cotton, cellulose, wool, polypropylene, rayon, fiberglass, acrylics, nylon, and fluorinated hydrocarbon polymers. In recent times, several specially designed synthetic polymers like polysulfone, polyvinylidene fluoride, polyethersulfone are used to prepare efficient microfiltration membrane. Membrane modules also come in different configurations such as spiral wound, flat sheet, tubular, disc-type, plates, and hollow fibers formats. The shelf-life of polymer-based membranes varies from 5 to 8 years depending upon the application, concentration of effluent such as dissolved solids and applied operation conditions. Also, common MF process applications mainly involve the separation of large suspended solids, macromolecules and biological contaminants like bacteria, cellular broths, fat from dairy industry, distillery industry, paper industry, sugar industry effluents and so on. [40, 41]
7. **Ultrafiltration:** Ultrafiltration (UF) is also a pressure driven membrane technique in which pore size regime is slightly reduced compared to microfiltration. UF process falls in between MF and NF processes among a family of pressure-driven membrane processes, with pore size ranging from 1-100 nm. This pore size range restricts the passage of medium to high molecular weight pollutants or solutes such as macromolecules, colloidal suspensions, proteins, bacteria, non-volatile organic compounds and so on. Unlike MF, ultrafiltration membranes are generally categorized with respect to their ability to reject particular solute- based on their molecular weight cutoff (MWCO) as loose UF closer to MF membrane pore regime and tight UF membranes, close to nanofiltration regime. Nevertheless, a wide range of pore size distribution in UF membranes regime allows it to be adopted for a wide variety of water and wastewater treatment from different industrial origins. Therefore, UF process is generally used for the concentration of dairy products, treatment of food and dairy industry effluents, beverage industries, automotive industry, domestic wastewater treatment from which high quality reusable water can be recovered for non-potable applications. [42]

8. **Nanofiltration:** Nanofiltration covers a larger range of molecular weight range solute than MF and UF due to its small pore size which also plays an important role in inducing charges to the membrane surface that helps in solute rejection. NF processes possess a pore size range between 0.1-10 nm. With this pore range, NF membranes can retain low molecular weight solutes, uncharged solutes, multivalent inorganic pollutants, sugars, emerging organic pollutants like dye molecules, and pharmaceutical waste. NF membranes also retain charged species, such as polyvalent ions and large monovalent ions, whereas smaller monovalent constituents like (Na^+, K^+) are allowed to pass through. In addition to this, due to its small pore size range NF process can substantially reduce hardness from ground and brackish, remove some of persistent contaminants, and dissolved organic matter (DOM) from various industrial wastewater sources. [43]
9. **Reverse Osmosis:** Among all pressure-driven liquid/liquid separation membrane processes, reverse osmosis (RO) membranes bear lowermost pore size regime capable of retaining almost all dissolved solutes including monovalent ions, dissolved particles, and small organic molecules within a feed stream or effluents. In the RO process, the separation of solutes or solvent, in most of the cases water is permeated selectively through semi-permeable membrane with less than 1 nm pore size. RO membrane is also referred to as dense membrane with hydrophilic surface nature largely used for the separation not only through size exclusion mechanism but also via diffusive separation mechanism as well. To achieve the efficient separation of solutes from liquid components, the applied pressure in the RO process is vital and adequately high to overcome osmotic pressure of feed or effluent. The RO process is most used for seawater desalination, demineralization, emerging pollutant treatment, and process water recovery and food industrial application like beverage concentration. [44]

   These types of membranes with different functionality, nature and diffusivity can also be adopted for the gas separation applications. Both RO and gas separation membranes possess extremely narrow pore size and require higher applied pressure to separate liquid or gaseous components from the mixture compared to all other pressure driven membrane processes.
10. **Electrolysis:** Electrolysis is an electrochemical effluent treatment technique that is popularly used for the effluent treatment. Even though electrolysis is a complex process that involves many reagents and adjusting physical parameters that use consumable electrodes useful for treating ionic species contaminated water. Electrolysis is a highly efficient technique to remove toxin contents in effluents like pharmaceutical wastewater, pump, electroplating industry, textile, and paper mills discards. This is also used as a complementary technique along with other electrochemical water treatment methods like electrochemical oxidation of organic pollutants, removal of cyanides, water disinfection,

and sulfides. Further, electrochemical reduction is applied for metal ion recovery and converting persistent or emerging organic pollutants into less toxic constituents.

11. **Electrodialysis:** Electrodialysis (ED) is a process in which electrically charged perm-selective films known as anion-exchange membrane (AEM) and cation-exchange membranes (CEM) are used to separate ions from an effluent or aqueous solution driven by an electrical potential applied across the oppositely charged electrodes. In ED, the process of ionic transport across is controlled by an applied electric field that allows the separation of pollutants in effluents and minerals in water from feed water solution to concentrate. The ion exchange process takes place when the dissociated ions selectively pass through ion-perm-selective membranes (AEM and CEM) from dilute (feed or effluent) to a concentrate (brine) tank. Electrodialysis is popularly used for concentrating trace contaminant treatment like brackish water treatment, electroplating industry wastewater, nitrate removal, fluoride removal and so on. Traditionally, ED has been considered an alternative technique to reverse osmosis (RO) process for desalination applications. [45]
12. **Solvent extraction:** Organic solvents that are immiscible with water however capable of dissolving pollutants are added to wastewater to remove or extract pollutants through the process known as solvent extraction. In this method, hexane, benzene, acetone, and other hydrocarbons are commonly used for solvent extraction purposes. Occasionally, a small or trace quantity of the solvent remains in contaminated product water, which is generally recovered using the distillation method. However, solvent extraction techniques are effective in removing oily contaminants, greases, and organic contaminants. Nevertheless, with specific effluents conditions and extraction parameters, actinide chemicals and other metal ions can also be removed simultaneously recycling high quality reusable water. [46]
13. **Adsorption:** Adsorption of pollutants on various high surface area materials such as activated carbon is often used as one of the tertiary water treatment techniques. In this route, contaminated water is purified using the high surface area carbaneous substrate by retaining organic micro-pollutants, heavy metals, COD, organic complexes, and sometimes relatively low concentrations of emerging pollutants from wastewater. Adsorption in traditional adsorbents takes place through two possible mechanisms, namely, physical adsorption also known as physisorption via van der Waals interactions between pollutant and adsorbent, whereas chemisorption involves surface functional groups to retain pollutants on activated surface. Physisorption is a reversible phenomenon in which physically adsorbed pollutants can be readily recovered to get recycled adsorbent. Here, the intermolecular attractive forces between a pollutant and the substance facilitate the adsorption of the process. In this process, adsorption capacity of substrate depends on various factors such as functional groups and nature of pollutant-

to-be-removed. This weak interaction can be reversed with suitable pH variations or change in the adsorption process parameters. On the other hand, chemisorption is irreversible in nature, this is due to the stronger chemical interactions between the pollutant and adsorbent. Adsorption is a surface mass transfer process in which sorption of solutes or pollutants is adsorbed on the solid substrate from mixture of liquid. Traditionally, carbon-based materials are widely used for absorption of contaminants for both domestic water treatment as well as effluent treatment. This approach helps in improving the quality of water by eliminating objectionable chlorine, removing undesirable tastes and odors. [47]

14. **Crystallization:** In the crystallization process pollutants from wastewater are removed by elevating their concentrations to a point at which they start to crystallize.. During this secondary treatment step, pretreated effluent condition is subjected to either by evaporation or lowering the temperature of the effluent by mixing other solvents. This method is particularly useful in crystallizing higher concentration total dissolved solids (TDS), other inorganic minerals and saturated organic pollutants. Among many inorganic minerals like ammonia, bicarbonate, sulfite, etc., can be crystalized from the effluents. Also, the crystallization process may require adjusting pH, temperature, and concentration parameters. Generally, the crystallization process is popularly used for the treatment of wastewater released by paper industry, dying industries, cooling towers, coal and gas fired boilers. [48]
15. **Precipitation:** Precipitation is a physicochemical technique traditionally used for separating mixture of pollutants dissolved in effluents. The solubility of a dissolved pollutant or toxic compound depends on the ionic strength of the effluent or origin of the wastewater and their physical parameters such as temperature and pH. This is the critical point in which manipulation of physicochemical parameters can be used to revert the solubility of a compound to precipitate out making more insoluble solid subsequently fall out of effluents or wastewater. In wastewater influent, precipitation process transforms a dissolved constituent (pollutant in case of wastewater) into an insoluble solid from a supersaturated solution. The solid is therefore separated as precipitate. Further, the clarified floating liquid can be separated as high-quality reusable water via centrifugal solid phase separation known as 'supernate' or 'supernatant'. Further, chemical precipitation is used to remove ionic pollutants from contaminated water by adding precipitating agents by reducing their solubility. [49]

2.3.4. Advanced water treatment techniques

Ever growing population, increased production and unregulated consumption of personal care pharmaceutical products are expected to cause serious threats to natural resources and the environment at large. On the other hand, advent in new water monitoring and analysis techniques have revealed the existence of new persistent contaminants and emerging pollutants in various water resources. The new revelations list over 3000 chemicals as persistent contaminants of emerging pollutants. This also forces many monitoring and regulatory authorities around the globe to develop and adopt novel water treatment technologies to address the impending water treatment crisis. On the other hand, developments in innovative approaches and leading-edge research have provided significant technological advantages in elevating efficiencies of conventional water treatment technologies. Also, a combination of scientific innovations and information technology related future solutions have enabled the growth of advanced wastewater treatment plants (WWTPs) that could enable handling of large volumes of wastewater to achieve high quality water recovery. These advanced water treatments have potential to achieve direct dischargeable quality treated water to natural water resources or ocean. Currently, most of the advanced approaches are characterized by high cost and limited access, however advantages in nano and materials science related innovation show increased potential for emerging pollutant treatment. On the other hand, use of information technology assisted components help in accurate detection of emerging pollutants. This approach can help in selecting suitable water treatment methods and protocol for specific contaminants. Nevertheless, creation of automated WWTPs with advanced treatment and improved discharge norms with better quality water to rivers or the ocean. This also helps in retrieving and segregating nutrients and organic matter from hazardous pollutants to fertilize and improve soil upon landfill. [50]

In any case, conventional separation technologies require serious attention to improve or replace some of the shortcomings such as the high energy consumption and to improve the life of a treatment. In this section, lists of advanced techniques being developed or adopted in place of traditional water treatment techniques are discussed in detail. However, in many cases stage hybrid technique approaches have been adopted to improve the quality of recovered water from traditional separation methods like primary treatment protocols, coagulation, sedimentation, distillation, and adsorption techniques. This is due to some of the physicochemical methods in combination are effective technologies for treatment of wastewater like domestic effluents.

(i) ***Fenton, Photo-Fenton's Process and Advanced Oxidation Process***: Fenton and photo-Fenton's process is one of the effective advanced treatment processes adopted in recent times. In this method, the combination of hydrogen peroxide (H_2O_2) and UV radiation with a Fe^{3+} or Fe^{2+} ion produces more hydroxyl radicals ($\cdot OH$) which increases the rate of degradation of organic and other emerging pollutants. Among several route, hydroxyl radical ($\cdot OH$) can be produced from the Fenton's reagents between aqueous ferrous ions and hydrogen peroxide (H_2O_2). Hydroxyl radicals produced from Fenton's reagents can destroy refractory and toxic organic as well as emerging pollutants in water and wastewater. However, increased complexities of wastewaters are forcing researchers and technologists to design new combinations of degradation catalysts. Also, the single oxidation reagent mentioned above is sometimes not sufficient and is ineffective for the total decomposition or degradation of conventional as well as emerging organic pollutants in various wastewaters.

On the other hand, the photo-Fenton process is presently considered an advanced oxidation process (AOP) that is extensively applied for the treatment of industrial wastewaters as well as domestic effluent. In this method, the photochemical methodology employs with the help of UV radiation as well as visible light to enhance the rate of reduction of Fe^{3+} to Fe^{2+}. Therefore, advanced oxidation processes (AOPs) involve the simultaneous use of more than one oxidation process, and include the accelerated creation of the extremely reactive hydroxyl (·OH) free radical. Further, recent studies have indicated the use of novel set of reactions to make AOPs more effective that include techniques like Fenton's oxidation, visible light assisted photolysis and sonolysis that can destroy the organic contaminants at ambient temperature and pressure. The main advantage of the AOP is that organic pollutants are commonly oxidized to CO_2 and to settle all degraded organic pollutants to the bottom in the form of sludge. In addition to this, a wide variety of AOP reagents are available like chemical oxidation processes using new Fenton's reagents, ozone, combined ozone, photocatalysts, peroxide, UV-enhanced oxidation such as UV-Fenton or photo-Fenton, UV-hydrogen peroxide, UV-ozone, UV-air wet air oxidation and catalytic wet air oxidation in which air is applied as the oxidant.

On the other hand, several novel photocatalysts have been developed covering both transition metal series and non-noble metal-based catalysts. Among all, photocatalysis is one of a series of AOPs applied successfully for organic and pharmaceutical pollutant degradation. In photocatalysis, light energy from a light source, traditionally UV-range, excites an electron from the valence band of the photocatalyst to the conduction band to initiate a series of reactions thereby causing the formation of hydroxyl radicals. AS produced hydroxyl radicals facilitate the high oxidizing potential attacking all accessible organic pollutants in wastewater through the process of oxidation. More recently, various chalcogenides, oxides like TiO_2, CeO_2, ZrO_2, ZnO, Fe/Al-oxyhydroxides, etc., or sulfides such as ZnS, CdS, etc., have been widely developed to suit the nature of wastewater and applied as photocatalysts in AOPs. Also, sonolysis process which makes use of ultrasonic waves has been adopted for degradation of organic molecules such as textile pollutants like dyes. In sonolysis-assisted AOP, the process of organic molecule degradation proceeds with the creation of short-lived radical species produced in aggressive cavitation events. [51-53]

2.3.5. Importance of advanced technology approaches

Historically, development of water treatment methods that trace back to the mid-nineteenth century that had limited purpose and end use such as sewage cleaning, and irrigation activities. However, industrial, and agricultural revolutions broaden the scope of water usage and regeneration. Further, with the advent of personal care products, petroleum products, nanomaterials and extensive use and discharge of pharmaceutical ingredients enforced renewed interest in filtration technologies. This led to the use of large-scale septic tanks, contact beds, sedimentation units and trickling filters with the help of natural materials and intermittent sand filters. Among other traditional sewage treatments Anoxic Oxic (AO) and Anaerobic-Anoxic-Oxic (A^2O) processes were extensively used for converting sewage into activated sludge. However, conventional treatment methods often resulted in poor effluent quality separation and clarified water due to unstable sludge in the secondary sedimentation tank. Lately, the membrane-based methods with hybrid approaches like membrane bio-reactor (MBR) forward osmosis (FO) have been reintroduced to address complex wastewaters. Therefore, advanced membrane separation processes are increasing sludge-water separation thereby improving the quality of recovered water greatly.

In the last decade researchers explored a series of innovative methods to address emerging challenges in the field of separation science and technologies. In addition to this, innovations in the field of nanotechnology and materials science have shown promising avenues for the water and wastewater treatment field. All these developments brought about the renewed curiosity and interest in high quality reusable water recovery from wastewater including the availability of reliable pressure-driven membrane processes like MF, UF, NFO and RO and recent renewed addition to this is forward osmosis (FO) which was earlier thought to be only suitable for pretreatment applications. Also, the innovation in ozone equipment designs have also enabled the adaptation of the technology as a hybrid system along with other efficient methods. Now, the use of ozone coupled tertiary water treatment has been developed as advanced technique such as biological filtration in combination with low, and medium UV doses. The high energy UV is also adopted for the direct disinfection of water sources. Similarly, the UV with high energy doses is used as a rate enhancing tool in advanced oxidation processes. Therefore, several such approaches can be now used to treat and selectively retain pollutants which cause acute toxicity such as microorganisms, heavy metals, and chronic toxicity causing contaminants like pharmaceutical traces, drug molecules, and personal product chemical constituents. [54]

2.4 Separation Emerging Contaminants

In recent years, increased concern related to extensive use of personal care products, pharmaceuticals, surfactants, pesticides, industrial chemicals, perfluorinated compounds, gasoline additives, water disinfection byproducts, synthetic nanomaterials, and industrial additive consistently being found in groundwater, surface water bodies, and wastewater. Another major concern is the direct discharge or partially treated industrial effluent to surface waterbodies and their potential entry into the food chain and larger ecosystem. To address this impending problem researchers have adopted advanced water treatment techniques. These innovations are practically used widely both at wastewater treatment plants as well as in fields or waste generation sites like pharmaceuticals, cosmetic industry, mining, food processing units and waste segregation industries. Also, most of these wastewaters often include small organic moieties and inorganic salts that may create serious hindrance in treatment. Due to this, wastewater treatment using adsorption, photocatalysis, membrane technology and thermal methods become ineffective. This necessitates simple technologies to be developed or adopted as hybrid technologies. However, more recently with the help of advanced membranes, RO and NF membrane processes have been effectively used for the selective separation of dyes and salts making use of the Donnan effect. Also, the advent of loose nanofiltration membranes have shown promising results in addressing salt contaminated industrial wastewater. [55]

At present there are two main preparation approaches of loose nanofiltration membranes (LNF) : (a) in situ nanomaterials induced polymerization, and (b) two-dimensional (2D) material-based nanofiltration membranes. These nomenclatures have been adopted based on the way of preparation of membranes. Also, based on the nanomaterials or 2D material agglomeration and their morphologies in the resulting membrane governs the wide range of applications of LNF. Currently, it is perceived that preparation of nanofiltration and reverse osmosis membrane via conventional phase inversion technique is relatively inexpensive. However, use of nanomaterials in the membrane matrix is expensive and due to their physicochemical interaction, controlling pore size and their distribution in the resulting membrane is vital. However, presence of nanomaterials or 2D materials in the membrane matrix have proven great advantageous. In order to enhance the membrane flux, it is often prefaced on the permeance and rejection trade-off. However, nanomaterials-induced LNF membranes have overcome the traditional trade-off limitations. [56, 57]

Consequently, it is vital to develop a simple, efficient, simple, and economical membrane targeting both specific emerging pollutants as well as conventional pollutants. Besides, traditional separation methods involving activated carbon adsorption, ozone oxidation, evaporation recovery, ion exchange, and so on will have to be only effective when they are used strategically in hybrid combinations for effective removal of emerging pollutants. Nevertheless, these conventional separation methods still exhibit many disadvantages such as adsorbent shelf-life, high energy consumption, and adsorption capacity against emerging pollutants. In addition to this, these secondary treatment techniques may also hinder the speed of the overall process. Therefore, it is crucial to develop simple hybrid treatment techniques with easy to recover saving energy technologies. Emerging hybrid-secondary and advanced water treatment technologies are expected to adopt both membrane-based processes and conventional secondary treatment methods as efficient candidates for the treatment of emerging pollutants with minimum energy footprint.

Considering the complexities of the emerging pollutants and their limited understanding in terms of their occurrence and form of existence, emerging membrane-based hybrid treatment technologies can resolve these problems. However, a widespread distribution of emerging pollutants and lack of their detection and analytical data limits the effectiveness of a particular pollutant treatment. For example, the membrane-based hybrid technologies may show some limitations because of several factors associated with membrane like membrane fouling pollution, flux resistance, temperature of the effluent which may affect the membrane, corrosive nature of pollutants may cause instability in long term. These factors may limit overall process efficiency. While there is a large enormous amount of research devoted to resolving aforementioned issues of the hybrid treatment techniques. Also, future research should concentrate not only on the flux performance of the membrane, but also must design, develop, and integrate a new hybrid membrane and suitable module from a comprehensive approach involving low-cost and sustainability.

Further, forthcoming sections of this chapter also consider elaborating modern water treatment techniques and water reuse practices. Upcoming sections will also consider subject matters including:

(a) advances in water reclamation,
(b) the significance of modern technology in treating effluent and water recovery,
(c) changing perspectives of water reclamation and reuse,
(d) quality of reclaimed water and their reuse in various applications,
(e) assessment reclaimed water use for domestic applications,
(f) improving quality of recycled water from non-domestic wastewater for reuse,
(g) systematic protocol development for quantifying unexpected and remote use of reclaimed water for potable purposes,
(h) human health, ecological and environmental issues associated with treated water, and
(i) overall review and expansion of water reuse criteria and standards.

2.5 Advances in Water Recovery

In a larger context, efficient wastewater treatment and recovery of large quantities of reusable quality water can bring economic benefits in almost every sector. Modern technologies have changed many things in the field of water reclamation and reuse. Water reclamation has a long history of water reclamation and reuse and contemporary water recycling efforts takes back over a century (1900 to present), first being use of recycled water for agricultural irrigation. However, implementation of modern technologies received increasing popularity in the last three decades. Among many efforts, one of the most relevant changes is the acknowledging importance of effluent treatment and recognizing the importance of reclaimed water in an integrated water resources treatment and management plan. Today most of the industries have implemented water treatment plants and domestic end users like medium to large colonies have also adopted treatment technologies. Further, reclaimed water sources have become a new resource for gardening purposes, agricultural applications in the form of alternative, additional, dependable water supply sources at the doorstep. This is gaining more importance at a larger level to access the recycled water at the doorstep of municipal water treatment plants for numerous applications in various environments. These individual as well community level efforts have even systematically studied, documented, and recognized by the United Nations. These efforts have now been been popularized through UN's World Water Development programmes (World Water Development Efforts Report, 2017). Furthermore, positive stories on water recycle and reuse have expanded the boundaries from irrigation, agricultural and domestic wastewater restricted municipal uses to a variety of applications including drinking purpose. [58-60]

Table 2.1. Typical sources of recycled water and their reuse in different fields or sectors and their limitations.

Source of waste water	Reuse in the field/sector	Limitations
Domestic wastewater	Agriculture, urban gardening	Point of use often far from the site
Sugar industry wastewater	Ferti-irrigation for agriculture	Requires suitable infrastructure
Paper industry wastewater	Crop irrigation, landscape irrigation & construction	Requires suitable infrastructure
Food industry wastewater	Non-food contact and cleaning	Limited demand
Textile industry wastewater	Farming and irrigation	Requires suitable infrastructure
Contaminated groundwater	Groundwater replenishment, seawater intrusion barrier	Requires intermediate reservoir
Agricultural overflows	Crop irrigation and nurseries	Seasonal demand
Low contaminated industrial effluent	Cooling water, boiler feed, electronic manufacturing	Site-specific demand
Municipal wastewater	Lake and pond rejuvenation, snow production	Transportation cost
Planned use of potable water	Drinking water supply	Lack of potential health risk assessment and high cost

The reuse types listed in Table 2.1 are ranked in comparative order of total usage as practiced worldwide. Reuse of treated water has expanded from earlier restricted agricultural and irrigation practices to industrial applications and in some cases potable use. This is also due to the improved effluent treatment technologies in recovered water quality. Also, there has been a worldwide trend to diversify water reuse practices in almost every sector. The reuse of recycled water from various wastewater sources is one of the main choices being considered a new source of water in areas where freshwater is scarce. However, reuse of recycled water can also be carefully assessed and linked with human and environmental health risks. Some of the major concerns related to recycling industrial wastewater is the presence of traces of toxic biological pollutants such as pathogenic bacteria, enteric viruses, and protozoa. On the other hand, some inorganic pollutants from wastewater like nitrogen, and phosphorus, may pose both positive as well as negative effects on vegetation, plant growth, crop yields, and the floral environment. However, higher salinity, suspended solids, heavy metal concentration and some of the emerging pollutants can be harmful for overall agricultural soils and agricultural infrastructure. Nevertheless, to reduce the danger and impact associated with the use of recycled water from various industrial wastewater on the overall environment, many countries, local governments, and regulatory authorities have adopted rigorous monitoring standards and guidelines that can help in assessing the agriculture sector. In addition to this, several monitoring and regulatory agencies are also planning, designing, and implementing innovative technologies for water reclamation and reuse. These strategies also consider the intended water reuse sector or specific application which dictate the level of contamination, standard of wastewater treatment, and distribution network required. [61]

Interestingly, regulations formulated for wastewater recycle and reuse in agriculture emphasis mainly on hygienic and environmental protection. These regulatory guidelines typically refer to setting reliable detection and analysis of wastewater, selection of appropriate wastewater treatment technology, sustaining reclaimed water quality, establishing recycled water distribution infrastructure, and identifying suitable crop or type where recycled water is used. In all cases, the fundamental requirements are based mainly on the extent of treatment required in context to the permissible contamination level of recycled water in concurrence with quantitative limits on turbidity, suspended solids, and bacteriological quality. Nevertheless, at present all the set standards and required norms for the safe use of recycled water and their type and amount are contentious. Therefore, the wastewater treatment and reuse of reclaimed water economy considering health and environmental safety measures need to be tailored to fit the balance between end user target, affordability, and risk. All these parameters are also assessed with the help and collaboration of regulatory authorities and various end users. Major factors associated with the water recycling and reuse will mainly depend on the following factors. Some of the other parameters include:

(1) assessing realistic environmental factors in relation to local conditions,
(2) affordability to larger segments of manufacturer or wastewater producers, and
(3) feasibility and enforceability. In all cases, economic constraints limit the level of wastewater treatment that can be provided for reuse by the community regularly.

Also, the quality of recycled water from industrial wastewater is required as standard disease-control norms to meet strict microbiological guidelines. Therefore, increased use of reclaimed water in the field of agricultural irrigation could possibly produce the greatest reimbursement to global water resources. In fact, recycled water from industrial or domestic wastewater sources can be widely used to complement agricultural irrigation purposes in largely every sector, particularly reclaimed water use in arid areas could be of great potential. With all initial success stories and studies, the WHO has framed guidelines for reuse of reclaimed agriculture in the irrigation sector which intends to serve as a standard for worldwide regulatory and monitoring institutions. These guidelines are particularly intended to guide stakeholders to be consulted when developing methods and protocols for agricultural reuse. These general approaches and protocols can include:

(a) replacing reclaimed water for applications that do not require drinking water,

(b) supplementing prevailing water sources and providing an additional source of water supply to assist in meeting both current and forthcoming water needs,

(c) conserving aquatic ecosystems by reducing the deviation of freshwater and reducing the extent of nutrients and other toxic pollutants entering waterways,

(d) stalling and dropping the need for water control structures, and

(e) complying with environmental regulations by better managing water consumption and wastewater discharges.

Typically, traditional wastewater treatment procedures and systems include multiple levels of physical, chemical, and biological treatment in individual or hybrid combinations in order to ensure the discharge of treated water. This ensures environmental protection and does not pose a significant risk to the ecosystem or human health. Treated wastewater at different competency is usually discharged to natural waterbodies and such recycled water sources are frequently used for downstream utility. Hence, many conventional water recovery systems and reclaimed water resources reuse wastewater inadvertently. Nevertheless, such unintended reuse and unregulated reuse are often rarely acknowledged due to their irregular practices. Thus, there is still some level of disapproval for direct reuse of reclaimed water from wastewater sources may be due to public resistance or hesitancy associated with health risks.

Therefore, it is important to devise advanced methods, protocols and technology involving integrated approaches to drastically impact on both the quantity and quality of the global water resources. Nevertheless, water reclamation from waste sources and reuse methods still need effective accommodation of water resource management courses in the face of stressors. Most importantly, sustainable, and advanced water reclamation and reuse methods should also contribute greatly to climate change efforts by allowing recycled water resources to be diversified and conserved as well. However, hindrances still exist for use of reclaimed water for potable applications. Therefore, advanced techniques would focus on using reclaimed water to permeable land surfaces or directly injected into the ground for the purpose of reviving groundwater aquifers and averting saline intrusion in coastal areas. Meanwhile, advanced treatment techniques are being integrated to recover value added products. In these approaches, alongside water recovery, the nutrients can be recovered from wastewater sources and are too treasured to waste in areas where resources are limited. These practices not only increase the bulk of water accessible for crop irrigation but alsoutilize the nutrients recovered from wastewater thereby contributing to better quality of human life by enhancing collective domestic water availability.

2.6 Advances in Waste Management, Disposal, and Economical Aspects

The direct impact of wastewater generation and discharge occurs on the environment contaminating natural habitats, wildlife and larger ecosystem exposing them to detrimental chemicals, substances, compounds, and complex systems that would otherwise may not be present naturally. Advanced wastewater treatment and reusable water techniques offer a multiple-value proposition. In addition to health and environmental benefits, advanced wastewater treatment techniques and approaches can fetch economic benefits through value added product and/or byproduct recovery and reuse of reclaimed water in different sectors. Generally, value added products or byproducts like nutrients, rare earth metals, minerals and biogas can be used for agricultural applications and energy generation. These approaches can also yield additional revenues by the sale and adaptation in other practices which can help to cover water management, operational and maintenance costs. Therefore, now onwards wastewaters will no longer be called waste and instead would be renamed as '*new resources*.' This is the core of a circular and sustainable economy, an economic structure or a system that aims at minimizing waste and making reuse of resources via recycle and reuse approaches.

As populations continue to grow and most of the population now migrating to cities and new booming economical hubs, the future urban and municipal development necessitate approaches that incorporate sustainability aspects by minimizing resource consumption and resource recovery following principles of circular economy. However, the ultimate target for wastewater treatment and recycling come from the economically viable applications like reducing stress on freshwater sources by replacing fresh water with reclaimed water in agricultural applications, municipal use, cleaning, and environmental restoration, toilet flushing, gardening, and industrial uses. Therefore, efficient advanced water treatment techniques for high quality reclaimed water can directly contribute to conservation of freshwater resources, pollution reduction and environmental protection.

On the other hand, a key advantage of wide acceptance of circular and sustainable economy principles in wastewater management is that resource recovery and reuse could change sanitation from an expensive service to self-sustaining low cost through value added products. These broader and long-term projections will help every country to bridge the resources and funding gap to achieve the United Nations Sustainable Development Goals (UNSDGs). As of early 2020s, resources to implement UN SDG agendas comprehensively by all member states is still a challenging task. In this context, the requirements of monetary and technical resources to implement water reclamation and reuse agendas will significantly depend on the type of value-added product recovery and application that is premeditated for the reclaimed water. Therefore, water resource monitoring, management, and distribution executives must fully comprehend the economics associated with developing and managing the specific water stock, management system, and projected water reuse system. All these efforts should consider the cost benefits for implementing water reclamation and reuse plans along with maintaining conventional water and wastewater managing approaches. A commercial analysis should be estimated in order to assess the cost of sustaining conventional approaches to develop supplementary water sources versus the cost of retrofitting current and building new infrastructure for reuse applications. Such evaluations should also reflect the number of economic paybacks associated with water reclamation and reuse methods such as reduced treatment costs and the recovery of valuable nutrients and other value-added products from wastewater.

On the other hand, creating a program implementing structure considers several key capacity building foundations that are essential to safeguard the quality of policy making decisions and performance monitoring. Generally, the United Nations Environmental Program (UNEP) issues broader guidelines on the planning and implementation of water reclamation and reuse programs. However, member countries may amend and improve the guidelines to suit the local conditions and larger situations. Also, advanced water treatment and reclamation program implementation requires well trained human resource and technical personnel for sustainable execution and operation. This would strengthen local water supply and wastewater treatment operations both technically and executive ability which is limited in current practices. In addition, these futuristic approaches also require utilizing potential benefits out of the reuse market of reclaimed water as well as the nurturing of their capability to implement new programs. All these efforts should always be supported by policy and regulatory framework. These policies and legal frameworks facilitate safe and suitable reclamation and reuse plans to either be formed or associated with larger UN frameworks in order to ensure the safety of human health and the environment.

2.7 Advances in Reclaimed Water Management and Reuse

Conventional waste treatment and disposal process involves steps of waste identification, analysis, collection, processing, and recycling or disposal of segregated waste materials following prescribed protocols. In a later stage, waste is categorized by their source, composition, and level of toxicity. Generally, segregated waste after treatment ends up in either concentrated liquid or solid sludge form (sewage, hazardous solid/liquid waste, plastic, and electronic waste) or their main constituents classified as hazardous or inert. Waste management in industrialized countries and developing economies are traditionally classified municipal wastewater as major source of environmental concerns and is funneled through standard sewage systems, where collected municipal wastewater is subjected to systematic treatment steps. Standardized wastewater treatment steps largely removes largely all or majority of impurities from wastewater before they are discharged to groundwater aquifers or fresh surface waters source like ponds, lakes, rivers, estuaries, and oceans.

However, increased threat of emerging pollutants and lack of their identification techniques and suitable treatments forcing almost every water management and monitoring agencies to look for advanced treatment and management techniques. As fully industrialized and developing nations are grappling with the problem of emerging pollutants, search for prompt, safe and sustainable waste disposal is increasing. On the other hand, non-biodegradable, complex and toxic wastes like radioactive remnants can possibly cause irreversible damage to the humans and environment, if not tactically disposed of. Though conventional waste disposal or clearance has always been a matter of alarming for several decades. Now, to make the situation worse, emerging pollutants are entering almost every freshwater resource untracked. This is going to increase to massive proportions due to ever growing population and industrialization. These two major factors need to be addressed while designing advanced water treatment, collection, and disposal strategies.

Nonetheless, several attempts are being made to adopt advanced waste disposal technologies, though these are still inadequate. However, emerging pollutants, semi-volatile organics and microplastics type newer pollutants pose a serious challenge in detection, quantification, treatment, and disposal. In this scenario, to create effective and efficient advanced waste segregation and disposal aiming at newer pollutants, a specific type of technological assessment involving financial implications need to be analyzed — known as life cycle cost (LCC) — to broadly evaluate circumstances under which waste treatment, water reclamation and reuse programs would be beneficial. This approach considers the economic considerations of a water reclamation and reuse program covering a complete life cycle, including technology design, instrumentation and equipment production, field installation, operations, maintenance, restoration, and disposal. There have been several attempts to apply the LCC approach in cities like Tokyo, where recycled water from wastewater sources was used instead of conventional freshwater in several office buildings. The LCC analysis found that if the reclaimed water volume from wastewater is more than 100 m^3 per day, the wastewater reuse option is more cost effective when compared to the freshwater. Such case studies and LCC analyzes data in large samples can be helpful in defining and estimating the economic feasibility of water reclamation and reuse programs.

Due to increased awareness and strict regulation, the future would see the institutionalized efforts that are involved in water supply, management, environmental and health protection likely to be involved in water reclamation and reuse projects. As the wastewater complexities increase, many advanced attempts are being made to coordinate among many institutions at the local, regional, and national levels. In this effort, it may be necessary to restructure and reorganize organizational duties into a general group that coordinates reclamation and reuse projects. To sustain long operational and economic benefits, it would be best to avoid conflict of interest for either the water supplier or the wastewater administrator to function as a monitoring and management agency that has enforcement accountability over all the stakeholders and partners involved in water reuse. Hence, it may also be obligatory to task an independent agency as an in-charge entity to protect public health or the environment. Recently several developing countries like Egypt, Tunisia, Morocco, and India have implemented such institutional arrangements to successfully adopt advanced infrastructure supported by policy framework for water reclamation and reuse programs.

However, it is important to note that all advanced water reclamation and reuse approaches employ the all-basic conventional treatment technologies including secondary advanced clarifiers, filtration modules with innovative designs, advanced membrane-processes, and post-disinfection steps. *Nonetheless, all possible perspectives discussed above discussed each and every advanced water reclamation treatment method, techniques, scheme systems and protocol will involve some degree of customization.* This requires a great amount of strategic plan and implementation to define appropriate treatment and reuse applications for water reclaimed by primary methods like sedimentation, secondary process like biological oxidation and disinfection, and tertiary advanced treatment techniques like chemical coagulation, membrane filtration, advanced oxidation, nanocatalysts-based Fenton-type treatment and disinfection processes. [62] On the other hand, regardless of method or technique used for wastewater treatment, recycled water is rarely recommended for direct potable reuse. These recommendations mainly consider two main technical reasons, namely:

(1) even with adaptation of all likely advanced technology to remove possibly all known pollutants during treatment steps from wastewater, there may be unknown and emerging contaminants may be present, and
(2) as it is evident from several research articles and reports that the identification, detection, quantification and effective removal techniques for emerging and unknown pollutants still to be realized and at present these mysterious pollutants may cause major risks to human health.

These two important factors still pose a serious potential threat to human health thereby restricting direct reuse of reclaimed water for potable applications. Even though globally, there is larger consensus on restricting direct use of recycled water for drinking water applications; however, continued efforts at Windhoek plant in Namibia, Singapore (NEWater) and Orange County, California (Water Factory 21) is the only case study that is showing affordable feasibility to directly use improved quality reclaimed water for drinking purposes on a long-term basis. [63] Conventionally, it has even been an unusual and rarely considered use of recycled water for rejuvenating drinking water reservoirs. Nevertheless, recharging of stressed freshwater reservoirs practice, generally known as indirect potable reuse is gaining increased popularity over the last decade and being effectively employed through several case studies around the world. In all cases, reclaimed water undergoes additional advanced treatment steps beyond traditional tertiary treatment steps such as advanced membrane processes like reverse osmosis, dissolved air flotation and UV irradiation. Nonetheless, reclaimed water largely finds more applications in non-potable reuse than as potable reuse in view of public health risks. [64, 65]

Recent times have seen greater coordination among water resource managers and innovation in treatment materials and processes. This vast information is accessible for water resource managers, regulators, and innovators to formulate further futuristic guidelines on water reuse, public health risks assessment, provide proper feedback for water reuse segment and technology developers. It is imperative to mention that most of the recycled water from wastewater treatment plants in developing countries are largely used for agricultural irrigation without adequate post-treatment. This approach may often result in high risk of contracting intestinal disease when harvested crops or vegetables cultivated using recycled water are consumed raw or undercooked. However, with appropriate post-treatment steps executed appropriately, water recycling and reuse can contribute to social and economic development by reducing environmental pollution while simultaneously recovering value added products.

2.8 Advances in Value Addition during and after Water Treatment

Value addition concept is increasingly gaining importance in the chain of green chemistry practices due to the important features involving prevention and reduction of waste generation. Extensive use of new substances, chemical and redundant products in everyday life, especially the domestic segment is characterized by large scale emerging pollutant contamination which is ending up as unchecked and unmonitored. On the other hand, the rapid population growth and urbanization makes it imperative to use recycled or second-hand products or sensibly use the existing ones. Its uncontrolled consumption or overexploitation of existing resources continues, there is a potential risk of people succumbing to the ill effects of toxic wastes.

Traditionally, so called lost resources in wastewater are disposed of consciously to minimize the environmental risk associated with their existence. Traditionally, four main methods are used to dispose of liquid or solid waste with minimum collateral damage to environment, namely:

(1) **Landfills:** Solid or highly concentrated liquid wastes are disposed of in landfills located at safe, designated, and isolated places.

(2) **Recycling:** In this process waste is analyzed for their constituents and useful recyclable products are extracted with minimum energy inputs. Recycling also refers to both the direct reuse of used products and quantifiable material recycling, that is, the recovery of raw materials from waste.

(3) **Composting:** Composting is a simple and natural biodegradation process that takes organic solid or highly concentrated wastes turn into nutrient-rich food and used as fertilizer in agriculture applications.

(4) Incineration: In this method combustible waste from waste treatment plants, households and industries that is not appropriate for recovering is subjected to thermal treatment using appropriate furnaces. In many cases, the heat generated during incineration of waste is used to generate electricity or heating buildings. However, highly concentrated waste having high calorific value and low contamination level can be directly used in industrial furnaces or boilers such as cement manufacturing units as an alternative to fossil fuels. On the other hand, highly concentrated liquid or solid waste that is characterized with organic pollutants is subjected to separate thermal treatment such as an incinerator with special design to capture toxic fumes (flue gas venting system).

However, recent innovations have facilitated specialized waste disposal companies, organizations, socio-environmental groups, and private individuals to treat the waste and extract value-added products in accordance with the requirements of international standards. These assurances that the waste treatment will be of a high quality and reduce the chance of risk. These institutions, companies and monitoring groups ensure that no undesirable alteration occurs when water is extracted. The organizations that are most likely to be involved in water recycle, recovery and reuse projects are those in charge of water supply, waste management, water resources supply, management, ecological conservation, and public health. This is because of the inherent complexity involved in the initiative that attempts to synchronize all stakeholder involving local, regional, and national levels thereby necessitating restructuring of administrative duties into a wide ranging group to coordinate in water management projects. Potentially, all solid waste disposal techniques can be effectively used for the recovery of value-added products. As discussed, wastewater is now considered a new resource and with appropriate treatment techniques can bring economic benefits through reuse in different sectors. Conventionally, several products and byproducts have been extracted from the concentrated mass. The byproducts, such as nutrients, heavy metals-free sludge, precipitated mass as carbon precursor and have been used for agriculture and energy generation, conversion, and storage devices.

2.9 Advances in Waste Segregation

Conventionally, there have been several socio-political barriers that repeatedly limited successful employment of water recycling, reclamation, and reuse programs. In many cases, public disapproval to the use of recycled water for any potential and possible application to which humans might be exposed can hamper progress. On the other hand, conventional and obsolete technologies have hindered successful employment of water recycling, reclamation, and reuse plans as well. This in the past had led to exorbitantly high costs that effectively limited the operation of such projects and programs. However, depleting natural resources and need for extracting high value solute materials in the wastewater necessitate adaptation of advanced waste indemnification, analysis, and segregation. To successfully achieve these targets, several important forecasts have been considered and adopted in the process. These involve factors like achieving:

(a) **Better Environment:** To achieve a clean and green environment to impart healthy resources utilization and reclamation in the interest of larger public and environmental demands for which waste segregation has a significant influence on the well-being and welfare of the environment and the society. This is initiated by placing multiple waste collection and clearance sites in cities to help the cleaner environment.

(b) **Reduce pollution:** Waste segregation does not merely condense waste from the environment but also eradicates the impact of harmful greenhouse gasses like carbon monoxide, carbon dioxide, and methane. Potentially, the effective waste segregation method or protocol directly decreases the dependence on conventional waste disposal methods like landfills that adversely affect the local ecology and environment at large.

(c) **Energy conservation:** Waste segregation is a significant step in the process of waste management. Appropriate waste identification and segregation at right place and form aids in conserving energy and plummeting carbon footprints.

(d) **Improve employment opportunity in the water treatment sector:** If a large section of industry, municipal corporation, and community adopt rigorous waste segregation practices, it will generate and initiate for an organized structure or institution to sell recycled and value-added products. These opportunities will lead to numerous employment opportunities in the waste management sector.

(e) **Create change in the mindset and approach among general public:** Although increased population and rapid growth of urbanization may not entirely get rid of waste, however, increased awareness can help in reducing the waste by segregation, recycling, and value addition through ecofriendly practices. This overall approach generates a change in the mindset of the general public significantly and imparts characteristic behaviors, adopt nature-friendly practices, habits and embrace a more sustainable and greener approach.

In addition to detailed waste segregation, simultaneous water recovery and proposing useful reuse strategies is paramount to achieve sustainability to the overall process. The last section of this chapter summarizes the potentials and possibilities of wastewater recycling and reuse for a beneficial purpose.

2.10 Advances in Process Water Recovery and Reuse

One of the main advantages of recycling reusable quality water is to preserve and protect freshwater resources by reducing disposal of polluted water into the environment. More than economic considerations, it is important to acknowledge the conserving natural resources post their utilization which helps in maintaining balance in natural habits. Every effort that is targeted to recycle and reuse should consider the selection of suitable modern technologies and their practices. These considerations are:

(a) the type and mode of modern technology,
(b) socio-economic awareness on water reclamation and reuse,
(c) defining specific water reuse application and sector,
(d) regular analysis of water quality and its monitoring,
(e) exploring water reuse sectors other than non-domestic sources,
(f) understanding possibilities of reuse in unplanned and emerging sectors,
(g) regular monitoring and regulation of health and environmental impact, and
(h) investing in research and technological understanding of emerging trends in water reuse and future challenges.

All these efforts are expected to shed light on promoting reclaimed water utility which can tremendously improve the additional revenue by selling treated wastewater thereby creating the parallel income that can cover operations and maintenance costs.

2.11 Conclusion

Increased awareness among policy makers, water resources managers, researchers' technologists and water end users about green and sustainable world has led to the adaptation of several water recycling technologies. This chapter has given detailed account on the wastewater assessment measure and the efforts made in various sectors to reduce, reclaim and recycle the water from waste sources. This chapter also has given an overview of policy efforts associated with water management and water recycling efforts being undertaken. Further this chapter has listed technologies developed and in practice for effective water recycling and treatment. It also enlists the advanced methods and modern efforts being developed for effective water recycling and identifying the new sectors for reuse of recycled water.

References

1. https://www.unwater.org/water-facts/scarcity/
2. M. Mekonnen Https and Arjen Y. Hoekstra, Four billion people facing severe water scarcity, Sci. Adv. 2016; 2 : e1500323.
3. Fiona Burlig via Barron's, Water Shortages Threaten the Economy and Our Food Supply, INSIGHT / OPINION & ANALYSES•OCT 04, 2021, https://epic.uchicago.edu/insights/water-shortages-threaten-the-economy-and-our-food-supply/
4. Dolan, F., Lamontagne, J., Link, R. et al. Evaluating the economic impact of water scarcity in a changing world. Nat Commun 12, 1915 (2021). https://doi.org/10.1038/s41467-021-22194-0.
5. David L Hoover, Brandon Bestelmeyer, Nancy B Grimm, Travis E Huxman, Sasha C Reed, Osvaldo Sala, Timothy R Seastedt, Hailey Wilmer, Scott Ferrenberg, Traversing the Wasteland: A Framework for Assessing Ecological Threats to Drylands, BioScience, Volume 70, Issue 1, January 2020, Pages 35–47.

6. Gouri Sankar Bhunia, Uday Chatterjee, Pravat Kumar Shit, Anil Kashyap, Chapter 5 - Wasteland reclamation and geospatial solution: existing scenario and future strategy, Editor(s): Gouri Sankar Bhunia, Uday Chatterjee, Anil Kashyap, Pravat Kumar Shit, Modern Cartography Series, Academic Press, Volume 10, 2021, Pages 87-113, ISSN 1363-0814, ISBN 9780128238950, https://doi.org/10.1016/B978-0-12-823895-0.00006-3.
7. Microbiology of Drinking Water Production and Distribution, First Edition. Gabriel Bitton. © 2014 John Wiley & Sons, Inc. Published 2014 by John Wiley & Sons, Inc.
8. Pooi, C.K., Ng, H.Y. Review of low-cost point-of-use water treatment systems for developing communities. npj Clean Water 1, 11 (2018). https://doi.org/10.1038/s41545-018-0011-0.
9. Alejandra de la Luz-Pedro, Efraín F. Martínez Prior, M. H. López-Araiza, S. Jaime-Ferrer, A. Estrada-Monje, and Jennifer A. Bañuelos, Pollutant Removal from Wastewater at Different Stages of the Tanning Process by Electrocoagulation, Journal of Chemistry, Volume 2019 | Article ID 8162931 .
10. Tortajada, C. Contributions of recycled wastewater to clean water and sanitation Sustainable Development Goals. npj Clean Water 3, 22 (2020). https://doi.org/10.1038/s41545-020-0069-3.
11. Michelle, Allison. "Why Is it Important to Recycle Water?" sciencing.com, https://sciencing.com/important-recycle-water-6454840.html. 12 August 2022.
12. M.T. Amin, A. A. Alazba, and U. Manzoor, A Review of Removal of Pollutants from Water/Wastewater Using Different Types of Nanomaterials, Nanotechnology for Energy and Environment Volume 2014 | Article ID 825910.
13. Malcolm J. Brandt, K. Michael Johnson, Andrew J. Elphinston, Don D. Ratnayaka, Chapter 7 - Chemistry, Microbiology and Biology of Water, Editor(s): Malcolm J. Brandt, K. Michael Johnson, Andrew J. Elphinston, Don D. Ratnayaka, Twort's Water Supply (Seventh Edition), Butterworth-Heinemann, 2017, Pages 235-321, ISBN 9780081000250. https://doi.org/10.1016/B978-0-08-100025-0.00007-7.
14. R.S. RAMALHO, Introduction to Wastewater Treatment Processes, Second Edition • 1983, ISBN: 978-0-08-092533-2, Academic Press, Elsevier Inc.
15. Y.J. Shao, Anmin Liu, Frank Wada, John Crosse, David Jenkins, Advanced primary treatment: An alternative to biological secondary treatment. The city of Los Angeles hyperion treatment plant experience, Water Science and Technology, Volume 34, Issues 3–4, 1996, Pages 223-233, ISSN 0273-1223.
16. https://www.lakeside-equipment.com/how-screening-removes-waste-during-water-treatment/
17. Yoshiaki Kiso, Yong-Jun Jung, Takashi Ichinari, Minsoo Park, Takane Kitao, Kazuyuki Nishimura, Kyung-Seok Min, Wastewater treatment

performance of a filtration bio-reactor equipped with a mesh as a filter material, Water Research, Volume 34, Issue 17, 2000, Pages 4143-4150, ISSN 0043-1354.

18. Emerging Technologies for Wastewater Treatment and In-Plant Wet Weather Management, Office of Wastewater Management U.S. Environmental Protection Agency Washington, D.C. EPA 832-R-12-011, Tetra Tech, Inc. Fairfax, Virginia. March 2013.
19. Cheremisinoff, P.N. (1995). Handbook of Water and Wastewater Treatment Technology (1st ed.). Routledge. https://doi.org/10.1201/9780203752494
20. Rasheed Ahmad, 1. Municipal Water Supply, Gravity Separation/ Sedimentation, Water Encyclopedia, First published: 15 July 2005 https://doi.org/10.1002/047147844X.mw181.
21. Hongmei Cui, Xing Huang, Zhongchen Yu, Ping Chen and Xiaoling Cao, Application progress of enhanced coagulation in water treatment, RSC Adv., 2020, 10, 20231-20244.
22. Chee Yang Teh, Pretty Mori Budiman, Katrina Pui Yee Shak, and Ta Yeong Wu, Recent Advancement of Coagulation–Flocculation and Its Application in Wastewater Treatment, Industrial & Engineering Chemistry Research 2016, 55(16), 4363-4389. DOI: 10.1021/acs iecr.5b04703.
23. Orescanin V, Kollar R, Halkijevic I, Kuspilic M, Flegar V. Neutralization/ purification of the wastewaters from printed circuit boards production using waste byproducts. J Environ Sci Health A Tox Hazard Subst Environ Eng. 2014; 49(5): 540-4. doi: 10.1080/10934529.2014.859034. PMID: 24410684.
24. Englande AJ Jr, Krenkel P, Shamas J. Wastewater Treatment & Water Reclamation. Reference Module in Earth Systems and Environmental Sciences. 2015: B978-0-12-409548-9.09508-7. doi: 10.1016/B978-0-12-409548-9.09508-7. Epub 2015 Jul 22. PMCID: PMC7158167.
25. George Z. Kyzas and Kostas A. Matis, Flotation in Water and Wastewater Treatment, Processes 2018, 6(8), 116.
26. J Rubio, M.L Souza, R.W Smith, Overview of flotation as a wastewater treatment technique, Minerals Engineering, Volume 15, Issue 3, 2002, Pages 139-155, ISSN 0892-6875.
27. Zouboulis, A.I., Matis, K.A., Stalidis, G.A. (1992). Flotation Techniques in Waste Water Treatment. In: Mavros, P., Matis, K.A. (eds) Innovations in Flotation Technology. NATO ASI Series, Vol. 208. Springer, Dordrecht. https://doi.org/10.1007/978-94-011-2658-8_25.
28. 13 Advanced Waste Water Treatment, Editor(s): Mircea Negulescu, Developments in Water Science, Elsevier, Volume 23, 1985, Pages 472-509, ISSN 0167-5648, ISBN 9780444995612.
29. Maulin P. Shah and Susana Rodriguez-Couto, Microbial Wastewater Treatment, 2019, ISBN: 978-0-12-816809-7, Elsevier.
30. Stahl N, Tenenbaum A, Galil NI. Advanced treatment by anaerobic process followed by aerobic membrane bioreactor for effluent reuse

in paper mill industry. Water Sci Technol. 2004; 50(3): 245-52. PMID: 15461419.

31. G. Lettinga, J. Field, J. van Lier, G. Zeeman, L.W. Huishoff Pol, Advanced anaerobic wastewater treatment in the near future, Water Science and Technology, Volume 35, Issue 10, 1997, Pages 5-12, ISSN 0273-1223.
32. Mark L. Brusseau, Ian L. Pepper and Charles P. Gerba, Environmental and Pollution Science, Third Edition, 2019, ISBN 978-0-12-814719-1, 2019.
33. Grazia Leonzio, Optimization of column distillation in a wastewater treatment plant, Journal of Environmental Chemical Engineering, Volume 5, Issue 6, 2017, Pages 5732-5745, ISSN 2213-3437.
34. Koczka, K., Mizsey, P. "New area for distillation: wastewater treatment", Periodica Polytechnica Chemical Engineering, 54(1), pp. 41–45, 2010. https://doi.org/10.3311/pp.ch.2010-1.06.
35. C. Calmon, Recent developments in water treatment by ion exchange, Reactive Polymers, Ion Exchangers, Sorbents, Volume 4, Issue 2, 1986, Pages 131-146, ISSN 0167-6989.
36. Vitor L. Martins, Mark D. Ogden, Mark R. Jones, Sam A. Trowsdale, Peter J. Hall, Henriette S. Jensen, Opportunities for coupled electrochemical and ion-exchange technologies to remove recalcitrant micropollutants in water, Separation and Purification Technology, Volume 239, 2020, 116522, ISSN 1383-5866.
37. Deng, Y., Zhao, R. Advanced Oxidation Processes (AOPs) in Wastewater Treatment. Curr Pollution Rep 1, 167–176 (2015). https://doi.org/10.1007/s40726-015-0015-z
38. Urs von Gunten, Oxidation Processes in Water Treatment: Are We on Track?, Environmental Science & Technology 2018 52 (9), 5062-5075. DOI: 10.1021/acs.est.8b00586.
39. Yang Su, Chapter 1:Current State-of-the-art Membrane Based Filtration and Separation Technologies , in Graphene-based Membranes for Mass Transport Applications, 2018, pp. 1-13.
40. Rathore AS, Shirke A. Recent developments in membrane-based separations in biotechnology processes: review. Prep Biochem Biotechnol. 2011; 41(4): 398-421. doi: 10.1080/10826068.2011.613976. PMID: 21967339.
41. Shaheen Fatima Anis, Raed Hashaikeh, Nidal Hilal, Microfiltration membrane processes: A review of research trends over the past decade, Journal of Water Process Engineering, Volume 32, 2019, 100941, ISSN 2214-7144.
42. Saif Al Aani, Tameem N. Mustafa, Nidal Hilal, Ultrafiltration membranes for wastewater and water process engineering: A comprehensive statistical review over the past decade, Journal of Water Process Engineering, Volume 35, 2020, 101241, ISSN 2214-7144.
43. A.W. Mohammad, Y.H. Teow, W.L. Ang, Y.T. Chung, D.L. Oatley-Radcliffe, N. Hilal, Nanofiltration membranes review: Recent advances

and future prospects, Desalination, Volume 356, 2015, Pages 226-254, ISSN 0011-9164.

44. Yang, Z.; Zhou, Y.; Feng, Z.; Rui, X.; Zhang, T.; Zhang, Z. A Review on Reverse Osmosis and Nanofiltration Membranes for Water Purification. Polymers 2019, 11, 1252. https://doi.org/10.3390/polym11081252.
45. Sajjad Al-Amshawee, Mohd Yusri Bin Mohd Yunus, Abdul Aziz Mohd Azoddein, David Geraint Hassell, Ihsan Habib Dakhil, Hassimi Abu Hasan, Electrodialysis desalination for water and wastewater: A review, Chemical Engineering Journal, Volume 380, 2020, 122231, ISSN 1385-8947.
46. Kiezyk, P.R. and Mackay, D. (1971), Waste water treatment by solvent extraction. Can. J. Chem. Eng., 49: 747-752. https://doi.org/10.1002/cjce.5450490607.
47. Ahmer Ali Siyal, Muhammad Rashid Shamsuddin, Aaron Low, Nurul Ekmi Rabat, A review on recent developments in the adsorption of surfactants from wastewater, Journal of Environmental Management, Volume 254, 2020, 109797, ISSN 0301-4797.
48. Haijiao Lu, Jingkang Wang, Ting Wang, Na Wang, Ying Bao, Hongxun Hao, Crystallization techniques in wastewater treatment: An overview of applications, Chemosphere, Volume 173, 2017, Pages 474-484, ISSN 0045-6535.
49. Wang, L.K., Vaccari, D.A., Li, Y., Shammas, N.K. (2005). Chemical Precipitation. In: Wang, L.K., Hung, YT., Shammas, N.K. (eds) Physicochemical Treatment Processes. Handbook of Environmental Engineering, vol 3. Humana Press. https://doi.org/10.1385/1-59259-820-x:141.
50. Sasan Zahmatkesh, Kassian T.T. Amesho, Mika Sillanpää, A critical review on diverse technologies for advanced wastewater treatment during SARS-CoV-2 pandemic: What do we know?, Journal of Hazardous Materials Advances, Volume 7, 2022, 100121, ISSN 2772-4166.
51. Rakshit Ameta, Anil K. Chohadia, Abhilasha Jain, Pinki B. Punjabi, Chapter 3 - Fenton and Photo-Fenton Processes, Editor(s): Suresh C. Ameta, Rakshit Ameta, Advanced Oxidation Processes for Waste Water Treatment, Academic Press, 2018, Pages 49-87, ISBN 9780128104996.
52. Marialy N. Sanabria, Milene M. Hornink, Valquíria G. Correia, and Leandro H. Andrade, Nontraditional Application of the Photo-Fenton Process: A Novel Strategy for Molecular Construction Using Formamide and Flow Chemistry, Organic Process Research & Development 2020 24 (10), 2288-2297. DOI: 10.1021/acs.oprd.0c00057.
53. Park JH, Cho IH, Chang SW. Comparison of fenton and photo-fenton processes for livestock wastewater treatment. J Environ Sci Health B. 2006; 41(2): 109-20. doi: 10.1080/03601230500364740. PMID: 16393899
54. João Peres Ribeiro, Maria Isabel Nunes, Recent trends and developments in Fenton processes for industrial wastewater treatment – A critical review, Environmental Research, Volume 197, 2021, 110957, ISSN 0013-9351.

55. S.K. Nataraj, Emerging Pollutant Treatment in Wastewater, ISBN 9781032103242, Published March 31, 2022 by CRC Press, 326.
56. Xiaoquan Feng, Donglai Peng, Junyong Zhu, Yong Wang, Yatao Zhang, Recent advances of loose nanofiltration membranes for dye/salt separation, Separation and Purification Technology, Volume 285, 2022, 120228, ISSN 1383-5866.
57. Bin Liu, Zhihao Xue, Zihao Wu, Hanxuan Zeng, Changrong Zhao, Lin Deng, and Zhou Shi, Layer-by-Layer-Assembled Loose Nanofiltration Membrane for Persulfate Activity Enhancement: Performance and Process Regulation, ACS ES&T Water Article ASAP. DOI: 10.1021/acsestwater.2c00267.
58. Devda V, Chaudhary K, Varjani S, Pathak B, Patel AK, Singhania RR, Taherzadeh MJ, Ngo HH, Wong JWC, Guo W, Chaturvedi P. Recovery of resources from industrial wastewater employing electrochemical technologies: status, advancements and perspectives. Bioengineered. 2021 Dec; 12(1): 4697-4718. doi: 10.1080/21655979.2021.1946631. PMID: 34334104; PMCID: PMC8806852.
59. Sathya, K., Nagarajan, K., Carlin Geor Malar, G. et al. A comprehensive review on comparison among effluent treatment methods and modern methods of treatment of industrial wastewater effluent from different sources. Appl Water Sci 12, 70 (2022). https://doi.org/10.1007/s13201-022-01594-7.
60. Philipp Kehrein, Mark van Loosdrecht, Patricia Osseweijer, Marianna Garfí, Jo Dewulf and John Posada , A critical review of resource recovery from municipal wastewater treatment plants – market supply potentials, technologies and bottlenecks, Environ. Sci.: Water Res. Technol., 2020,6, 877-910.
61. Jebin Ahmed, Abhijeet Thakur and Arun Goyal, CHAPTER 1:Industrial Wastewater and Its Toxic Effects , in Biological Treatment of Industrial Wastewater, 2021, pp. 1-14 DOI: 10.1039/9781839165399-00001.
62. (The UNEP report on water and wastewater reuse (UNEP and GEC, 2004), UNEP and GEC, 2004; US EPA, 2004.
63. Metcalf & Eddy, Inc. an AECOM Company, Takashi Asano, Franklin Burton, and Harold Leverenz. 2007. Water Reuse: Issues, Technologies, and Applications. Written. New York Chicago San Francisco Lisbon London Madrid Mexico City Milan New Delhi San Juan Seoul Singapore Sydney Toronto: McGraw-Hill. https://www.accessengineeringlibrary.com/content/book/9780071459273.
64. Daniel A. Okun, Distributing reclaimed water through dual systems, First published: 01 November 1997, Volume 89, Issue 11, November 1997, Pages 52-64.
65. Jimenez, B. and Asano, T. (2008) Water Reclamation and Reuse around the World. In: Jimenez, B. and Asano, T., Eds., Water Reuse: An International Survey of Current Practice, Issues and Needs, IWA Publishing, London, 648.

CHAPTER

3

Membrane Processes: Advances in Novel Membranes

— ABSTRACT —

Membrane-based separation and purification technologies have been known to both industry and academia for several decades. Membrane-based separation and purification technologies have been successfully used for several industrially important applications like process intermediate purification, separation, product separation, and segregation as well as in waste management systems. These membrane-based separation processes have seen several phases of their evolution. Among many processes, pressure-driven membrane processes have shown greater potential in water management and water-based waste treatment. Technologically, the membrane separation process has shown several advantages in water-based waste treatment and management. Over several decades of their existence, inorganic, polymer, composite, and nanocomposite-based membranes have shown greater potential in addressing the real-time industrially important applications. However, the rise of new challenges and inherent limitations in membranes materials, chemistry followed in preparing membranes, processes restrictions and traditional engineering modules design have led innovators, researchers, and industrial partners to rethink strategies in developing advanced membranes and engineering modules, and plant design and specialized applications. This chapter evaluates the innovations and advances in the field of membrane-based separation applications.

3.1 Introduction

Issues related to water scarcity are widely discussed and debated due to ever-increasing pollution and access to limited freshwater resources. Water scarcity is also increasing due to the lack of water treatment and recycling infrastructures in many cities. Creating appropriate water recycling facilities could have led to the reuse of high-quality reusable water for secondary purposes releasing much of the stress on freshwater resources. Nevertheless, increasing human activities and rapid industrial activities consume billions of liters of fresh water every day. According to some estimates more than 80% of consumed freshwater end up as wastewater after partial or minimal use in all sectorss. In addition to this, wastewater or industrial effluents carry a variety of contaminants during their use and end of their usage. Millions of people around the globe are potentially exposed to such contaminated water sources having inorganic pollutants like heavy metals, fluoride, nitrides, organic pollutants like dye molecules and organic matter, emerging organic pollutants involving semi-volatile compounds, active pharmaceutical ingredients, and so on. These pollutants in water attain various toxic forms in the long run. For instance, according to the World Health Organization (WHO) report, long-term assimilation of heavy metal-contaminated water may cause risk to lungs, kidney, bladder, skin damage, and cancers. Various types of contaminants including oil residue, drug molecules, pesticides, and so on originated from different chemical and anthropogenic sources are causing irreparable threats to the environment and human health which require innovations in membrane technology. [1, 2]

Membrane technologies provide a technological advantage in producing direct high-quality water from a contaminated source. Such a situation also leads to elevated water stress and environmental crisis. Therefore, the remediation of polluted water with undesirable constituents is given the highest priority, both at the industrial scale and on groundwater resources. Several technologies have been tried and tested in the past for the removal of heavy metals, organic contaminants and so on from groundwater as well as industrial wastewater and have shown certain definite results like: 1) partial, full or substantial contaminant destruction and degradation requires further investigations to understand nature of degraded components or completely remediate the contaminants from water sources, 2) pollutant sampling via extraction for additional processing or disposal, and 3) preventing unstable contaminants to make water treatment more reliable, stable and sustainable. It is noticed that more of the treatment techniques either use a multi-step treatment system or require additional chemicals to either extract, degrade or adsorb the pollutant. On the other hand, membrane-based technologies are relatively clean and do not require further addition of chemicals or pre- or post-treatment techniques. Therefore, membrane-based technologies can be used in combination with other conventional technologies to rectify mysterious problems related to water contamination, complex wastewater, and industrial effluents where no other conventional technique offers better solutions. [3]

Membrane-based purification and separation technologies over the years have shown increased demand from various sectors of the industries as well as domestic purposes. Recent trends in membrane materials, preparation techniques as well as plant design have led to adopting technologies for newer applications. Therefore, progress in polymer development suitable for membrane manufacturing technology and new avenues in the application of membrane technology cemented the ways to adopt membrane technology for water treatment, wastewater treatment, and drinking water reclamation. Recent advances have revealed that both direct membrane separation as a standalone technique, as well as a membrane-assisted hybrid process like biological reactors, adsorption, thermal processes and so on, have their own merits in water and wastewater treatment.

Water is a precious commodity and essential for life on the earth. a Water is overexploited in various sectors. Overexploitation has also led to the unregulated discharge of extra water from various steps and processes. Most of the over-exploited and discharged waste sources have been characterized by different types of contaminants such as organic pollutants, inorganic contaminants like heavy metals, emerging or persistent contaminants, etc. Most of the wastewater from industries, agriculture, and household discharges is also characterized by their high concentration of total dissolved solids (TDS), total dissolved solids (TDS), and high chemical oxygen demand (COD), biochemical oxygen demand (BOD), and so on. Nevertheless, increased awareness among the industrial community, the general public as well as water managers shifted the focus on water treatment technologies. In this regard, every effort is being made to monitor, regulate, and manage the water quality at various levels. The absence of suitable water treatment techniques leads to the degradation of water leading to increased accumulation of contaminants which may grow above the permissible limits thereby worsening the wastewater contamination level. Conventional wastewater treatment systems such as adsorption, flocculation, coagulation, and distillation have been effectively used to partially or fully remove various contaminants. Most of the conventional techniques have been effectively applied, however, modern contaminants pose serious limitations to conventional water treatment techniques. However, some have proven ineffective and most of these methods have shown several limitations due to the use of additional chemicals, sludge formation, large energy consumption, a large carbon footprint, and moreover, most of the processes leave behind a generation of secondary pollutants in the ecosystem and environment at large. Nonetheless, modern water treatment methods involving physical treatment techniques have shown promising potential. [4]

One such technique with additional advantages in overcoming the limitations in conventional methods is membrane-based technique. Among many membrane processes, pressure-driven processes like microfiltration (MF), ultrafiltration (UF), nanofiltration (NF), and reverse osmosis (RO) have been recently adopted for water and wastewater treatment applications. In several cases, membrane processes have proven highly effective and efficient in obtaining high-quality reusable water. Unlike other conventional water treatment techniques, membrane processes offer instant purified water, uses no added chemicals, does not require additional post-treatment step, and offer a wide range of options for treating a variety of contaminants ranging from ionic species to organic contaminants, heavy metals to biological contaminants like bacteria, virus, and other pathogens. However, membrane technologies also suffer from several limitations such as membrane instability due to phenomena like fouling, reduced permeability over the period, membrane durability, etc., which cause a drastic decline in product flux, decrease in separation efficiency and overall reduces shelf-life of a membrane. Nonetheless, recent advances in the field of membrane science and technologies have led to the design and development of new material stock and the development of novel membrane preparation techniques. These developments are showing improved results in the preparation of membrane and module with superior chemical resistance, antifouling, thermal resistance, permeability, durability, along with high permeability and rejection which make the overall process cost effective. [5]

In this chapter, advances in pressure-driven membrane processes and technological innovations are discussed. This chapter will also evaluate recent advances in nanotechnology solutions used for the preparation of high-performance membranes and in all supporting materials that go into module design and testing. Also, new analytical parameters associated with membrane characterization, evaluation, and rating will also be carefully assessed. The traditional wastewater treatment approach to achieve reusable quality water recovery has shown mixed results in the past. The conventional treatment methods such as carbon adsorption, chemical oxidation, coagulation, multimedia filtration, ozonation, etc., on secondary effluents have shown important advantages in producing quality water. However, these traditional technologies suffer from certain disadvantages such as cost, large treatment plant footprint which requires a specific site, operational problems, unreliable product water quality due to varying load of treatment reagents, variability in treatment site conditions, etc. However, membrane technology is considered stable and reliable in the set of conditions provided in use. Therefore, advances in membrane material design, manufacturing technology, and applications are hovering the potential of membrane technology as an effective alternative to wastewater reuse applications.

3.2 Potential Membrane Technologies

Membrane-based separation processes have been mainly classified as advanced water treatment techniques. However, membrane processes have also been adopted both standalone as well as in combination with other traditional processes based on their requirement. Nonetheless, the membrane process in both direct steps and in combination with a conventional technique such as biological treatment has its own advantages in wastewater treatment to obtain reusable quality product water. Pressure-driven membrane processes produce high concentration retentate while producing quick good quality water. Even though this is an advantage with the membrane process, however, resultant high concentration retentate residue generated is a serious concern. In such cases, the membrane process is coupled with biological treatment to preferentially extract residual water from the concentrate. This indirect complete treatment process approach has been widely applied for treating complex industrial wastewaters. For example, complex industrial wastewaters like distillery and tannery wastewater are difficult to handle in direct membrane processes.

In such complex wastewater systems, bio-origin pollutant separation in cross-flow membrane filtration is difficult and therefore preferred to treat using secondary sedimentation processes. In such situations, the membrane process coupled with a secondary method like the activated sludge treatment step is the most popularly used to achieve the extraction of high-quality water. In some cases, the membrane process is also used in combination with secondary treatment techniques like the membrane bioreactor (MBR) step. However, a major disadvantage of this approach is the high cost involved in the overall management of wastewater. Nevertheless, a combination of MF or UF processes in cross-flow modules with MBR substantially reduces the potential of membrane fouling or concentration polarization. Initially, researchers like Yamamoto et al. have adopted submerged membrane modules in the aeration tanks instead of a crossflow MF/UF system to avoid membrane clogging due to high crossflow velocity. On the other hand, Yamgiwa (et al.) used the high-pressure retentate produced in the cross-flow filtration for direct aeration in a specially designed liquid plunging set up to use as an air jet in an aeration unit which makes it an energy-efficient process. On the other hand, Visvanathan (et al.) used the direct backflushing technique to flush the membrane surface at high speed on submerged membrane modules which helped in maintaining the stable flux in a long-term operation. All these approaches have consumed higher energy inputs for producing quality water from complex wastewater sources. Nevertheless, recent advances in innovative approaches have led to adopting various energysaving options to make the overall process sustainable. [6, 7]

Among many approaches, advanced methods consider comprehensive tactics to extract quality water from waste sources. The advantages of direct membrane separation can be described as follows:

(i) contaminants from wastewater can be eliminated to obtain clean water of high standard either using a direct membrane process or hybrid-membrane system coupled with other secondary or other advanced wastewater treatment techniques,

(ii) direct method can also use multiple-membrane processes steps to attain reusable quality water recovery such as use of microfiltration as pretreatment technique for screening large size pollutants like colloids or suspended solids followed by nanofiltration or reverse osmosis to remove fine pollutants like organic matter, and ionic monovalent contaminants,

(iii) for the treatment of biologically intense wastewater, membrane processes can be coupled with biological treatment techniques in which membrane process like micro- or ultrafiltration can be used for bio-solids or macromolecules contamination separation and further fine pollutants from NF or RO process, further,

(iv) since membrane processes are considered advanced physical treatment techniques and no biological or chemical sub-step involved during the treatment, the overall operation can be shut down at any time readily and can be restarted whenever required without loss in water quality,

(v) since membrane processes operate as closed systems, intermediate complications associated with odors and obnoxious smell can be eliminated,

(vi) since the membrane process is largely controlled with operating parameters like applied pressure, cross-flow velocity and feed volume, the process automation can be easily achieved even for remote controlling and monitoring,

(vii) when treated using membrane processes, the product water quality does not suffer from feed water quality, and

(viii) unlike the traditional primary and secondary wastewater treatment technique such as screening, aeration processes, multimedia filtration, adsorption where relatively larger space is required to effectively operate process, membrane-based separation processes are very space friendly and takes minimum footprint.

3.3 Advancements in Membrane Materials and Sources

Membrane technology has been in practice for several decades now, however, several disadvantages with the existing processes and limitations with membrane materials have led to several innovations in the separation and purification sector. However, dependency on membrane technology is increasing at an unprecedented rate. The membrane technology market size was estimated at USD 19.6 billion in 2020 and is forecasted to grow multifold and the global membrane science separation technology market is expected to grow at a compound annual growth rate of 12.2 % from 2020 to 2028. [8] As conventional membrane materials are known to have low sustainability due to their fabrication involving majorly synthetic polymers. Another major disadvantage associated with membrane materials in engineering modules disposed of post-active use is categorized as one of the main plastic wastes which is impacting the environment globally. In addition to this, the growing environmental concerns associated with water pollution are starting point for researching alternative potentials of natural polymers that can substitute the conventional synthetic polymer for the membrane preparation. [9, 10]

3.4. Synthetic Polymer Membranes

Conventionally the selection of suitable synthetic polymeric materials has been a key component for tuning the pore characteristics in membranes by effectively using their physical and chemical properties. These properties have been further optimized and improved to overcome some of the disadvantages of surface characteristics due to the nature of the synthetic polymer and specific reactivity. These modifications are targeted to obtain the enhancement in permeability, selectivity, and durability of the membrane. Therefore, these polymeric membranes, also known as organic membranes, include a family of technology for liquid separation which is an industry leader in efficiency and performance. These membranes in porous form have been generally categorized on their pore size, pollutant rejection pattern, and separation characteristics, namely,

(i) microfiltration,

(ii) ultrafiltration,

(iii) nanofiltration, and

(iv) RO/FO processes.

Further, polymers used for the fabrication of membranes in these categories are carefully selected to ensure excellent high permeate flux or product water rates at specific operating operations and process environments. Further, post-membrane preparation filtration modules also comprise a series of engineering materials that need to have compatibility with the main polymer precursor. Also, membrane surface and pore characteristics are carefully controlled to particularly attain favorable affinity towards specific pollutant or fluid(s) to be separate and cleaning materials required to help maintain hygiene and stability in performance. Therefore, the most used polymers in membrane preparation are cellulose and its derivatives like cellulose acetates, cellulose triacetate, nitrates, and esters, polysulfone, poly (ether sulfone), polyetheretherketone, polyacrylonitrile, polyimides, polyamides, polyethylene and polypropylene, polytetrafluoroethylene, poly(vinylidene fluoride), poly(vinyl chloride) have been commercially used. Among these polymers, polysulfone (PSf), polyethersulfone (PES), polyurethane (PU), polysulfone/PS/PS, polypiperazine, polyamide as well as polyvinylpyrrolidone (PVP) also adopted in large scale membrane plants. Further, a detailed list of some of the synthetic-origin polymers widely used or various types of membranes has been listed in Table 3.1. [11-13]

Table 3.1. Synthetic polymers used as precursors for the preparation of membranes and their structures.

No.	Synthetic polymer	Structure	Ref.
1	Polysulfone (PSF)	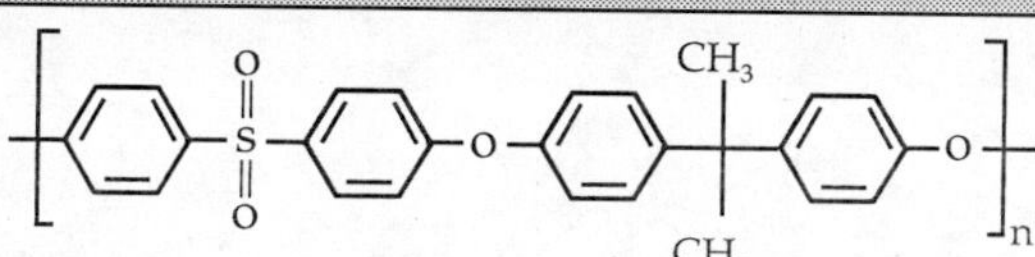	[29]
2	Polyvinylidene fluoride (PVDF)	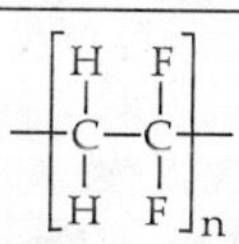	[31]
3	Polyether sulfone (PES)	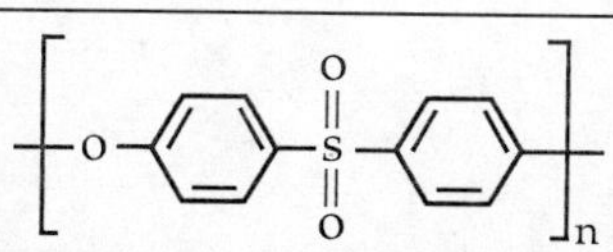	[33]
4	Polyamide (PA)	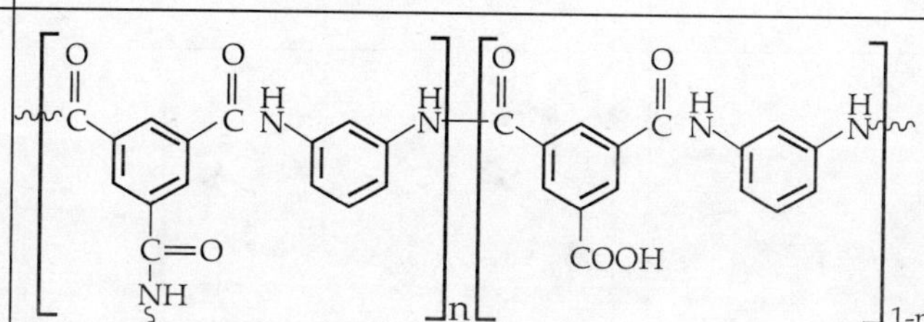 Crosslinked structure Linear structure	[35]

Contd...

5	Cellulose acetate (CA)	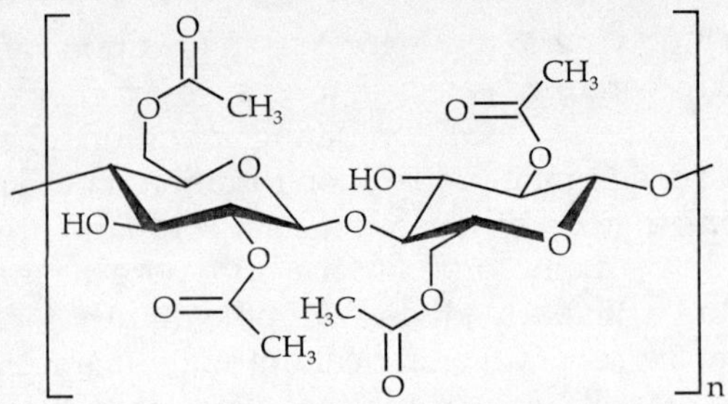	[37]
6	Polyacrylonitrile (PAN)	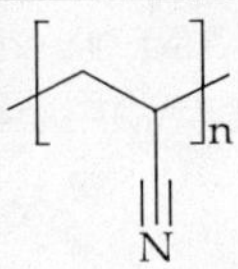	[39]
7.	Polyvinyl alcohol (PVA)	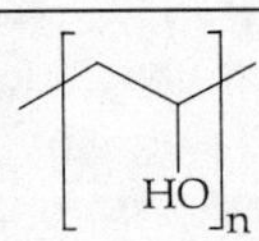	[30]
8.	Polycarbonate (PC)	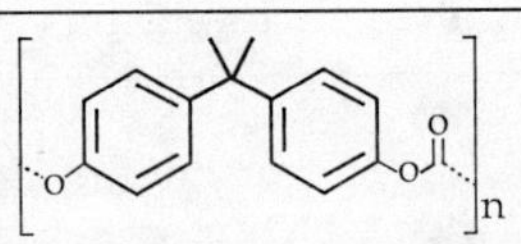	[32]
9.	Polyurethane (PU)	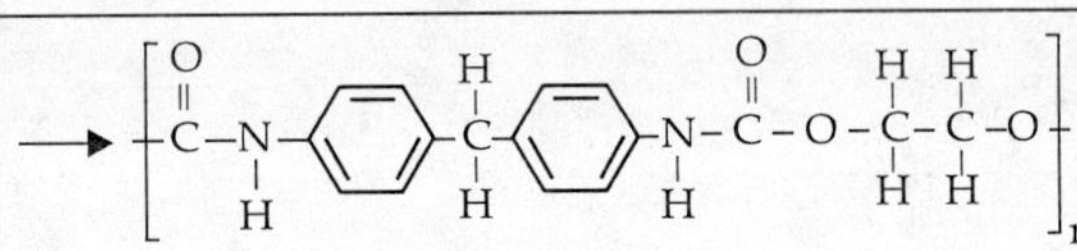	[34]
10.	Polyimide (PI)	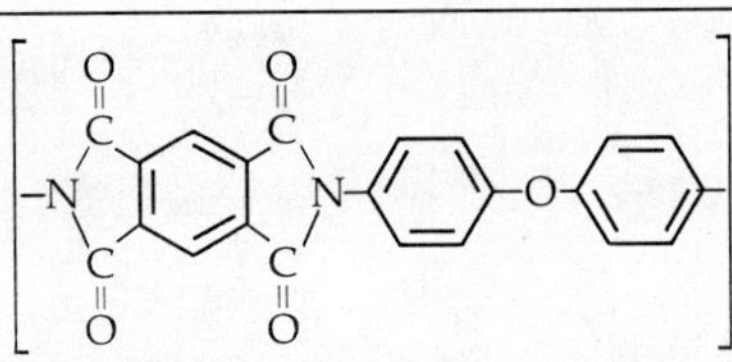	[36]
11.	Polyvinyl-pyrrolidone (PVP)	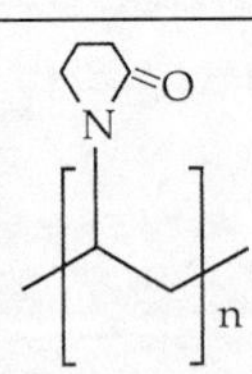	[38]
12.	PIM-1	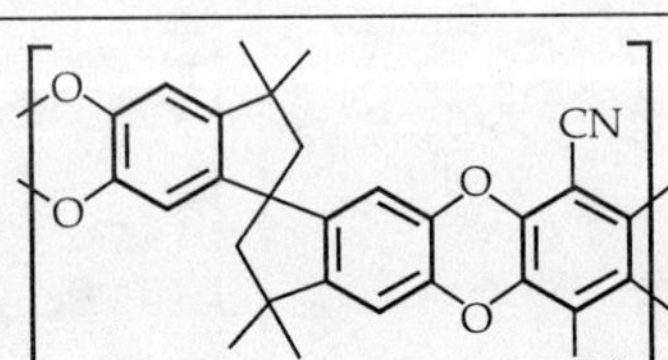 PIM-1	[40]

Broadly, every membrane which is used successfully at commercial level mainly consists of organic and synthetic polymers such as cellulose acetate, polysulfones, sulfone polyether, poly(vinyl diameters), polyacrylonitrile, and so on. Among pressure-driven membrane processes, ultrafiltration and microfiltration with large pores and different membrane preparation conditions with adopted parameters frequently use identical polymer materials precursors as a dope solution. Also, during MF/UF membrane preparation process solution casting and coagulation determine the pore size and their distribution and is usually followed to produce different pores and morphologies.

Also, advanced innovations in the field of Materials Chemistry also provide various material stocks and options for the preparation of inorganic membranes such as ceramics or metal oxides/hydroxides. Also the preparation of composite membranes using inorganic materials as one or major constituent in a polymer matrix is yielding advanced membrane structures. Further, blending two or more polymers and preparing composite membranes using various types of indifferent precursor materials is seen as a futuristic potential to induce important membrane properties both in bulk and on the surface which helps in the separation and permeation of feed constituents. In addition to membrane performance parameters, polymer systems have also used suitable composite materials to improve their thermal stability, and chemical resistance, thereby membranes with suitable composite precursors have proven to be an advantage over only polymer membranes. On the other hand, there are also various disadvantages with ceramic, or inorganic membranes processed for implementation in MF such as a high cost and ceramics being thermally fused face mechanical fragility that prevented their widespread use. Similarly, metallic membranes are often made-up of stainless steel and can be prepared with fine porosity with controlled pore characteristics. However, mainly metallic membranes have been broadly used for gas separation, nonetheless, but recent advances have also been tried to use metallic membranes as both support as well as individual separation media for water filtration at high operating temperatures. Nevertheless, the major disadvantage with metallic membranes comes from their scale-up difficulties in which large-scale use of metallic membranes poses serious issues associated with design, and module preparation and their inflexibility create difficulties associated with plant footprint as well as logistical problems. The present and near-futuristic tendencies of advanced membrane development are to use nano-functionalized membranes with tailor-made separation and purification characteristics. Polymer-nanomaterials composite membranes with active nanomaterials besides are an example of advanced membranes that are forecasted as extremely important separation media due to their various proven advantages over only polymer membranes. [14-15]

Furthermore, synthetic-polymer-based membranes can be classified as organic and inorganic. Inorganic membranes that include different constituents such as zirconia, alumina, titanium, zeolites, palladium, silver, and oxides of metals are used to prepare ceramic membranes. Organic membraneprecursors are mainly precursors like polytetrafluoroethylene, polystyrene, polyamide, and other macromolecular systems that are used for most commodity membranes for both domestic as well as industrial applications. In order to achieve enhanced separation and permeation properties in membranes, the improved film-forming characteristics with high tensile strength available at cost-effective precursor sources need to be found. Further, to maintain the conventional scale-up protocols and minimize the damage to existing membrane preparation infrastructure, the use of synthetic polymers as one of the dope constituents is needed. Inorganic-origin membranes, even though highly reliant on the separation mechanism via charge exclusion, promote retention of pollutants via electrical repulsion and mass transport. Charged surface properties in inorganic-based membranes can be easily adjusted further to ensure optimum pollutant rejection and separation through various mechanisms which are also influenced by various parameters, like ionic strength, feed solution pH, and operating temperature conditions. With the optimum load of filler in the membrane matrix which alters the surface properties such as charge and pore characteristics, the overlap effect of the double-layer would be avoided during the ionic pollutants rejection. Therefore, the main advantages of synthetic polymer membranes are that macromolecular-based filtration matrices offer extensive opportunities to alter the constituents for tailor-made end-user via their composites and blends and polymers can be adaptable in any form, such as flat sheet and hollow fiber. Over the years, progress in polymer science has offered various innovative polymers for membranes with special characteristic features that can be used through various morphologies.

Nonetheless, bio-origin from the plant, animal, and seaweed such as cellulose-based polymers, alginate, polyisoprene, starch, polylactic acid, polyhydroxyalkanoates, polybutylene succinate, and agarose/agar, respectively has been well researched. Also, several new entrants to this list are bacterial fermentation products like collagen, sericin, chitin, and chitosan, which fascinated the research along with the growing worldwide trend towards sustainability. The use of biopolymers for membrane-based separation and purifications is very well documented in the literature. There have been several reports of advanced biopolymers that have been tried and tested at lab scale and pilot scale and several other biopolymers have been well documented even if their large-scale applications are still a challenge. Moreover, recent studies exploring the use of biopolymers in membrane applications with critical analysis in terms of their origin, extraction, synthesis, properties, membrane-forming properties, and applications with specific emphasis on the main results of bio-gen membranes during the last decades in the different fields of membrane technology including micro- and ultrafiltration, nanofiltration, reverse osmosis, forward osmosis, pervaporation, gas separation and so on.

3.5. Biopolymer Membranes

Natural materials have been identified as one of the major sources for extraction of biopolymers and other bio-gen nanomaterials useful for water treatment applications. A new class of biopolymers extracted from renewable resources includes a variety of macromolecular systems having the main backbone as cellulose and its derivatives which find direct applications in water treatment are listed in Table 3.2. On the other hand, biomacromolecules or biopolymers are generally known as substances produced through plants, live organisms, and so on. Biopolymers also possess unique characteristics and highly functionalized long chains with several active groups induce robust activity in the overall polymer. More significantly, active functional groups and their repeating chemical blocks make them biocompatible and biodegradable which easily depletes these bio-origin macromolecule resources in the environment. Therefore, biopolymers carry many repeat-active units in their repeating units with cellulosic backbone, and their derivatives are useful for further functionalization and preparing tailor-made end products.

Table 3.2. Bio-genic polymers used for membrane application.

No	Biopolymers	Structure	Ref.
1.	Chitosan (CS)	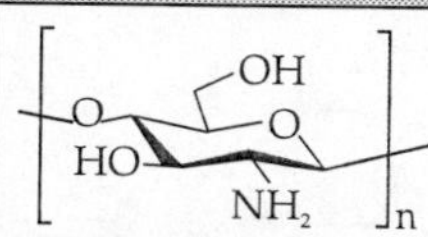	[16]
2.	Sodium alginate (SA)	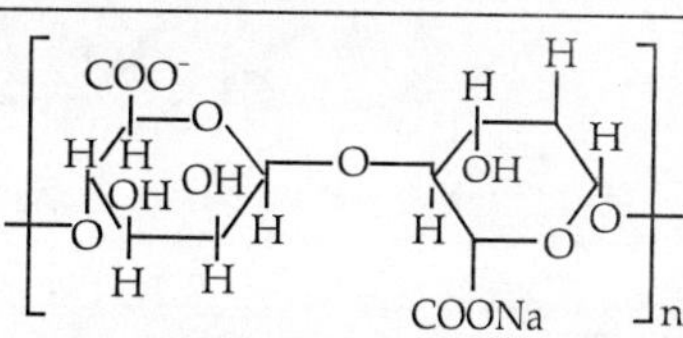	[17]
3.	Agar (AG)	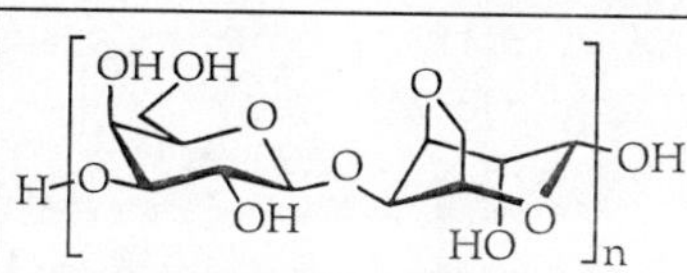	[18]
4.	Agarose (AGR)	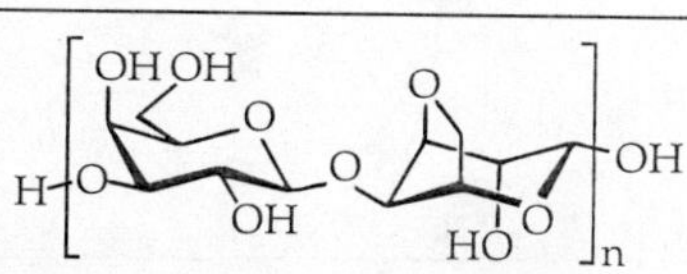	[19]
5.	Chitin (C)	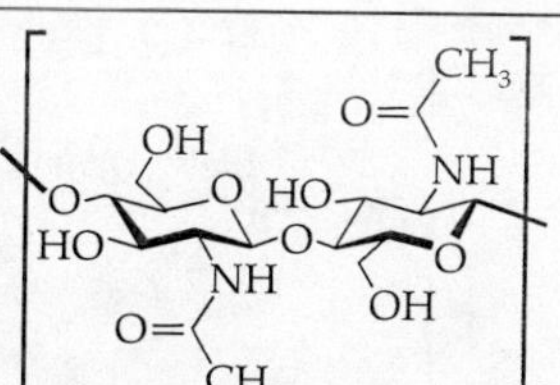	[20]
6.	Cellulose	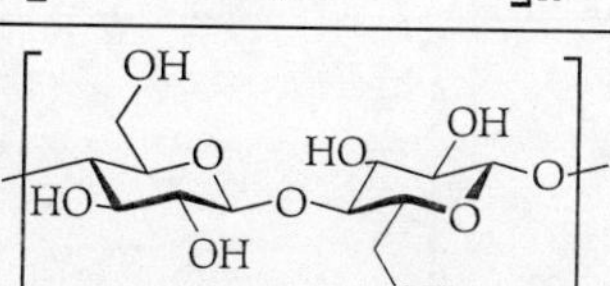	[21]
7.	Carrageenan (CAR)	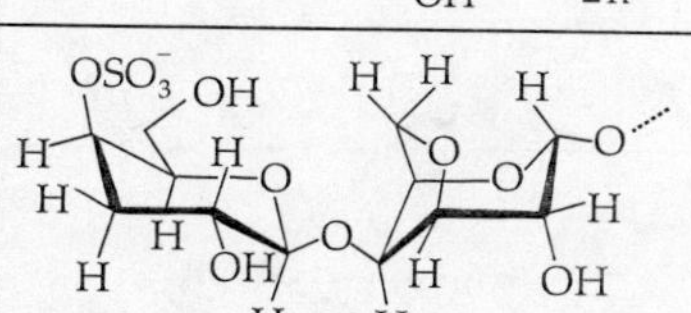	[22]

8.	Polylactic Acid (PLA)	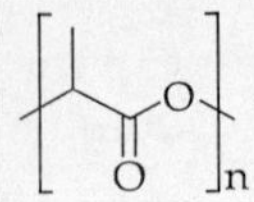	[23]
9.	Polyhydroxyalkanoates (PHAs)	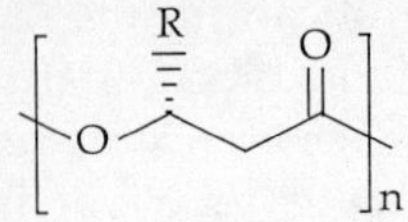	[24]
10.	Polyhydroxybutyrate (PHB)	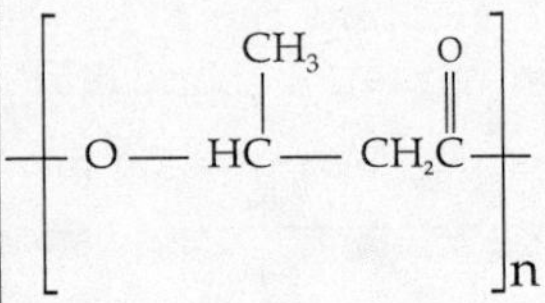	[25]
11.	Pullulan	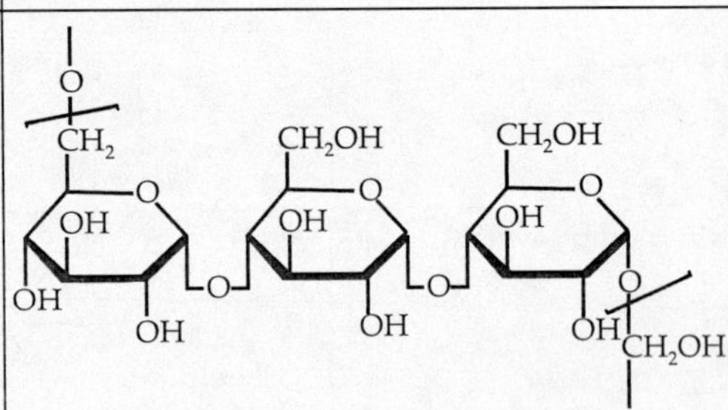	[26]
12.	Nitrocellulose (NC)	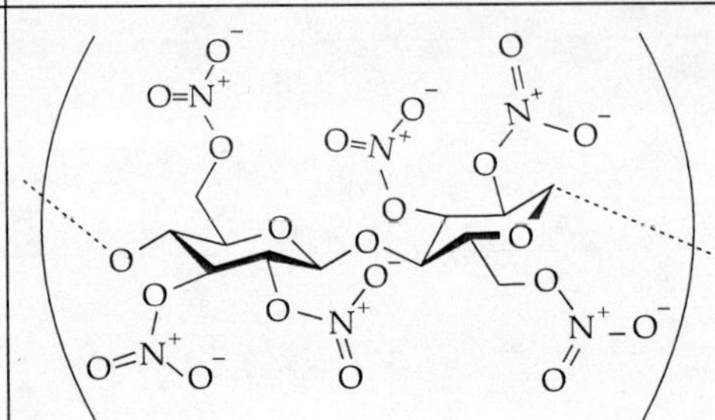	[27]
13.	Cellulose acetate (CA)	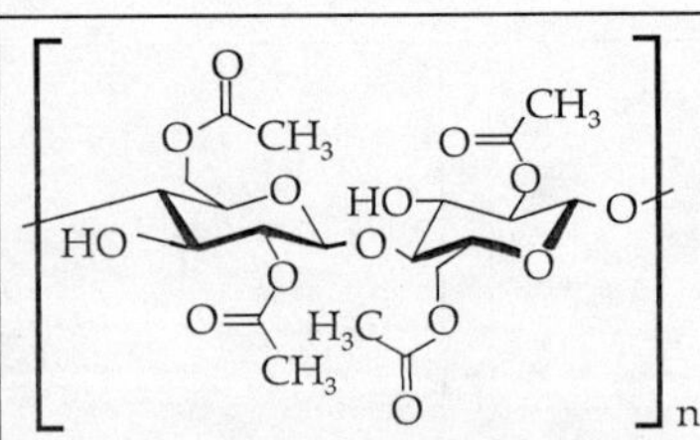	[28]
14.	Cellulose triacetate (CTA)	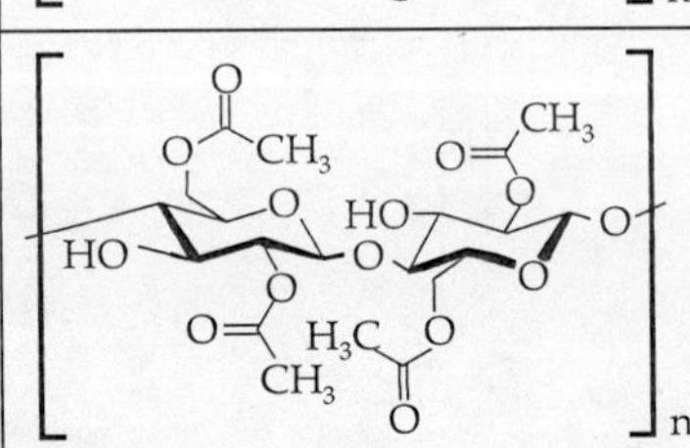	[29]

On the other hand, until recently biopolymers have not been extensively used for membrane applications even though bio-macromolecules possess excellent film-forming properties. Even though the researchers have used biopolymers for the preparation of membranes for selective separation of organic solvent traces from water in pervaporation applications. However, many biopolymers continue to be untapped, uncovered, and under-utilized. Now, biopolymers with the rapid development of extraction technologies, increased understanding of basic biosynthetic pathways and bio-processing protocols have been widely applied to convert biopolymers into different types of membranes. Also, biopolymer backbones having lipids, polysaccharides, polyphenols, proteins, and specialty macromolecular chains offer a series of advanced separation characteristics. The use of these specific repeating units helps to glue, induce functionality, and tune the morphologies of the final membrane architecture.

Advanced approaches including use of biopolymer in combination with synthetic polymers at commercial level have shown increased interest to impart green, sustainable, and durable end products with suitable modifications. This advanced approach would add value to environmental compatibility, enhanced shelf-life, and protection in large-scale membrane modules. Also, biopolymers carry remarkably important inherent membrane properties like hydrophilic and biocidal characteristics that are significant for water treatment applications of both domestic as well as industrial wastewater nature. Higher hydrophilicity also induces unique wettability in biopolymer membranes which enhances the affinity of membrane surface towards water component in the contaminated water. Combined characteristics in biopolymers such as hydrophilicity, wettability, biocompatibility, and biocidal properties in biopolymer membranes are drawing considerable research interest in addressing challenges associated with commercially successful synthetic polymers which lack both biocompatibility and biodegradability. In addition to this, several advancements in blending and composite thin-film membrane preparation with detailed insight in the scientific understanding and potentials of functional groups also provide remedy in conventional membrane materials such as fouling, solute scaling, and biofouling which thereby control the concentration polarization. This increases the durability of a membrane along with drastically reducing the maintenance and operating costs. Based on these reasons, researchers, innovators, and technologists have intensified their approach in using biopolymers in membrane-based separation processes.

Therefore, biopolymers with biocompatibility, biodegradability, hydrophilicity, wettability, portability, and ecological compatibility can be an extremely important alternative to the conventional synthetic polymers-based petroleum origin. On the other hand, advanced approaches have shown great potentials of biopolymers in water treatment and wastewater treatment applications via their various modification processes involving physical, chemical, and physicochemical properties such as pH, temperature, mechanical and thermal stress, etc. More significantly, biopolymers also offer low-toxicity for human health with high biocompatibility and biodegradability. Future would also use large scale application of various biopolymers such as cellulose, cellulose acetate (CA), chitosan (CS), polylactic acid (PLA), polyhydroxyalkanoates (PHA), polyhydroxybutyrate (PHB), and cellulose triacetate (CTA) are now substituting conventional and commercially successful synthetic polymers like polysulfone (PSF), polyethersulfone (PES), polyimide (PI), etc. Nonetheless, other important properties of biopolymers also come from their inherent film-forming characteristics, high affinity towards water molecules and can be easily adopted for the composite formation potential with different nanomaterials. However, several drawbacks remain significantly unexplored in membranes prepared with biopolymers such as long shelf-life and low mechanical strength. Nevertheless, advanced, and innovative membrane preparation techniques have shown great potential in preparing enhanced mechanical characteristics using various nanomaterials that can potentially be of immediate prosperity to future membranes. [16-29]

3.6. Advances in Anti-fouling Membrane Development

One of the major challenges in sustaining modern society and lifestyle is an inadequate supply of safe and clean water. Recent advances in the development of sustainable materials process are opening new avenues for futuristic technologies due to their cost-effective and energy-saving features. Membrane-based technologies have become an integral part of modern society in the form of platform technologies for water purification, both brackish water and seawater desalination as well as domestic wastewater or industrial wastewater treatment. However, membrane fouling, scaling and concentration fouling that arise from the non-specific and multiple interaction between membrane surface and foulants from contaminated feed streams. The scaling, fouling and concentration polarization significantly imped the efficiency of membranes in separation and purification performance. Therefore, preparing anti-fouling membranes is a fundamental strategy to deal with pervasive fouling associated issues from a variety of foulants in contaminated waters. Recent major advancements have been made in preparation of anti-fouling membrane techniques and further elucidating the anti-fouling mechanisms of membrane processes such as microfiltration, ultrafiltration, nanofiltration, reverse osmosis and forward osmosis. This section will further discuss the list of modern foulants and the principal mechanisms involved in membrane fouling, highlighting deposition and growth of the fouling layer and its development, and present status in developing inherently anti-fouling membranes. We will also discuss advanced strategies and mechanisms of understanding involving membrane fouling during practical applications. [30, 31]

Fouling on membranes occurs from the deposition of unwanted solutes or pollutants and their gradual adhesion on the surface which significantly hinders the performance of the overall process as shown in Figure 3.1. In recent advances, researchers have innovated new materials, combinations, processes, and designs of novel membrane preparation techniques to suit the nature treatment applications. The recent studies scrutinize the fouling phenomenon through the fundamentals of membrane properties and their affinity towards foulants in association with classical theories. Therefore,

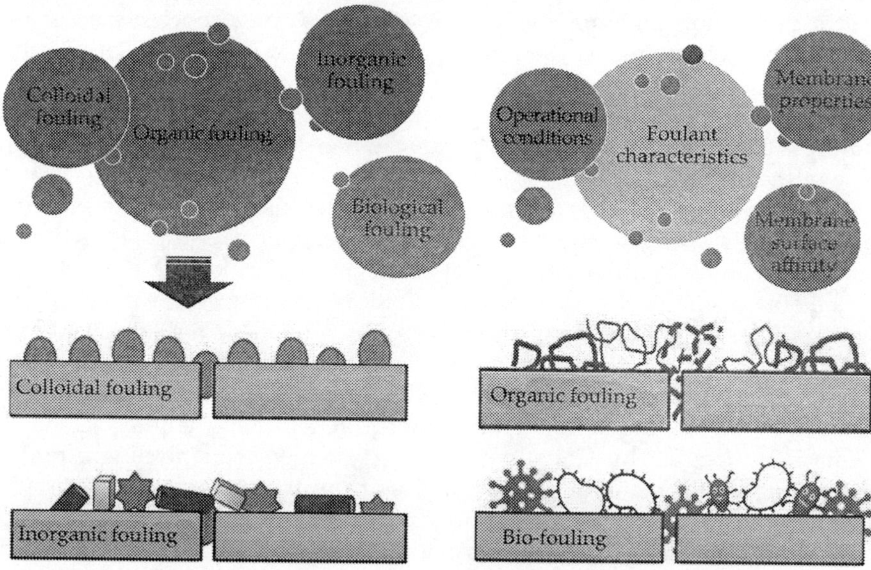

(a) Types of fouling (b) Factors influencing fouling

Figure 3.1. Schematic representation of (a) different types of fouling, and (b) factors influencing various types of fouling on membrane surface.

lately antifouling membrane surfaces are being widely prepared inspired from nature which offer a series of speciality properties such as super-wetting surfaces, anti-microbial characteristics, and biosynthesis strategies. [32]

In desalination, fouling of membranes occurs as a result of undesirable deposition or accumulation of inorganic species, organic compounds, colloids foulants, and microorganisms on/in the membrane's surface or within the bulk of membrane during the filtration process. Fouling due to deposition of various types of foulants damages the performance of membranes by drastically affecting permeate output, increases trans-membrane pressure, and decreases shelf-life and the effective operation time of membrane. As it is understood that the fouling is quite a complex process that is influenced by various factors via different mechanisms in its formation on surface or within the membrane bulk, it should be then studied to understand application of suitable innovative strategies to mitigate, minimize, and clean the fouled membrane.

Major foulants can be categorized into colloidal, organic, inorganic and/or biological as shown in Figure 3.1(a). Colloidal fouling is largely caused by the convective depositions and accumulation of colloidal systems like silica, clays, paint pigments, etc., on the surface of the membrane. The colloidal deposition generally tends to create additional layers or cake layers due to gradual scaling formation on the surface of the membrane. Inorganic fouling on the other hand is usually initiated with the dispositive accumulation of inorganic precipitates like metal oxides, metal hydroxides, and other ceramic particles on the surface on the membrane. Continuous deposition of foulants typically denotes the formation of inverse-solubility salts such as calcium sulfate, calcium carbonate and calcium phosphate that are known to cause serious concern for RO and NF membranes owing to their rejection pattern to inorganic species that gradually build up a concentrated layer interfacing with membrane surface which further facilitates complete precipitation. On the other hand, organic fouling is caused by the buildup of organic matters of various sizes and nature from influent process streams such as humic acid in surface and groundwater, dye molecules, oily constituents, macromolecules constituents like proteins, carbohydrates, semi-volatile organic compounds, pharmaceutical ingredients, anti-foams and so on. Among many biomacromolecular constituents like proteins have one of the most complex fouling patterns and features due to their labile and dynamic nature. For example, proteins being large and long molecules in the process stream tend to fold together known as coiling after interacting with each other resulting in colloidal aggregates which subsequently concentrate on the membrane and further precipitate at the interface of the membrane surface. This leads to severe organic membrane fouling. Another major concern in membrane fouling arises from the biological or microbial fouling which is also considered more dangerous among all types survival of membrane in a long run due to the formation of the undesirable microbial layers on membrane surfaces in the form of bio-film subsequent microbial gel-layer originated from highly contaminated influent sources. Bacterial, pathogenic and combination of organic moieties which facilitate a conducive atmosphere for the growth of microbial pollutants in contaminated water systems. Increase in the COD and BOD conditions of contaminated sources further leads to rise in the temperature of wastewater due to fermentation-type microbial activities. It is considered a promoter of biofouling in membrane processes. During this process bacterial cell activities also lead to the formation of several by-products of complex nature and some are in soluble nature generally characterized with high-molecular weight products comprising mainly large molecules like lipids, proteins, and polysaccharide constituents. Further, with gradual increase in the concentration of biological constituents, growth of biofilms occurs which catalyzes in inorganic precipitation via enhanced nucleation, and crystallization kinetics. Nevertheless, recent understanding of fouling phenomenon reveals that there are no single fouling mechanism which proceeds over other in real membrane processes. However, it is realized that a combination of various fouling mechanisms take place due to combination of pollutants and co-pollutant materials making it more complex to deal with as illustrated in Figure 3.1(b). [33, 34]

On the other hand, major factors that affect fouling phenomenon on surface and in the bulk of the membranes can be mainly categorized based on their, (a) foulant characteristics such as molecular size, concentration, charge on foulant, solubility, diffusivity, polar nature, etc.; (b) membrane properties such as hydrophobicity, surface charge, surface roughness, nature of functional groups on membrane surface, pore size and their distribution; (c) operational parameters such as flow velocity, pH, ionic strength, temperature, flux, and solution chemistry, and the concentration of organic/inorganic matters in feed stream. Even though, types and extent of fouling that occur on the surface of the membrane and inside pores ilargely depends on the types of foulant, concentration, and their physicochemical properties of influent contaminated water. The nature of fouling is also determined by the presence of co-pollutants, solution chemistry, interactions between the co-foulant and the affinity of membrane surfaces that could greatly increase the fouling propensity. Therefore, membrane surface properties play an important role that largely influence the severity and nature (reversible or irreversible) fouling. On the other hand, operational conditions such as the cross-flow velocity, applied pressure, temperature and washing pattern can also affect the extent of fouling in the membrane process.

Nevertheless, as water scarcity continues to haunt researchers and technologists to discover efficient treatment technologies and to explore various pathways to minimize the water consumption or utilization cycle at the nexus of water, energy, and food systems. In this direction, there is a paradigm shift required towards realizing advanced and efficient water treatment technologies such as membrane processes to sustain the circular economy. These strategies are important in modern membrane techniques considering deteriorating conditions of municipal as well as industrial wastewater qualities as one of the valuable sources for extracting clean nutrients and water thereby drastically reducing dischargeable effluents to the environment.

Nevertheless, several technologies have been used to extract treated water as a new and alternative source to augment the safe water supply and address the issues caused by a lack of freshwater in different parts of the world. This is more so important in the arid regions. Further, recent advances in the reuse of reclaimed water from wastewater sources has spurred new research goals to convert wastewater treatment plants into both water recovery as well as value added product recovery facilities. Even though several conventional methods are being continuously improved to apply for resource recovery and the extraction of high-quality water. Among many existing techniques water and resources can be achieved using physical, chemical, chemo-physical and/or biological processes such as precipitation, adsorption, coagulation/ flocculation, ion exchange, advanced oxidation process, electrochemical methods, and membrane filtration. However, advanced membrane-based separation processes have proven as an important and viable immediate option for water-scarce regions.

Even though membrane processes provide a simple physical barrier for contaminants to be retained and selectively allow high quality water. The use of MF, UF, NF, and RO depending on the pore size and molecular size allows the rapid rejection of various pollutants such as organic, inorganic, colloidal and biological constituents from various effluent streams. Furthermore, FO, MD and electrodialysis have also been reinvented to address the demerits associated with processes to overcome in desalination solutions. Therefore, membrane technologies are evolving rapidly and steadily improving largely based on the realization of their potential in almost every water-intensive sector.

Nonetheless, conventional polymeric membranes are still being employed as main separation media for water treatment and desalination owing to their remarkable pollutant retention property, mechanical characteristics, permeability, and relative cost-effectiveness. However, increased concerns over membrane fouling related parameters involving scaling and concentration, polarization is one of the greatest bottlenecks for membranes during the water treatment process. Nonetheless, advanced efforts are being made to find the anti-fouling, mechanically robust and thermally resistant new materials stock to be utilized in combination with polymer matrices during membrane fabrication. In this direction, one of the main strategies has been the fabrication of mixed-matrix membranes (MMMs) having super-hydrophilic characteristics and use of speciality additives like crosslinking agents to improve the shelf-life of membrane materials to achieve the realistic water treatment goals. [35]

To overcome the challenges associated with fouling, researchers have adopted various innovative approaches. Among many, developing anti-fouling membrane is known to yield favorable and encouraging end results from many study reports. In anti-fouling membrane preparation approaches, physical disruption of fouling layer existed due to the presence of bio-material matrices, control by biocides using quorum quenching and phage-based decomposition approaches and chemically modification of membrane surface with anti-fouling behavioral fillers and surface modifying agents such as coatings, imbedded approach or thin-film formation of active anti-fouling nanomaterials, etc., are prominent approaches so far.

On the other hand, membrane surface modifications using suitable anit-fouling agents are mainly classified into two categories, passive or active modification approach. Passive approaches include the modification of membrane surfaces to impede foulant adhesion which facilitate easy removal of foulants and cost-effective cleaning. For example, fouling resistance can be enhanced in conventional membranes by increasing membrane surface hydrophilic which directly influences the repulsion of foulants from active surface interfaces. Furthermore, breaking the scaling pattern and foulant accommodation can be controlled with the tailor-made surface modification approaches such as creating amphiphilic surfaces, which help in driving the pollutants away from the membrane surfaces. Even relatively high hydrophilic membrane surface can effectively repel and resist the accommodation of foulant which is non-migratory in nature such as biomacromolecules, dissolved organic matter and normal organic matter. However, such surfaces often fail to resist fouling of spreadable foulants like oils and greasy pollutants. Therefore, membrane fouling, particularly organic type, cannot always be avoided or suppressed by hydrophilic membrane surfaces or hydrophilization. Therefore, recent reports list the limitations of passive approaches due to their non-general application and emphasie to formulate innovative and advanced antifouling membranes with wide-ranging and universal applicability. This is also recommended as an effective approach in curbing the fouling issues in membranes and an active approach is greatly anticipated in the last few years. In comparison to passive approaches discussed above, active anti-fouling approaches have an emphasis adopting novel routes by removing the foulants by destroying, degrading, and deconstructing them via membrane surface contact agents or the release of various agents to actively participate in the process. Interestingly, since one ant-ifouling mechanism and unique approach may not help in dealing with combination of foulants or the use of a single antifouling agent may only have limited effect on a particular range of foulants, multi-component approaches are being followed to effectively induce the surface activity. This can be achieved by incorporating anti-fouling agents having multi-mechanism effects such as active–active, passive–passive, or/and passive–active approaches in anti-fouling membrane preparation protocols. With these novel membrane modification approaches it is expected to achieve robust anti-fouling properties for sustainable water treatment and purification application. [36]

In this direction, recent studies have used a series of nanomaterial-based fillers to induce tailor-made anti-fouling properties to achieve efficient retention of targeted foulant from contaminated sources. This is also achieved by following novel strategies like surface functionalization of membrane polymers which may also induce synergic effects during the separation process. Further, advances in 2-dimensional nanomaterials (2D-NMs) have opened a new avenue for overcoming the several issues associated with membrane fouling during water treatment due to their unique morphologies, structural characteristics, and superior sieving properties. Among many advanced materials, variety of 2D-NMs have been formulated, designed, and developed to act as anti-fouling agent or constituent involving metal

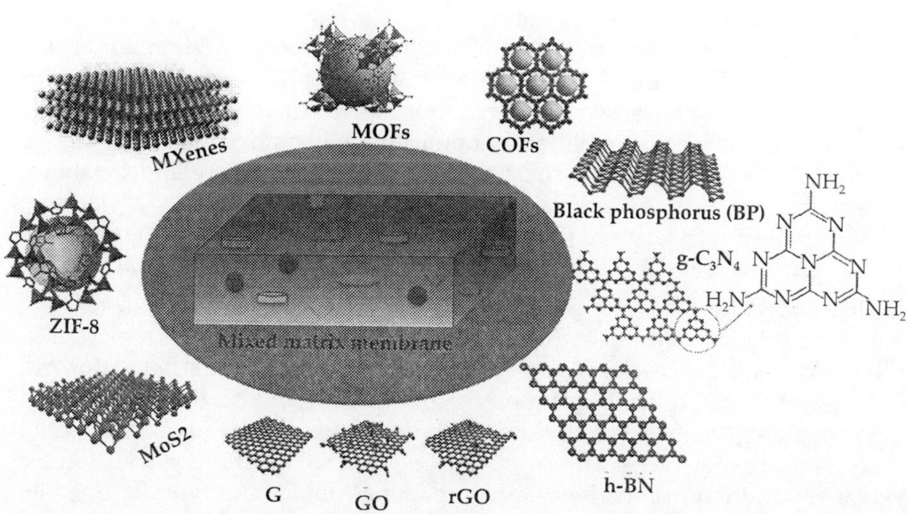

Figure 3.2. Mixed matrix membranes having various emerging 2D-nanomaterials like MOFs, COFs, BPs, h-BN, Graphenes, MOS2, ZIF-8 and MXene incorporated within polymeric membranes used in water treatment applications.

organic frameworks (MOFs), covalent organic frameworks (COFs), graphene, MXenes, transition metal dichalcogenides (TMDs), graphitic carbon nitride (g-C_3N_4), transition metal oxides (TMOs), and so on. Therefore, advanced 2D nanostructures such as MOFs, COFs, ZIFs, MoS_2, graphene, boron nitride (h-BN), MXenes, g-C_3N_4, and black phosphorus (as shown in Figure 3.2) are one of the main constituents of membrane dope solution. They have demonstrate a remarkable surface-to-volume ratio, which promises ultra-low quantity material use, ultra-fast processing time, and ultra-high treatment efficiency during wastewater treatment applications. [37-40]

3.7. Advances in Selective Layer Tuning in Membranes

In the last two decades, thin-film composite membranes have revolutionized the water treatment applications in desalination of seawater. Initially, a thin film composite (TFC) membrane, developed for reverse osmosis applications, composed of mainly 3 layers in a unique engineering design and manufacture protocols. In TFC membranes all 3 layers are bonded together to achieve mechanically robust and high performing separation media at high applied pressure ranges. Among these, the 2-base layers consist of support matrix and asymmetric porous ultra-filtration porous support and the top-most layer is known as selective skin layer. Conventionally, in TFC membranes the top-most selective layer is made of polyamide on porous ultra-filtration and charged polysulfone support. The top-most selective skin layer is generally used for rejecting ionic species or salt , where, bottom two-layer base provides a porous asymmetric structure whose principal function is to provide mechanical strength to overall thin-film composite architecture (Figure 3.3).

Since, the introduction of the first polyamide TFC membrane prepared using trimesoyl chloride and piparizine monomers via interfacial polymerization has gained enormous attention among academia and industrial sectors. Thus, the unique architecture of TFC has shown great potential in achieving the two most important membrane parameters simultaneously, namely, flux and salt rejection when used in desalination applications. Another major advantage of using TFC membrane is associated with their adaptability and flexibility in selecting polymer precursors in optimizing for the formation of selective layer, availability of series of micro-porous middle-substrate layer and can be further modified selective thin-film layer with nanomaterials. [41, 42]

In this architecture, TFC membranes have been widely applied in NF, RO, and FO mainly for seawater and brackish water desalination. Nevertheless, lack of freshwater resources is causing serious concern to the industrial sector and endangers the lives of people around the globe. Additionally, the uncontrolled and unregulated generation and discharge of wastewater from the industrial sector into rivers and freshwater sources is also destroying the aquatic ecosystem and environment at large impacting human and wildlife directly. On the other hand, limitations with conventional TFC membranes are also undoubtedly forcing innovators and researchers to design advanced versions of membranes to suit the futuristics applications. Also, successful application

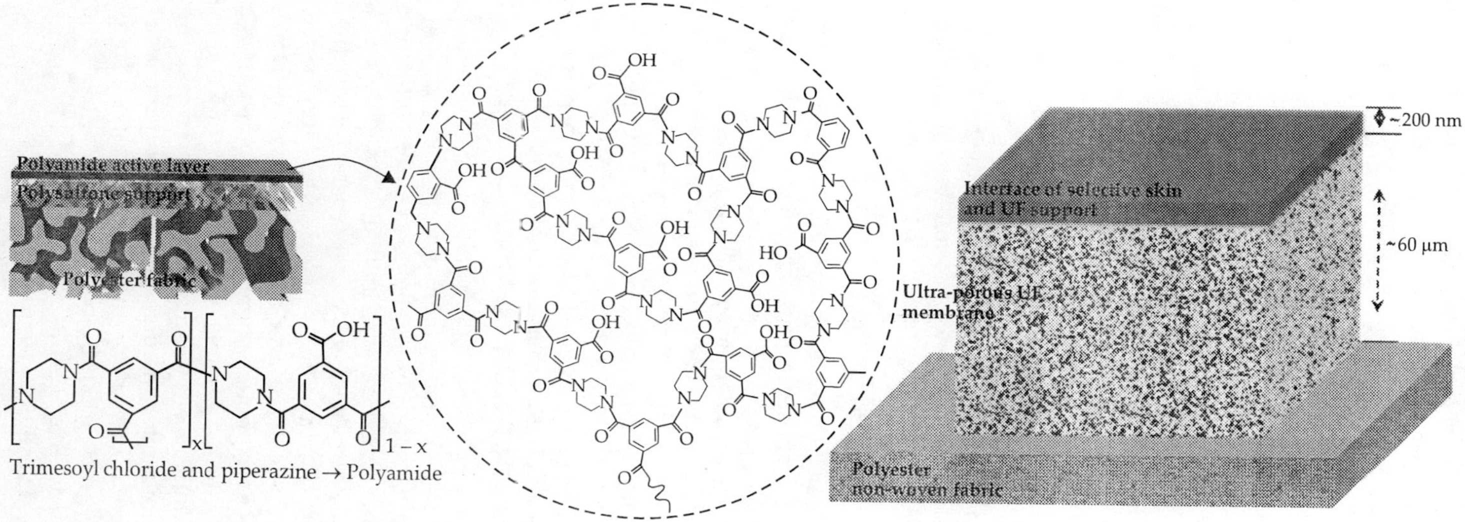

Figure 3.3. Schematic representation of a typical TFC membrane and its architecture with structural specifications prepared using polyester support, polysulfone ultraporous mid-layer and polyamide prepared following interfacial polymerization of trimesoyl chloride and piperazine monomers.

of TFC-polyamide membranes has also been adopted to industrial wastewater treatment owing to their chemical and mechanical robustness in the field of desalination applications at high operating pressure conditions. Therefore, a TFC membrane is regarded as a practical solution in eliminating the toxic pollutants and compounds from the wastewater sources and simultaneously reclaiming reusable quality water thereby value addition purposes would be highly served. Therefore, recent studies have already started adopting new polymeric middle-support structure other than conventional polysulfone (PS) such as relatively more semi-hydrophobic polymer such as polyethersulfone (PES), poly(ether ether sulfone) (PEES), polyvinylidene fluoride (PVDF), polyacrylonitrile (PAN) and so on. Interestingly, these polymer materials are not only available at cost-effective prices but also provide good mechanical and chemical stability and most importantly they show excellent compatibility and membrane forming properties via a simple phase inversion technique.

Nevertheless, in overall TFC membrane performance, compatibility of the ultra-porous polymeric middle-substrate layer plays a crucial dual functionality role in achieving a stable term goal. In the first place, a polymer substrate must provide a better interfacial compatibility to the selective skin layer polymer precursors to initiate interfacial polymerization. Secondly, middle layer polymers need to have sufficient interpenetrating capacity to fuse into the polyester fabric to attain maximum mechanical stability. In all these cases, primarily the dope solution design, formulation and interfacial polymerization conditions in synergetic combination are of utmost importance. This approach yields a TFC membrane with a sponge-like texture and architecture on top of the substrate layer and finger-like channeled porous structure throughout the middle layer of the membrane to ease the water transportation. Further, to convert the membrane into a working module, TFC offers great flexibility maintaining every barrier property. Also several other studies have also clearly explained the importance of 3-layer composite structure to yield both constant and reliable water permeability at high operating pressure maintaining mechanical robustness.

Traditionally, the selective skin layer of TFC membrane is made up of a relatively dense and cross-linked polyamide layer following interfacial polymerization of various suitable monomer precursors that can allow preferrable water transport but hinder the passage of pollutants. However, recent advances have adopted several innovative approaches in creating novel selective skin layers. These include preparation of thin-layer TFC membranes that can be fabricated following simple solution casting, in situ graft polymerization, dip-coating, layer-by-layer coating, interfacial polymer deposition and so on. Even though there are many innovative approaches that have been tried and tested as some of the feasible routes to prepare thin-film layer formation, until recently interfacial polymerization has been the most widely used technique for the preparation of polyamide selective skin layer. Nevertheless, recent breakthroughs have re-adopted simple polymer crosslinking approaches for various biopolymer systems to stabilize them on support in thin-film architecture with improved hydrophilicity and tailor-made permeability and selectivity towards pollutants.

In addition to their high-performance characteristics, polyamide-based TFC have been widely adopted in the desalination market due to their relatively low-cost monomer precursors and simple polymerization technique. Also, several combinations of monomers have already been explored for the crosslinked aromatic polyamide TFC membranes preparation with limited success which also showed less interest in the desalination market due their chemical natures. This also led to the drastic decline in the exploration of new synthetic origin monomer materials for TFC selective layers. At the same time, increased awareness in greener and sustainability concerns have forced researchers to look for biogenic and environmentally friendly alternatives. Therefore, recent research and development efforts are diverted on the sourcing, selection, and optimization of the environment-friendly and easy to handle and polymerizable precursors. [43, 44]

In this direction, the double skinned concept in membrane preparation is being adopted with a sandwiched structure with two-selective layers mainly used for forward osmosis (FO) applications. With this architecture, an active polyamide layer is used for the feed side and a secondary active layer is exposed to draw solution. This approach results in a porous support layer sandwiched between two thin selective layers one on the top and another onbottom part of the polymeric substrates. It can also use conventional polyamide selective layers that can be prepared following interfacial polymerization and further be added with a secondary layer through tailor-made procedure with the introduction of additional functionality at the bottom part. The secondary layer can be thus added to the substrate following various innovative techniques like hydrophilization, grafting, crosslinking, masked-dip coating and layer-by-layer assembly. For this purpose, several novel polymers can be directly used as feasible options for secondary selective layers such as charged polymers. In this direction, positively charged poly(allylamine hydrochloride) (PAH) assembling and negatively charged poly(sodium 4-styrene-sulfonate) (PSS) polyelectrolytes can be easily coated as a secondary layer via self-assembly techniques. Further, the mussel inspired hydrophilic and ionic polymers like polydopamine (PDA) can also be used as a secondary rejection layer through surface hydrophilization. In another approach grafting technique is used to prepare polyelectrolyte brush type selective layer using zwitterionic (poly(3-(*N*-2-methacryloxyethyl-*N*, *N*-dimethyl)ammonia propanesultone) (PMAPS). Through all these approaches, the double skinned membrane acquires unique barrier properties which allows the FO unit operation to be more efficient. Recent studies also reveal that the polyamide thinselective layer facing the draw solution has been successfully tested with high intrinsic permeate flux, lower fouling, and internal concentration polarization. Further, advanced approaches focus on tuning the bottom selective layers with various customized polymer materials design and fabrication that induce surface hydrophilicity and surface functionality with controlled charges to favor selective pollutant separation, chlorine resistant and anti-fouling propensity. [45-47]

As discussed in detail, the present membrane science and technology sector adopted advanced research targeting wastewater treatment and high-quality water reclamation from contaminated sources. However, major concern associated with wastewater treatment is lack of suitable anti-fouling membranes which necessitate designing various advanced strategies modifying membranes following surface hydrophilization approach, incorporating anti-fouling nanomaterials, coating zwitterionic polymers on conventional polymer surfaces, photocatalytic decomposition of catalysts like MOFs, and electrically active or switchable antifouling agents.

3.8. Advances in Development of Nanocomposite Membranes

Recent advances in nanoscience and nanotechnology have attracted the attention of membrane scientists to adopt novel nanomaterials for enhancing the membrane characteristics in their real-world applications. A nanocomposite is a multi-phase material in which one of the phases in 1D, 2D or 3D nanomaterial is less than 100 nm, or in nanocomposite systems, phases would have nano-scale distances between embedded fillers. Advanced thin-film nanocomposite membrane (TFNCM) is generally configured inducing nanomaterials within either on surface or in bulk of the polymer matrix or/and polymeric substrate. Advanced nanocomposite membranes are prepared by incorporating inorganic or organic nanomaterials into polymeric membrane matrix. Some of the nanocomposite membranes reported include inorganic nanoparticles such as metal oxides, hydroxides, oxyhydroxides such as zeolites, TiO_2, Al_2O_3, SiO_2, ZnO, Si, Ag, ZrO_2, $Mg(OH)_2$, $TiSiO_4$, MIFs, COFs, etc. On the other hand, organic nanomaterials include carbon nanotubes, graphene oxide (GO), fullerenes, etc., Recent nanocomposites membranes also use hybrid inorganic/organic nanomaterials, viz., GO-TiO_2, GO-SiO_2, SiO_2-TiO_2, Ag-SiO_2, etc. Therefore, nanocomposite membranes with nanomaterials of inorganic, inorganic and/or hybrid inorganic/organic nature such as metal oxide, MOFs, zeolites, carbonaceous 2D nanomaterials, and grapheneous family are commonly introduced to overcome major intrinsic limitations of conventionally composite membranes. Unlike conventional approaches, the nanocomposite membrane fabrication process can directly use the nanomaterials either by adding to the monomers or polymer dope recipe during the interfacial polymerization of selective layer, bulk of the membrane simply by mixing physically these nanomaterials in polymer dope solutions. [48-50]

3.9. Advances in Functional Materials-based Membrane

The demand for futuristic and sustainable materials stock encourages researchers to greatly improve materials properties by introducing highly functional active sites. Water treatment technologies are also facing their limitations in providing desired end results. In this condition, the development of the sustainable materials for separation technologies is considered a high priority task to reduce waste load to the environment and achieve "zero discharge" target through water treatment. In this direction, functional membrane materials development is an immediate approach to satisfy the growing demand in separation technologies. Among many approaches, the development of a functional porous or non-porous material as a selective barrier for target molecules or ions from a wastewater stream is known to yield encouraging results. Therefore, functional materials are generally characterized as particular native properties and can be further tuned to acquire added functional properties via different functionalization processes. Largely, functional materials were earlier designed to have optical, electronic, magnetic, and piezoelectric properties to be used in the applications like energy conversion, harvesting, storage, memory, and communication devices. [51]

On the other hand, functional membranes offer multiple features and series of advantages, in helping separation of pollutants or contaminants from contaminated water sources. These membranes with suitable functionalities on the surface also contribute to degrading, adsorbing, and/or disabling contaminants. Polymer-based functional membranes initially designed with specific active sites on their surface and bulk of the structure for targeted applications. For example, initial functional polymer membranes were used for the separation of organic and inorganic contaminants such as dyes, emerging pollutants, and heavy metals, etc.

However, recent advances in functional polymer production using highly active functional groups on polymer backbone involves processing and the subsequent formation of active membranes following innovative methods in order to achieve stable membranes. These processes maintain original intensity properties of parent polymers while inducing additional novel properties like bio-compatibility, hydrophilicity, anti-static characteristics, and performance. Among many methods, simple route of physically inducing nanomaterials into polymer matrix such as natural clays, modified clays, colloidal suspension of various nature and speciality polymer nanomaterials are also used to enhance the functionality in conventional polymer membranes such as swelling patterns, surface affinity, adsorption properties, physico-chemical interactions, and rheological/colloidal phenomenon useful for complex wastewater treatment applications. [52]

Owing to the diverse architecture features in structure and materials, water treatment membranes follow two main mass transport mechanisms: pore flow through size, exclusive solute rejection mechanism and solution-diffusion. In conventional membranes, the size exclusive or pore flow mechanism follows a simple filtration process that separates solutes or pollutants or molecules by their size, which is largely known as size-sieving effect or size exclusive principle. It has been evaluated that the transportation mechanism in dense membranes follow the solution-diffusion transport mechanism and it has been largely the same as for gas separation membranes, except that the feed and permeate phases are different. Among many processes, MF, UF, MD and dialysis membranes follow the pore flow mechanism whereas RO, ED and FO have been known to follow the solution-diffusion mechanism. The loose or porous membrane structure allows the preferential transport of relatively smaller solute or pollutant molecules from feed through the pores and simultaneous rejection of larger components. The functionalized polymer membrane follows multiple transportation mechanisms involving ion-exchange, weak electrostatic interactive, hopping mechanism and so on. However, relatively complex and novel nanofiltration (NF) membranes are evaluated to follow a combination of several mechanisms as the pore characteristics are known to induce charge to the surface of the membrane under different operating and feed conditions. Overall, solute or pollutant separation mechanisms from contaminated feed waters can effectively remove solute ranging from 10^{-4} to 10^{2} μm, including organics, colloids, macromolecules like proteins, pathogenic pollutants viruses, bacteria and so on , as shown in Figure 3.4.

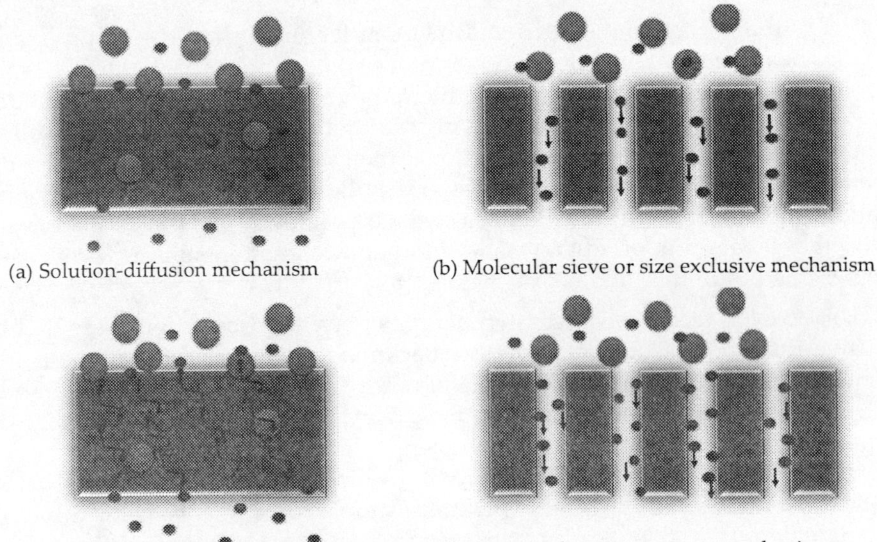

(a) Solution-diffusion mechanism

(b) Molecular sieve or size exclusive mechanism

(c) Functional group transport mechanism

(d) Functional group transport mechanism

Figure 3.4. Transportation mechanisms established in different types of membranes.

Therefore, the commercially most used polymers for RO and NF membranes have been so far polyamide thin-film composite membranes and cellulose acetate membranes both in dense as well as asymmetric configurations. Typical MF membranes are prepared using poly(vinylidene fluoride) (PVDF), polysulfone (PS), polyamides (PA), and polyethylene (PE). However, UF membranes use commodity polymer materials like (PVDF), (PS), poly(ether sulfone) (PES), CA, poly(ether imides), polyacrylonitrile (PAN), aromatic polyamides, and poly(vinyl pyrrolidone). Both MF and UF membranes are mainly prepared following solvent non-induced phase inversion technique for single component membranes and selective layer in nanosized thin-film on porous substrate is prepared following interfacial polymerization to optimize pore sizes less than 10 nm useful for NF applications. [53-64]

Even though last decade, Advances in the last decade in water membranes have shown great potentials in improving performance characteristics in real-world applications compared to conventional following novel materials stock and innovative fabrication protocols, however continued efforts in also yielding improved selectivity and permeability characteristics in recent developments. In recent advancements polymer membranes have prepared the following track etching process that irritates polymeric films gaining momentum with well-defined pore architecture to suit the application. In this method, charged particles in polymer membrane matrices are dispersed and the whole system then passes them through an etch solution to induce uniform and regular pores, albeit in a low volume. Among many recent innovative methods, stretching is another membrane fabrication technique used widely for the preparation of membrane in which crystalline polymer film is extruded up to 300%. In this process, polymers are extruded at a temperature near to their melting points and then rapidly cooled. This polymer stretching process is repeated in the perpendicular direction to induce slit-like pores which can create pores ranging from 20 to 250 nm. Other advanced membrane preparation techniques like grafting, coating, crosslinking, and spin coating have been continuously innovated to achieve precise pore control and induce desired surface functionality even in conventional as well as economically viable polymers.

Among many advanced techniques, use of 2D materials can induce fine and sub-nano sized pore characteristics with controllable pore size distribution. 2D nanomaterials can also induce layer-by-layer architecture within polymer matrices with specific channel-like voids that can be subsequently used as pores with improved mechanical strength. Many of these approaches in developing novel membranes to overcome all the limitations associated with fouling, low permeation, declining rejection patterns, relatively low mechanical strength, and overall shelf-life in each targeting specific applications.

Recently, membrane-based separation processes have been widely applied for oil contaminated wastewater due to their increased importance for ecological as well as environmental concerns. The oil is referred to the various compounds that originate from animal and vegetable origin having greasy constitutes, surfactants, fatty acids, petroleum hydrocarbons, naphthenic acids, phenolic compounds, etc. Cil is considered a significant environmental pollutant generated and discharged from various wastewaters from industries like food manufacturing or processing, steel, textile, aluminum, textile, petrochemical, leather, metal finishing as well as different types of domestic discharges. The oil in water takes the forms like free droplets, dispersion or emulsified droplets characterized by characteristic size distribution. Oil contamination comes in both colloidal size as well as thermodynamically stable emulsion form having droplet sizes as larger as 150 μm to dispersed sizes ranging from 20–150 μm. On the other hand, emulsified oil-droplet sizes can be less than 20 μm, even in some cases reaching nano-sized droplets. Conventionally, various simple physical methods have been used for the treatment of oil contaminated water such as grit removal, gravity separation, adsorption, and skimming. These methods are more useful and efficient for contaminated systems with free oil, while unstable oil–water emulsions require more advanced chemical or mechanical methods where physical methods fall short. However, all these traditional approaches suffer from various shortcomings such as use of toxic compounds, higher-cost, requiring large unit installation space, and more importantly producing secondary pollutants.

However, membrane processes have been successfully used for oil contaminated wastewater of various origin. Nevertheless, the main limitation in membranes for the treatment of oil contaminated water is due to fouling caused by the deposition of colloids oil deposits, particulates accumulated oil granules, or microbial adsorbed droplets onto the membrane surface. Here, oil droplets act as substrates for secondary pollutants providing a conducive atmosphere to grow or deteriorate the wastewaters. These conditions also make wastewater complex to treat in conventional membrane processes. To avoid this issue, conventional membranes are being modified to enhance their surface hydrophilicity. Therefore, increasing the membrane surface's hydrophilicity avoids a fouling phenomenon creating an aqueous barrier between the foulants and the membrane surface.

Therefore, oil/water separation from contaminated water typically uses MF and UF membrane processes. Therefore, commonly used oil/water separation membranes are made of rarely asymmetric single-component or multi-component nanocomposite with a thin selective layer on top supported by microporous substrate membranes largely used in cross-sectional configuration. Nonetheless, the UF process has been the most efficient membrane process for oil/water separation due to nominal molecular weight cut-off (MWCO), which refers to over 90% rejection of the smallest molecular weight component from the membrane barrier. The MWCO of UF membrane ranges between 1,00, 000 and 2,00, 000 daltons (Da) which is suitable for the rejection of stable emulsified oil droplets. On the other hand, MF membrane processes have shown a higher flux rate for oil components in contaminated sources but they have shown relatively low oil rejection efficiencies. Nevertheless, UF and MF processes have been the efficient media for oil/water separation of various origins following simple pore flow mechanisms.

However, the main challenge associated with UF and MF is membrane fouling due to adsorption of surfactant and subsequent pore clogging of membrane by the oil droplets. Therefore, recent advanced studies have invested significant research efforts for estimating and understanding the mechanism of membrane fouling and developing membranes with anti-fouling strategies targeting oily pollutants. Therefore, a wide range of membranes are being tried and tested as filler materials to tune hydrophilic/oleophobic, hydrophobic/oleophobic, or amphiphilic nature in polymer membranes prepared using most prevalent material like PES, PS, PAN, PVDF, and CA. Recent advances have also used polymer anti-fouling grafting agents having hydrophilic properties such as polyethylene glycol (PEG) and polyvinylpyrrolidone (PVP) via blending or physical and chemical surface modification. Recent innovation in surface and interfacial chemistry has also led to the discovery of a series of super-wetting and anti-wetting properties as emerging approaches for oil contamination separation from contaminated sources. These super-wetting and super-absorbent materials are used to selectively absorb oil or water components from the mixture while altogether leaving behind the other. Recent studies have also used ceramic membranes prepared using alumina, titania, zirconia, and silica. Overall, use of advanced techniques and innovative materials can make sustainable barriers for oil/water separation wastewater of complex nature.

Organic solvent nanofiltration (OSN) is gaining greater importance for the treatment or separation of organic components in the molecular range from different mixtures and contaminated water sources. In the recent past, organic solvent separation has been carried out mainly using OSN membranes which reject particles and dissolved molecules smaller than 2 nm. At the industrial scale, plate and frame, hollow fiber, and spiral wound membrane modules have been used for large scale treatment of contaminated or process sources. The OSN process depends on molecular interactions between the solute/solvent, solvent/membrane, and molecular size of solute and solvent molecules. However, molecular diffusion in conventional OSN membranes is limited by their size sieving effect and the physicochemical interactions between the membrane material and feed mixtures. As OSN membranes follow the characteristic solution–diffusion and pore theory transport mechanisms that also exploit various physicochemical properties of solutes including solute charge, size, and hydrophilicity/hydrophobicity in effective separation. [65]

Nevertheless, conventional OSN membranes have been prepared from organic or inorganic materials or combinations of organic/inorganic systems. As synthetic polymers offer economically viable options to convert them into membranes for effective separation of solutes. However, polymer membranes suffer from aging, mechanical deformation, and compaction issues. Thus, ceramic membranes have been included as compaction resistance, leaching resistance, less swelling, lower feed pressures, and easy cleaning. However, ceramic membranes are known to be brittle and face more challenges in scale-up activities than polymeric membranes. Some ceramic materials used include silicium carbide and zirconia, titania, and alunina. In advanced approaches, crosslinking and polymers with post-processing techniques are employed such as drying by solvent exchange, wet or dry annealing, and conditioning treatments to enhance their stability, selectivity, and permeability characteristics of OSN membranes.

In recent advances, OSN membranes are prepared using polymers like (i) PAN, PI, for supports, and (ii) polybenzimidazole (PBI), PSF, polyaniline (PANI), or sulfonated poly(ether ketone) (SPEEK) as NF selective layer applications and (iii) UF support have been largely prepared using polypropylene (PP), and poly(ether ketone) (PEEK). In addition to this, OSN membranes in thin-film composite form utilize solvent-resistant synthetic polymers and inorganic support. Further, all conventional polymers are modified using tailor-made monomers via interfacial polymerization or coating such as acyl chlorides, diamines, and grafting polymers such as poly(ethylene imine) and polydimethylsiloxane (PDMS). Therefore, advanced mixed matrices with zeolites, nanotubes, clay, and fullerenes as filler have been widely applied to OSN applications.

Nevertheless, membrane technologies have lately shown several limitations due to their chemical stability when used under organic solvents conditions. Another major disadvantage with OSN membrane is low permeability and issues associated with reproducibility. Therefore, advanced studies are now focusing on development of:

(i) chemical robustness of the membrane materials in harsh conditions,
(ii) mechanical robustness to control swelling and leaching,
(iii) enhance the multi-functionality to suit wider applications,
(iv) enhance the selectivity for molecular separation,
(v) understand the fundamentals of membrane characteristic using advanced modeling tools, and
(vi) most importantly substantially reduce the capital investment for installation membrane treatment units.

3.10. Block Copolymer Membranes

Block copolymers (BCs) are considered an important section of thermoplastic polymers which are considered viable alternatives in water treatment applications. In advanced approaches, novel block copolymers offer a unique self-assembly feature which helps in tuning polymer chain arrangements through which polymer voids can be utilized for separation and permeation purposes. In this process, isoporous block copolymers are largely used as they offer an easy self-assembling mechanism. The BCs are highly ordered macromolecular architectures formed by the self-assembly of two different or dissimilar monomer species. Due to their pre-defined molecular weight and well-defined nature of blocks BC-membranes can offer narrow high porosities, pore size distributions, and sharp MWCO. PIMs are an advantageous material with a high free volume content and resistance to swelling. The major disadvantage with conventional membrane preparation techniques like solution casting, phase inversion techniques is the uncontrol over pore size distribution in which external parameters play a crucial role. [66, 67]

Interestingly, the first BCP membranes were reported five decades ago as charge-mosaic membranes. The first BCP membrane was prepared using poly(styrene-b-2-vinylpyridine) on a stable substrate. In 1975, Odani et al., also created an ordered porous structure using poly(styrene–butadiene–styrene) (PSBS). In both the cases, researchers followed two main approaches to create ordered porous films, namely: 1) Film casting and followed by selective block degradation, and 2) self-assembly of BCPs through non-solvent-induced phase separation (NIPS). These porous and dense types of BCP films have been applied in the field of ultrafiltration, nanofiltration, gas separation, pervaporation as well as in energy conversion devices like fuel cells and batteries. Recent advances also used promising application of self-assembled copolymers having block polymers PEG and fluoro-blocks. These approaches have been the part of innovative ideas to control the porosity with ordered nano-structures with specific anti-fouling characteristics as inherent features.

Many advanced polymer membranes have been widely used and accepted as effective and environment-friendly separation media. Many inorganic membranes have also been studied as alternatives to polymeric membranes due to their chemical/thermal stability and mechanical strength. Even though many of these commercial membranes offer high performance for solute rejection and solvent recovery, many suffer from high energy consumption and surface fouling problems due to uneven distribution of pores. For all commercial asymmetric RO and NF membranes, a high operating pressure is required to maintain concentration gradient across the membrane whilst sustaining high product flux. In many cases, the operating cost at constant high pressure can exceed the capital cost of the system.

Therefore, there are several basic requirements for a high-performance nanoporous filtration material to tackle issues associated with modern complex wastewater. To achieve this, a high degree of accessible pores, which by implication leads to maximizing the available surface area per unit volume (voids) in the membrane. Another is the intrinsic strength, again related to the connectivity (or crosslinking) of the dense (non-void) part of the membrane. However, in realistic terms, block copolymer materials must be prepared as quickly as possible at low-cost. There are several key technologies where nanoporous media have been critical in enhancing membrane performance. In both fuel cell and dye-sensitized solar cells, this is seen by increasing the local surface area of the platinumized and titanium oxide electrodes. Also, controlling pore sizes whilst maintaining a satisfactory continuity from top-to-bottom in separation membranes will enable the application of these systems for nanofiltration. Doing so in a practical way would enhance the viability of membrane separation to replace conventional, expensive, and cumbersome distillation-based desalination technologies with smaller, simple membrane-based solutions.

Nanoporous polymeric membranes with good interconnectivity can be produced using several methods that invoke microphase separation. Thus, demixing of two polymers can be dynamically controlled to vary the size scale of pore sizes and the extent of their connectivity. A related dynamic method that in principle produces connected nanopores is the so-called breath-figure method via the condensation of water vapor into a drying polymer film. Microporous polymeric membranes are often prepared from rigid materials with large voids that create interconnected networks. This is as true of membranes formed by Loeb-Sourirajan type phase inversion methods as block copolymers. Typically, the thickness over which the nanoporous structure performs its separating function is a few hundred nanometers, which constitute the selective portion of the membrane. The remainder of the membrane material is mainly there to provide geometrical and mechanical support. Hence, there is a requirement to ensure that the selective portion of the membrane is fully open and accessible from the more macroporous sections of the membrane. However, it is well known that BCP films might form the selective layers that tend to have one component of the BCP preferentially wetting the bounding surfaces of the film. Moreover, depending on the architecture of the BCP, its orientation may not be conducive to direct diffusing species across the membrane. Some attention has been given to creating BCP films with a cylindrical morphology (later to become cylindrical pores) with an orientation that runs across the films. This has been achieved by several methods involving external fields or by the dynamic evaporation of selective solvents. However, there is a possibility that such methods can be hard to scale up or be difficult to control. A much more elegant method makes use of the thermodynamically stable double gyroid phase that is observed in some block copolymers since the two components of such BCP films are intrinsically connected throughout the film. However, the relative polymer compositions and molecular weight window where the double gyroid system exists is quite sensitive and the architecture has only been noted for a handful of polymer systems. [68-71]

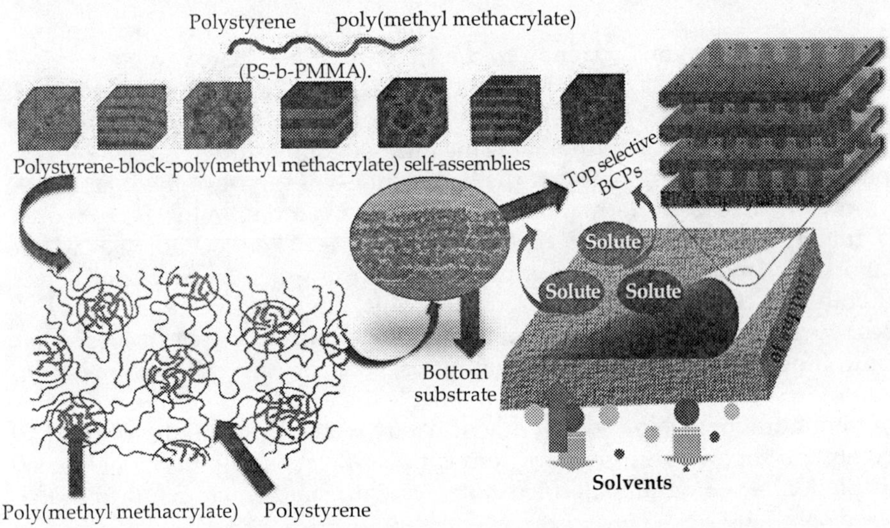

Figure 3.5. Block copolymer (BCP) self-assemblies obtained from diblock copolymer with different volume fractions in the varying molecular weight distribution block chains. These BCP patterns can be prepared beyond conventional equilibrium morphologies to obtain specific self-assemblies to obtain specific fractions to be selectively removed to create porous structures in thin-selective layers.

The preparation of the porous structure from the block copolymer itself takes place by the selective crosslinking and degradation of the two polymer phases simultaneously. For instance, well-studied polystyrene (PS) block copolymer systems with poly(methyl methacrylate) (PMMA) are widely used for preparing highly ordered porous nano-architecture on various substrates to yield well defined morphologies as shown in Figure 3.5. Similarly, porous thin can be prepared by degrading one block from the diblock in the polyisoprene and polybutadiene system with a combination of UV/ozone and subsequent dialysis of the degraded phase. In other systems, simple hydrolysis of one component is sufficient to create the nanoporous network. One of the assumptions that is implicit in literature is that the UV-induced crosslinking and degradation of the BCP leads to creating a replica of the original BCP architecture. Indeed this is also documented experimentally for cylindrical and lamellar systems of PS-*b*-PMMA as well. In the former case, PMMA cylinders have been degraded from the PS matrix and subsequently refilled. However, in the case of perpendicular lamellae, the degradation of the interlayers of PMMA leads to a collapsed structure of cross-linked PS sheets. The assumption, that seems to be implicit in the literature, is that the same UV process will have no benefit to a system of PS-*b*-PMMA spherical micelles, with PMMA as a minor phase. This is because the resulting degraded phase, if one can remove it at all, remains separated from other similar voids. It has been our observation that this assumption is quite wrong and, in fact, it is possible to generate nanoporous bicontinuous thin films from spherical block copolymers; and their remarkable utilization in filtration membranes is discussed in this chapter. It is demonstrated here that these materials can filter with 100% efficiency small dye molecules with molecular weights in the range of 200-900 whilst further tuning the BCP-based membranes allows their application for purification of organic solvents, removal of urea, biological contaminants as well as colloidal pollutants.

Therefore, molecular manipulation of a porous structure in combination with surface chemistry is now a well-known approach to the creation of well-ordered pore architecture. This is a need of hour as the emerging pollutants are posing a serious threat to the water treatment technologies. This is also required to create pores physically to increase the features of all membranes in certain applications. In this regard, use of block copolymer self-assembly has shown great potential. Block copolymers have now paved the way to overcome the several disadvantages of the conventional membranes. These BCPs can easily be used in the form of thin-selective film either porous or dense films on flat or hollow fiber substrates. With a self-assembled BCP route different chemically stable structures can be obtained by keeping one polymer block with a specific porosity and composition can also be altered in combination with other nanomaterials which suits huge applications in both gas and liquid separation applications such as pervaporation, gas separation, nanofiltration and fuel cells.

Besides diblocks, triblocks, and pentablock copolymers are also equipped with micro-domain islands, which are complex and their periodic geometry offers unlimited nanoscale engineering potentials for porous membrane applications. As it is carried out in the diblock copolymers, selective degradation of specific blocks by deterioration of one-block in spheres, lamellae, cylinders and complex nano-bicontinuous pores. Similarly, selective degradation can also be achieved in the triblock copolymers with more complex structures with numerous morphological components unknown to the science and technology membrane community. Nevertheless, once realized BCPs could revolutionize the properties of fluid transport and can offer a range of high-performance membranes for advanced treatment applications in near future.

3.11. Polyelectrolyte Complex Membranes

Polyelectrolyte complexes (PECs) are the association polymer complexes formed between oppositely charged polymer systems when combined in specific composition having unique loading as well as surface properties which are useful for applications like water purification and treatment. Polyelectrolytes have already been used in the past as chemical flocculants conventionally in water treatment in which they act as colloidal breaking agent in coagulation/flocculation stage. Polyelectrolytes have been traditionally used as colloidal particle stabilizers or breaking the colloidal stability with slightly varying their concentrations and pH condition which leads to conditioning or thickening leading to sludge formation. Advanced polyelectrolytic systems in simple charged agents as well as complex forms that are either available in branched or linear polymer backbone carrying ionic charge along their chain. Polyelectrolytes occur in either anionic, cationic, or non-ionic polymers. Conventionally, these charged flocculants neutralize surface charges of pollutants breaking the neutrality of contaminants leading to agglomeration of solute together through physical or chemical forces. On the other hand, PEC-based membranes have already shown promising materials, mainly in pervaporation separation of different contaminated mixtures and have shown encouraging separation performance in nanofiltration applications.

Interestingly, polyelectrolytes have a strong affinity to adsorb charged solute or pollutants on their surfaces in an aqueous suspension. The active groups in ionic polymers with characteristic zeta potentials are called polyelectrolytes that can be tuned to their tailor-made surface properties altering pH and hydrolyzed conditions. Recent advanced research is aiming to make polyelectrolyte complexes cost-effective, quick, high performance, efficient, thick, and chemically controlled to predict their end properties. In this direction, several approaches reveal that the innovative and advanced methods made PECs an easy-to-use composites in both membranes as well as non-membrane configurations. In terms of polymer molecular structure, PECs can be tuned to attain weak and/or strong electrolyte activity. This can be achieved by changing molecular weight, concentration, and charge density.

Some of the major advantages, before and after the treatment process, PECs can be separated following different mechanisms following their electrostatic repulsion forces. This type of repulsive forces is largely maintained on polymer chains due to their open structures and high charging density. In water treatment applications, PECs-based membrane can be extremely thin, however can act as an effective barrier for selective separation of pollutants. Also, as filler, incorporation of polyelectrolytes on the conventional membrane surface can also act as a highly effective and tailor-made separation barrier resulting in efficient separation of contaminants. In most of the cases, charge on polyelectrolyte can effectively be used to separate ionic pollutants. Therefore, salt ion penetration can be made to reject via polyelectrolyte-modified membranes. Recent approaches have also shown great potential in using polyelectrolytes that can be used for inducing uniform repulsive charges on membrane surfaces with excellent control over their distribution even on a series of conventional as well as commercially successful membrane surfaces. Therefore, polyelectrolyte complex membranes can be innovatively designed, fabricated, and adopted on various substrates using numerous stocks including biocompatible, and biodegradable resources like biopolymers. With these PECs-based systems having highly film forming properties an effective, sustainable, non-toxic, biocompatible, and cost-effective alternative in water treatment applications where conventional membranes fail to provide effective outcome. [72, 73]

3.12. Future Perspectives

Detailed analysis of conventional membranes and broader understanding of recent advances in various membrane material stocks as well as membranes reveal their ongoing relevance in the water treatment sector. On the other hand, TFC membranes have been so far widely used for various types of water and wastewater treatment applications with varying end results. Even though TFC membranes have been widely used for different industrially important applications like process water recovery, wastewater treatment as well as desalination of seawater. Nonetheless, TFC membranes have received fair share of criticism due to their inherent limitations associated with their fouling proficiency and low mechanical stability in the long run. TFC membranes have shown limited success when applied for complex wastewater treatment applications. Extensive studies reported on to evaluate the performance of TFC membranes for complex feed systems like leather industry wastewater, sugar industry waste, oily/greasy wastewater treatment and so on reveal that the separation and purification mechanism in TFC greatly depend on the type and concentration of wastewater sources.

Even though there have been numerous studies detailing the various separation properties of conventional membranes, however, the separation mechanisms in multilayer TFC membranes as well as the changes in the membrane-active layer and their transport behavior due to physicochemical properties during and after the separation process reveals the limited information. This complex multi-layer membrane architecture is a complex understanding as TFC membrane while producing product water cumulative effect is still to be analyzed. This is also due to the proecess of a range of inorganic and organic components in TFC as well as thin-film nanocomposite membranes that may affect the overall membrane performance and integrity, particularly for a long-term operation. In addition to this, the quantitative analysis and characteristic analysis of various conventional membrane material stocks as well as their fabrication techniques are important to assess keeping several factors that govern the membrane performance in mind such as anti-fouling properties.

On the other hand, nature and complexities of contaminated water and wastewater constituents also play a major role in overall performance of membranes applied. For instance, constituents in the oil-contaminated wastewater may interact with the conventional polymeric substrate and polyamide active layer in TFC have had limited understanding and further advanced studies are warranted. The complex pollutants in wastewater when exposed to selective layers of TFC as well as conventional membranes eventually lead to surface degradation and emergence concentration polarization enforces new organic contaminants in the product water or permeate. Therefore, there needs to be a series of studies to be constituted to better understand the nature, type of membrane materials, and change in the membrane preparation protocols can also yield change in the barrier properties. This would lead to the altering separation efficiency in terms of permeate flux, solute or pollutant rejection, and fouling resistance. These cumulative properties can also propel engineering of the membranes into working modules and their advancement in the water treatment sector. Besides more reliable and standard advanced analytical techniques are required to accurately quantify the composition membrane as well as composition of dissolved species complex wastewater. Further studies also warrant the understanding of product water quality that has complex nature due to the presence of a broad range of pollutants such as pharmaceuticals, dispersed oil, heavy metals, organic and inorganic compounds, as well as series of emerging pollutants in different concentrations.

This chapter also attempted to address the issues associated with intrinsic limitations of membrane technology, particularly the nature of membrane material, their preparation techniques as well as membrane fouling. Among these, membrane fouling is the primary concern which is mainly considered before membrane materials and modules are decided to use for practical application. In recent times, one of the most required efforts in this field is finding the suitable membrane material stocks, selection of combination of materials, and designing suitable membrane preparation protocols to achieve development of high anti-fouling membranes to improve the separation performance and enhance the sustainability of membranes during their practical application at industrial scale. In this regard, the fabrication of double-skinned thin-film composite membranes has been tried and tested as a promising strategy to address the issue associated with membrane fouling. For such efforts, fouling- resistant layers such as zwitterionic polymers, hydrophilic biopolymers, brush-like copolymers as well as nanomaterial-based antifouling agents have been successfully used. On the other hand, advanced studies have also critically evaluated the mechanism and severity of membrane fouling. Accordingly, the fouling takes place not only on the membrane surface but also critical evaluation of solute deposition in the interior of the pore walls. This information has also helped in minimizing overall solute scaling through incorporating antifouling fillers within the polymeric substrates.

Nevertheless, despite the improved efforts made in designing the innovative membranes, it is also important to understand the interaction between the membrane and the contaminants at molecular level and potential intermediate formation in order to have better insights of membrane fouling mechanisms in wastewater treatment. This chapter also discussed several factors that are worth mentioning on effective fouling mitigation strategies usually achieved via the synergetic effects obtained from the combination of membrane materials design, optimization of hydrodynamic conditions, module development, and redefining the membrane cleaning protocols post-water treatment applications. Among many other parameters, energy is required to operate membranes and economical aspects associated with commissioning of plants mainly determines the sustainability of the membrane process. Membrane-based technologies have always been considered capital intensive and substantial price goes in maintaining the overall process in operations. Nevertheless, membrane fouling and high maintenance cost remains the main deterrent for membrane technologies to be widely adopted for process applications in industries and highly contaminated water treatment.

Moving forward, the potential application of advanced membranes overcoming all traditional limitations at commercial scale wastewater treatment poses a series of critical questions on:

(a) economic trade-offs between lower permeability in asymmetric membranes maintaining negligible fouling propensity,

(b) innovating novel membrane modules to minimize treatment unit footprint, and

(c) adopting high temperature and mechanically stable membrane precursors to withstand harsh working environment and so on. Despite several innovative efforts made in the design and development of highly efficient and cost-effective membrane processes, a significant breakthrough is yet to be achieved to improve the shelf-life of the membranes.

In summary, membrane technologies have been largely used in many commercial-scale applications for both industrial applications as well as domestic water treatment purposes. Even though membrane technology is greatly contributing to the environmental cleanliness and recovering large numbers of resources from various waste sources, membrane processes and overall technology itself will have to become sustainable when the waste produced during preparation of membranes, during their use, maintenance, operation as well as post-use disposal is reduced. On the other hand, the growing environmental concerns associated with treatment technologies has also prompted the exploration of greener and bio-origin membrane materials to tackle the issues related to the sustainability of membrane processes. As of today, membrane module disposal post-use seems great into the dueto its synthetic polymer origin and use of several engineering materials which are again polymeric in nature. These discarded membranes and membrane modules have been considered waste. Therefore, use of sustainable, biocompatible, and biodegradable membrane materials stock as well as engineering materials in module preparation is highly anticipated. In this direction, several efforts have also been made to use biopolymers as effective alternatives for synthetic polymer or fossil-based polymer barriers derived from vegetable sources, plants, animal, seaweed-based macromolecules, and bacterial fermentation products and so on.

3.13 Conclusion

Membrane technologies have been in use for decades for effective water purification and wastewater treatment applications. The pressure driven membrane processes like MF, UF, NF, and RO have been successfully used for wide range of applications related to water purification and wastewater treatment. However, with increased water consumption and generation of uncontrolled amounts of wastewater from various sources, membrane processes are facing newer challenges. Even though membrane technologies have proven effective against conventional pollutants, but modern pollutants and emerging contaminants are posing serious threat to the membrane for their effective treatment. Nevertheless, modern membrane preparation methods, protocols and techniques have substantially improved the membrane stability and performance against modern pollutants and emerging contaminants. However, researchers and innovators need to pay deeper attention to tackle futuristic challenges from harsh and severely polluted water sources to make the overall process relevent.

References

1. Shams Forruque Ahmed, P. Senthil Kumar, Mahtabin Rodela Rozbu, Anika Tasnim Chowdhury, Samiha Nuzhat, Nazifa Rafa, T.M.I. Mahlia, Hwai Chyuan Ong, M. Mofijur, Heavy metal toxicity, sources, and remediation techniques for contaminated water and soil, Environmental Technology & Innovation, Volume 25, 2022, 102114, ISSN 2352-1864.
2. Luisa Lucattini, Giulia Poma, Adrian Covaci, Jacob de Boer, Marja H. Lamoree, Pim E.G. Leonards, A review of semi-volatile organic compounds (SVOCs) in the indoor environment: occurrence in consumer products, indoor air and dust, Chemosphere, Volume 201, 2018, Pages 466-482, ISSN 0045-6535.
3. Rajendran, S.R.C.K., Mason, B. & Doucette, A.A. Review of Membrane Separation Models and Technologies: Processing Complex Food-Based Biomolecular Fractions. Food Bioprocess Technol 14, 415–428 (2021). https://doi.org/10.1007/s11947-020-02559-x.
4. Kamran Ghasemzadeh, Abbas Aghaeinejad-Meybodi, Angelo Basile, Chapter 4 - Separation Theory of Silica Membranes, Editor(s): Angelo Basile, Kamran Ghasemzadeh, Current Trends and Future Developments on (Bio-) Membranes, Elsevier, 2017, Pages 65-95, ISBN 9780444638663.
5. Sweta Parimita Bera, Manoj Godhaniya, Charmy Kothari, Emerging and advanced membrane technology for wastewater treatment: A review, Journal of Basic Microbiology, 2021, Volume 62, Issue 3-4 p. 245-259.
6. Sameer Al-Asheh, Marzieh Bagheri, Ahmed Aidan, Membrane bioreactor for wastewater treatment: A review, Case Studies in Chemical and Environmental Engineering, Volume 4, 2021, 100109, ISSN 2666-0164.
7. Li, L., Visvanathan, C. Membrane technology for surface water treatment: advancement from microfiltration to membrane bioreactor. Rev Environ Sci Biotechnol 16, 737–760 (2017). https://doi.org/10.1007/s11157-017-9442-1.
8. https://www.grandviewresearch.com/industry-analysis/membrane-separation-technology-market.
9. Chernyshova et al.; Multidisciplinary Advances in Efficient Separation Processes ACS Symposium Series; American Chemical Society: Washington, DC, 2020.
10. Zakhar Maletskyi, Advances in Membrane Materials and Processes for Water and Wastewater Treatment, Multidisciplinary Advances in Efficient Separation Processes. January 1, 2020 , 3-35.
11. Dong X, Lu D, Harris TAL, Escobar IC. Polymers and Solvents Used in Membrane Fabrication: A Review Focusing on Sustainable Membrane

Development. Membranes (Basel). 2021 Apr 23; 11(5): 309. doi: 10.3390/membranes11050309. PMID: 33922560; PMCID: PMC8146349.

12. Ajith J. Jose, Jincymol Kappen, Muthukaruppan Alagar, 2 - Polymeric membranes: Classification, preparation, structure physiochemical, and transport mechanisms, Editor(s): Sabu Thomas, Preetha Balakrishnan, M.S. Sreekala, In Woodhead Publishing Series in Biomaterials, Fundamental Biomaterials: Polymers, Woodhead Publishing, 2018, Pages 21-35, ISBN 9780081021941.
13. Arumugham T, Kaleekkal NJ, Gopal S, Nambikkattu J, K R, Aboulella AM, Ranil Wickramasinghe S, Banat F. Recent developments in porous ceramic membranes for wastewater treatment and desalination: A review. J Environ Manage. 2021 Sep 1; 293: 112925. doi: 10.1016/j.jenvman.2021.112925. Epub 2021 Jun 7. PMID: 34289593.
14. Thanigaivelan Arumugham, Noel Jacob Kaleekkal, Sruthi Gopal, Jenny Nambikkattu, Rambabu K, Ahmed Mamdouh Aboulella, S. Ranil Wickramasinghe, Fawzi Banat, Recent developments in porous ceramic membranes for wastewater treatment and desalination: A review, Journal of Environmental Management, Volume 293, 2021, 112925, ISSN 0301-4797.
15. Muhammad Bilal Asif, Zhenghua Zhang, Ceramic membrane technology for water and wastewater treatment: A critical review of performance, full-scale applications, membrane fouling and prospects, Chemical Engineering Journal, Volume 418, 2021, 129481, ISSN 1385-8947.
16. A. Ghaee, M. Shariaty-Niassar, J. Barzin, T. Matsuura, Effects of chitosan membrane morphology on copper ion adsorption, Chemical Engineering Journal, 165 (2010) 46-55.
17. A. Moriya, T. Maruyama, Y. Ohmukai, T. Sotani, H. Matsuyama, Preparation of poly(lactic acid) hollow fiber membranes via phase separation methods, Journal of Membrane Science, 342 (2009) 307-312.
18. R.Y.M. Huang, R. Pal, G.Y. Moon, Characteristics of sodium alginate membranes for the pervaporation dehydration of ethanol–water and isopropanol–water mixtures, Journal of Membrane Science, 160 (1999) 101-113.
19. P. Tomietto, P. Loulergue, L. Paugam, J.-L. Audic, Biobased polyhydroxyalkanoate (PHA) membranes: Structure/performances relationship, Separation and Purification Technology, 252 (2020).
20. D. Chen, M.A. Hickner, E. Agar, E.C. Kumbur, Optimizing membrane thickness for vanadium redox flow batteries, Journal of Membrane Science, 437 (2013) 108-113.
21. J. Guo, Q. Zhang, Z. Cai, K. Zhao, Preparation and dye filtration property of electrospun polyhydroxybutyrate–calcium alginate/carbon nanotubes composite nanofibrous filtration membrane, Separation and Purification Technology, 161 (2016) 69-79.
22. P.-J. Lin, M.-C. Yang, Y.-L. Li, J.-H. Chen, Prevention of surfactant wetting with agarose hydrogel layer for direct contact membrane distillation used

in dyeing wastewater treatment, Journal of Membrane Science, 475 (2015) 511-520.

23. H. Pang, K. Tian, Y. Li, C. Su, F. Duan, Y. Xu, Super-hydrophobic PTFE hollow fiber membrane fabricated by electrospinning of Pullulan/PTFE emulsion for membrane deamination, Separation and Purification Technology, 274 (2021).
24. P. Samoila, A.C. Humelnicu, M. Ignat, C. Cojocaru, V. Harabagiu, Chitin and Chitosan for Water Purification, Chitin and Chitosan 2019, pp. 429-460.
25. Y. Yang, A. Raza, F. Banat, K. Wang, The separation of oil in water (O/W) emulsions using polyether sulfone & nitrocellulose microfiltration membranes, Journal of Water Process Engineering, 25 (2018) 113-117.
26. L. Manjarrez Nevárez, L. Ballinas Casarrubias, O.S. Canto, A. Celzard, V. Fierro, R. Ibarra Gómez, G. González Sánchez, Biopolymers-based nanocomposites: Membranes from propionated lignin and cellulose for water purification, Carbohydrate Polymers, 86 (2011) 732-741.
27. L.A. Goetz, B. Jalvo, R. Rosal, A.P. Mathew, Superhydrophilic anti-fouling electrospun cellulose acetate membranes coated with chitin nanocrystals for water filtration, Journal of Membrane Science, 510 (2016) 238-248.
28. A. Prasannan, J. Udomsin, H.-C. Tsai, C.-F. Wang, J.-Y. Lai, Robust underwater superoleophobic membranes with bio-inspired carrageenan/laponite multilayers for the effective removal of emulsions, metal ions, and organic dyes from wastewater, Chemical Engineering Journal, 391 (2020).
29. T.P.N. Nguyen, E.-T. Yun, I.-C. Kim, Y.-N. Kwon, Preparation of cellulose triacetate/cellulose acetate (CTA/CA)-based membranes for forward osmosis, Journal of Membrane Science, 433 (2013) 49-59.
30. D. Rana and T. Matsuura, Surface Modifications for Antifouling Membranes, Chemical Reviews 2010 110 (4), 2448-2471.
31. Runnan Zhang, Yanan Liu, Mingrui He, Yanlei Su, Xueting Zhao, Menachem Elimelech and Zhongyi Jiang, Antifouling membranes for sustainable water purification: strategies and mechanisms, Chem. Soc. Rev., 2016,45, 5888-5924.
32. Ibrar, I.; Naji, O.; Sharif, A.; Malekizadeh, A.; Alhawari, A.; Alanezi, A.A.; Altaee, A. A Review of Fouling Mechanisms, Control Strategies and Real-Time Fouling Monitoring Techniques in Forward Osmosis. Water 2019, 11, 695. https://doi.org/10.3390/w11040695.
33. Mohamed, Bassyouni., Mohamed, Helmy, Abdel-Aziz., Mohamed, Helmy, Abdel-Aziz., M., Sh., Zoromba., M., Sh., Zoromba., S.M.S., Abdel-Hamid., Enrico, Drioli., Enrico, Drioli. (2019). A review of polymeric nanocomposite membranes for water purification. Journal of Industrial and Engineering Chemistry, 73:19-46. doi: 10.1016/J.JIEC.2019.01.045.
34. Tahseen Kamal, Nauman Ali, Abbas A. Naseem, Sher B. Khan and Abdullah M. Asiri, Polymer Nanocomposite Membranes for Antifouling Nanofiltration, Recent Patents on Nanotechnology, ISSN (Print): 1872-2105 ISSN (Online): 2212-4020, Volume 10, Issue 3, 2016, p. 189 – 201.

35. Ashwin R. Kamble, Chetan M. Patel, Z.V.P. Murthy, A review on the recent advances in mixed matrix membranes for gas separation processes, Renewable and Sustainable Energy Reviews, Volume 145, 2021, 111062, ISSN 1364-0321.

36. Wen, Y., Yuan, J., Ma, X. et al. Polymeric nanocomposite membranes for water treatment: a review. Environ Chem Lett 17, 1539–1551 (2019). https://doi.org/10.1007/s10311-019-00895-9.

37. Wang J, Zhang P, Liang B, Liu Y, Xu T, Wang L, Pan K (2016) Graphene oxide as an effective barrier on a porous nanofibrous membrane for water treatment. ACS Appl Mater Interfaces, 8(9): 6211–6218.

38. Jabbari V, Veleta J, Zarei-Chaleshtori M, Gardea-Torresdey J, Villagrán D (2016) Green synthesis of magnetic MOF@ GO and MOF@ CNT hybrid nanocomposites with high adsorption capacity towards organic pollutants. Chem Eng J, 304: 774–783.

39. Abdi J, Vossoughi M, Mahmoodi N, Alemzadeh I (2017) Synthesis of metal-organic framework hybrid nanocomposites based on GO and CNT with high adsorption capacity for dye removal. Chem Eng J, 326: 1145–1158.

40. Torad N, Hu M, Ishihara S, Sukegawa H, Belik A, Imura M, Yamauchi Y (2014) Direct synthesis of MOF-derived nanoporous carbon with magnetic Co nanoparticles toward efficient water treatment. Small 10(10): 2096–2107.

41. Lu Elfa Peng, Zhe Yang, Li Long, Shenghua Zhou, Hao Guo, Chuyang Y. Tang, A critical review on porous substrates of TFC polyamide membranes: Mechanisms, membrane performances, and future perspectives, Journal of Membrane Science, Volume 641, 2022, 119871, ISSN 0376-7388.

42. Feng Liu, LanLan Wang, Dawei Li, Qingsheng Liu and Bingyao Deng, A review: the effect of the microporous support during interfacial polymerization on the morphology and performances of a thin film composite membrane for liquid purification, RSC Adv., 2019, 9, 35417-35428.

43. Lau WJ, Gray S, Matsuura T, Emadzadeh D, Chen JP, Ismail AF. A review on polyamide thin film nanocomposite (TFN) membranes: History, applications, challenges and approaches. Water Res. 2015 Sep 1;80:306-24. doi: 10.1016/j.watres.2015.04.037. Epub 2015 May 6. PMID: 26011136.

44. Jaydevsinh M. Gohil, Paramita Ray, A review on semi-aromatic polyamide TFC membranes prepared by interfacial polymerization: Potential for water treatment and desalination, Separation and Purification Technology, Volume 181, 2017, Pages 159-182, ISSN 1383-5866.

45. Lijo Francis, Oluwaseun Ogunbiyi, Jayaprakash Saththasivam, Jenny Lawler and Zhaoyang Liu, A comprehensive review of forward osmosis and niche applications, Environ. Sci.: Water Res. Technol., 2020, 6, 1986-2015.

46. Blandin G, Ferrari F, Lesage G, Le-Clech P, Héran M, Martinez-Lladó X. Forward Osmosis as Concentration Process: Review of Opportunities and Challenges. Membranes (Basel). 2020 Oct 14; 10(10): 284. doi: 10.3390/membranes10100284. PMID: 33066490; PMCID: PMC7602145.

47. Wafa Suwaileh, Nirenkumar Pathak, Hokyong Shon, Nidal Hilal, Forward osmosis membranes and processes: A comprehensive review of research

trends and future outlook, Desalination, Volume 485, 2020, 114455, ISSN 0011-9164.

48. Wen, Y., Yuan, J., Ma, X. et al. Polymeric nanocomposite membranes for water treatment: a review. Environ Chem Lett 17, 1539–1551 (2019). https://doi.org/10.1007/s10311-019-00895-9.
49. Sivasubramanian Manikandan, Ramasamy Subbaiya, Muthupandian Saravanan, Mohanadoss Ponraj, Masilamani Selvam, Arivalagan Pugazhendhi, A critical review of advanced nanotechnology and hybrid membrane based water recycling, reuse, and wastewater treatment processes, Chemosphere, Volume 289, 2022, 132867, ISSN 0045-6535.
50. Abedalkader Alkhouzaam, Hazim Qiblawey, Functional GO-based membranes for water treatment and desalination: Fabrication methods, performance and advantages. A review, Chemosphere, Volume 274, 2021, 129853, ISSN 0045-6535.
51. Eds. Inamuddin, Tariq Altalhi, Mohd Imran Ahamed, Mohammad Luqman Advanced Functional Membranes, Materials and Applications, Materials Research Foundations Vol. 120 Publication Date 2022, 228 Pages Print ISBN 978-1-64490-180-9 (release date March 2022) ePDF ISBN 978-1-64490-181-6.
52. Shoji Kimura, Transport Phenomena in Membrane Separation Processes, The Society of Chemical Engineers, Japan 1992 Volume 25 Issue 5 Pages 469-480.
53. J.-H. Choi, J. Jegal, W.-N. Kim, Fabrication and characterization of multi-walled carbon nanotubes/polymer blend membranes, Journal of Membrane Science, 284 (2006) 406-415.
54. L. Deng, T.-J. Kim, M.-B. Hägg, Facilitated transport of CO2 in novel PVAm/PVA blend membrane, Journal of Membrane Science, 340 (2009) 154-163.
55. X. Cao, J. Ma, X. Shi, Z. Ren, Effect of TiO 2 nanoparticle size on the performance of PVDF membrane, Applied Surface Science, 253 (2006) 2003-2010.
56. J.-Y. Lai, M.-J. Liu, K.-R. Lee, Polycarbonate membrane prepared via a wet phase inversion method for oxygen enrichment from air, Journal of Membrane Science, 86 (1994) 103-118.
57. J.-F. Li, Z.-L. Xu, H. Yang, L.-Y. Yu, M. Liu, Effect of TiO2 nanoparticles on the surface morphology and performance of microporous PES membrane, Applied Surface Science, 255 (2009) 4725-4732.
58. M.B. Karimi, G. Khanbabaei, G.M.M. Sadeghi, Vegetable oil-based polyurethane membrane for gas separation, Journal of Membrane Science, 527 (2017) 198-206.
59. J. Yin, G. Zhu, B. Deng, Graphene oxide (GO) enhanced polyamide (PA) thin-film nanocomposite (TFN) membrane for water purification, Desalination, 379 (2016) 93-101.
60. H.H. Yong, H.C. Park, Y.S. Kang, J. Won, W.N. Kim, Zeolite-filled polyimide membrane containing 2, 4, 6-triaminopyrimidine, Journal of Membrane Science, 188 (2001) 151-163.

61. M. Elimelech, Z. Xiaohua, A.E. Childress, H. Seungkwan, Role of membrane surface morphology in colloidal fouling of cellulose acetate and composite aromatic polyamide reverse osmosis membranes, Journal of Membrane Science, 127 (1997) 101-109.

62. M. Han, Thermodynamic and rheological variation in polysulfone solution by PVP and its effect in the preparation of phase inversion membrane, Journal of Membrane Science, 202 (2002) 55-61.

63. Y. Xiuli, C. Hongbin, W. Xiu, Y. Yongxin, Morphology and properties of hollow-fiber membrane made by PAN mixing with small amount of PVDF, Journal of Membrane Science, 146 (1998) 179-184.

64. J.M. Luque-Alled, A.W. Ameen, M. Alberto, M. Tamaddondar, A.B. Foster, P.M. Budd, A. Vijayaraghavan, P. Gorgojo, Gas separation performance of MMMs containing (PIM-1)-functionalized GO derivatives, Journal of Membrane Science, 623 (2021).

65. Patrizia Marchetti, Maria F. Jimenez Solomon, Gyorgy Szekely, and Andrew G. Livingston, Molecular Separation with Organic Solvent Nanofiltration: A Critical Review, Chemical Reviews 2014 114 (21), 10735-10806, DOI: 10.1021/cr500006j

66. Leiming Guo, Yong Wang and Martin Steinhart , Porous block copolymer separation membranes for 21st century sanitation and hygiene, Chem. Soc. Rev., 2021, 50, 6333-6348

67. Rini John, Kaushik Pal, Jitha S Jayan, Saritha Appukuttan, Kuruvilla Joseph, New emerging review on advances in block copolymer based water purification membranes, Journal of Molecular Structure, Volume 1231, 2021, 129926, ISSN 0022-2860.

68. Feng H, Lu X, Wang W, Kang NG, Mays JW. Block Copolymers: Synthesis, Self-Assembly, and Applications. Polymers (Basel). 2017 Oct 9;9(10):494. doi: 10.3390/polym9100494. PMID: 30965798; PMCID: PMC6418972.

69. Zili Li,Zhiqun Lin, Perspective: Self-assembly of block copolymers for biological applications, Polymer International, Volume 71, Issue 4 p. 366-370 First published: 16 November 2021 https://doi.org/10.1002/pi.6327.

70. Suzana P. Nunes, Chapter 11 - Block Copolymer Membranes, Editor(s): Gyorgy Szekely, Andrew Livingston, Sustainable Nanoscale Engineering, Elsevier, 2020, Pages 297-316, ISBN 9780128146811.

71. S.B. Darling, Directing the self-assembly of block copolymers, Progress in Polymer Science, Volume 32, Issue 10, 2007, Pages 1152-1204, ISSN 0079-6700.

72. Venkata S. Meka, Manprit K.G. Sing, Mallikarjuna R. Pichika, Srinivasa R. Nali, Venkata R.M. Kolapalli, Prashant Kesharwani, A comprehensive review on polyelectrolyte complexes, Drug Discovery Today, Volume 22, Issue 11, 2017, Pages 1697-1706, ISSN 1359-6446.

73. Lankalapalli S, Kolapalli VR. Polyelectrolyte Complexes: A Review of their Applicability in Drug Delivery Technology. Indian J Pharm Sci. 2009 Sep; 71(5): 481-7. doi: 10.4103/0250-474X.58165. PMID: 20502564; PMCID: PMC2866337.

CHAPTER

4

Advances in Forward Osmosis: An Alternative Membrane Process

— ABSTRACT —

Forward osmosis (FO) process has shown greater potential among the many membrane-based separation processes for wastewater treatment and high-quality reusable water recovery. Decades ago, taking inspiration from natural phenomena like osmosis, researchers and technologists designed engineered osmosis in the name of forward osmosis (FO). Earlier versions mainly focused on designing standalone FO processes. Today, researchers have adopted various configurations in both standalone as well as hybrid process forms which are widely used for both potable water treatment and wastewater treatment. Even though FO technology has been rated as one of the promising technologies, it still suffers from drawbacks like lack of suitable draw solution, efficient draw solution recovery, membrane fouling and internal concentration polarization (ICP) which restricts its widespread applications in long-term use. Therefore, this chapter attempts to assess the advances in FO technology performances in considering parameters like sustainability indicators for wastewater treatment and reusable water recovery.

4.1 Introduction

Presently, availability of fresh and safe water for drinking purposes as well as domestic applications has become a point of serious discussion. On the other hand, identification, collection, and treatment of abundant industrial wastewater has also become a global matter, particularly in developing nations where an enormous amount of industrial and municipal wastewater is produced daily and discharged into the freshwater bodies and environment. In this situation, there is enormous demand for speedy and efficient treatment like membrane-based separation processes. However, the major part of the membrane-based separation process is the availability of all-purpose membranes which can also be tailor-made to suit the particular contaminants in the wastewater. This task also necessitates the selection of sustainable and highly adoptable polymer with enhanced properties for membrane preparation to deal with the highly polluted effluents. Membrane processes are designed to withstand the high concentration and contamination level such as leather industry wastewater, distillery wastewater, sugar industry effluent, pharmaceutical wastewater and so on. Most of these effluents are designated as 'red category' wastewater due to their extreme contamination level including high BOD, high COD, and organic contamination. Therefore, to treat such extreme category contaminated water membrane technologies have been effectively adopted. As discussed in the previous chapter, a series of membrane-based treatment techniques have been successfully used. Recent advances in membrane materials and processes have also shown great deal of consideration due to the increased demand for characteristic advantages to effective treatment compared to other physicochemical methods. In this direction, numerous innovations at laboratory scale, pilot plants, several start-up activities and medium to large-scale studies on the tuning membrane materials, process and working plants are underway for utilization in water purification and treatment. All these developments are undergoing in compliance with rigorous environmental norms and regulations to satisfy the future needs for clean and safe water resources. [1, 2]

An enormous amount of wastewater, produced from industrial segments and domestic activitie,s is the major environmental and health concern. This is due to rapid industrialization that has led to the production of huge amounts of highly polluted effluents including conventional pollutants, biowastes, and emerging pollutants of diverse nature posing serious health and environmental threats. Most of these effluents are either partially treated or disposed of to the environment without treatment. However, serious efforts are being made to adopt advanced membrane technologies to effectively treat and recover maximum out of contaminated sources. In several cases, untreated and discharged toxic liquid waste of various origins and category along with the emerging and persistent contaminants are released into the freshwater bodies is a great cause of concern. On the other hand, large quantities of freshwater are consumed in manufacturing processes such as dying (textile), automobile cleaning and maintenance, tanning or leather industry effluent, ore extract and processing in mining industry, sugar industry, electroplating, etc., with approximately 90% of it is composed of unused raw materials, intermediate compounds and byproducts in them is directly discharged into the environment. This is majorly due to the absence of appropriate wastewater treatment technologies. This negligence is leading to severe problems pertaining to the discharge and disposal of highly contaminated effluents and difficulty in designing appropriate recycling methods According to the United Nations (UN) World Water Development Report published in 2017, nearly 80% of the global industrial and domestic wastewater is discharged and disposed of without proper treatment. Though this percentage ratio varies in different income group countries, it is likely that >90% wastewater in these countries goes untreated or partially treated. [3, 4]

Therefore, it is high time to devise technologies to overcome such problem. Nevertheless, as discussed in previous chapters various physical, physicochemical and tertiary treatment techniques like biological treatment methods have been adopted and are being implemented in several segments of industries to treat highly contaminated wastewaters before being discharged into the environment. In this segment among many advanced techniques, membrane-based separation processes have been promising and sustainable compared to other methods. Among all membrane separation processes, forward osmosis (FO) has attracted more attention due to its noteworthy potential in water reclamation of high quality from wastewater sources. The FO process in both standalone and hybrid configuration has shown greater potential as an emerging alternative to all other pressure-driven processes such as nanofiltration (NF) and reverse osmosis (RO). [5, 6]

Further, there have been numerous studies and reports featuring important observations pointing towards the successful implementation of technology. Also, there has been increased interest in designing effective FO membrane, manufacturing, process development, plant design and operational parameter optimization to demonstrate the potentials of large-scale considerations. These approaches are carefully optimized in various approaches keeping important sustainability criteria in mind. Forward osmosis process comprises three main important components (a) membrane, (b) draw solution, and (c) efficient draw solute/solution recovery and reuse mechanism. Efforts are being actively made to have a better understanding on the dynamics of separation process at the membrane interface with the emphasis on draw solution recovery and reducing energy consumption both for sustainable water and draw solute/solution recovery. In pressure-driven membrane processes like NF and RO processes in which unidirectional drift drives the mass transportation, however in the FO process the water diffusion and mass transportation in the membrane is a complex phenomenon. There are several parameters which can stimulate the osmosis process and can influence overall efficiency of the FO process in water treatment. Influential factors arethe nature of polymer used for membrane preparation, its morphology, pore characteristics, membrane thickness, wettability, and the solubility and diffusibility. [7, 8]

In addition to material parameters, operating and process parameters like low fouling propensity, low concentration polarization, high flux, good and rejection rates need to be considered for endorsing the FO process for large-scale potentials. The feature required to be an efficient FO membrane is the high concentration solute tolerance as the membrane is typically exposed to both high concentrations of feed solution as well as extreme concentrations of draw solute or draw solutions across the membrane. On the other hand, finding high osmotic pressure inducing draw solutes or solutions is also an important task. Several studies have highlighted these conditions in many combinations of materials. Some of the important case studies have also revealed the major issue related to severe internal concentration polarization (ICP), which hinders the membrane performance in the long run. This also necessitates developing washing or membrane maintenance steps which are typically cost intensive. Absence of this step may also impart inferiority to the overall FO process and long-term integrity under wastewater treatment conditions. [9]

Therefore, researchers have established that the concentration polarization effect mainly depends on factors related to membrane properties, nature of draw solute and the process design aspects. In addition to this, adopting ingenuous treatment configuration to suit the nature of effluent or nature of feed is of utmost importance to achieve separation of contaminants from reusable water. Indigenous plant design helps to attain the sustainable processes that considers all the criteria like technological feasibility, process economic, environmental benefits, and social aspects. [10-12]

In the past, FO has been widely used in standalone configuration for the treatment of shale wastewater recovery, multiple resource recovery and energy production [24], hyper-saline wastewater desalination, and in specialized applications of biological contaminant treatment and other fields like value addition. However, like any other new technologie, the FO process has also faced serious competition from the prevailing pressure-driven membrane processes and other modern technologies. At this stage, to overcome the drawbacks of FO process on various characteristic technological parameters, a systematic case study and critical evaluation of the process is also done thereby understanding the technology for futuristic benefits. [13-19]

Hence, in this chapter, attempts have been made to critically evaluate the technology for water treatment applications in both its FO standalone mode and FO-based hybrid processes configuration for wastewater treatment. Further, sections have been dedicated to give detailed analysis of FO process sustainability aspects along with the challenges and future opportunities.

4.2 Recent Advances in Forward Osmosis Process

Forward osmosis (FO) is one of the membrane-based separation processes which uses nature inspired osmosis as driving force to separate contaminants from water resources. In this technique selective water separation is achieved using a semipermeable membrane as filter media and the natural osmotic pressure as energy to achieve water separate from dissolved solutes. In the FO process, a high osmotic energy solution known as draw solution across the membrane against feed solution to achieve water separation. Here, the draw solute or draw solution employed creates an osmotic gradient to achieve effective separation of the clean or reusable quality water. Typically, inorganic salts such as ammonium salts, Typically, inorganic salts such as ammonium salts, sodium chloride, fertilizers and so on have been widely used as draw solutes to create high osmotic gradient across the semi-permeable membrane.

In this process, the osmotic pressure drives the transportation water across the semi-membrane while retaining all the dissolved solutes on the feed side. Conventionally, reverse osmosis has been ideally used for dissolved salt as RO membrane has smallest pore size ranges. Similarly, the FO process uses a similar semi-permeable membrane as separation media to achieve efficient separation. Also, the FO process can also be applied to treat complex industrial wastewater. The major issue in sustaining membrane in FO process is to retain clean membrane surface, the low concentration feed solution may not affect surface morphologies of FO membrane and are less sensitive to fouling, concentration polarization and scaling. However, emerging membrane materials and technologies are expanding the scope of FO membranes to a wide range of potential water treatment applications including desalination. On the other hand, the osmotic pressure gradient is used to pass water across the membrane barrier while holding all the dissolved solutes on the feed side back.

The FO process is designed to suit the specific application with the selection of a particular draw solute or solution against feed or contaminant to be separated. However, to make the process sustainable it is vital to recover from the diluted draw solution and reuse for further separation cycles. Once the draw solute is recovered from the diluted draw solution, it is then re-circulated in the forward osmosis to extract water from the feed side continuously. The selection of suitable draw solute for an envisioned FO application following specific guidelines such as (a) nature and type of feed solution, (b) concentration of feed, (c) type of membrane suitable for contaminants, (d) required osmotic gradient to induce osmotic gradient, (e) type of draw solute, (f) concentration of draw solute, (g) importance of draw solute recovery, and (i) extent and potentials of draw solute recovery and reuse. [19-21]

Several studies have listed sustainability criteria for various water treatment technologies at industrial scale as an important aspect. This mainly includes capital and operational costs, process feasibility and targets to achieve zero liquid and zero solid disposals to the environment at the end of the treatment process. Any new technology must go through the economic considerations which is also a main selection criterion to replace the existing alternatives. Along with FO technology, several novel methods have also been introduced for water treatment applications, however, many of the novel technologies serve small and medium scale purposes. For example, nonmaterial-based technologies have shown greater potential at laboratory scale for different contaminant treatment for both groundwater and wastewater treatment. However, there are several inherent limitations in adopting nanomaterial at scale-up levels. Even though several efforts have been made to overcome the limitations of nanomaterial scale-up, still detailed studies require deeper understanding. Nonetheless, nanocomposite-based separation media is effectively used in combination with polymers and bulk substrates for water treatment applications. However, the FO process has shown great potential to be a process capable in large scale applications like seawater desalination and wastewater treatment. Interestingly, nanocomposite FO membranes have been successfully used in extreme operational conditions with minimum internal concentration polarization and tolerance to the range of pH and draw solution conditions. In addition to this, scaling-up prospects for the prevailing lab-scale modules have been assessed and the plausible consequences such as environmental footprint are being addressed.

This chapter assesses the most sustainable types of FO technologies to deal with the contaminated wastewater in real applications and emerging versatile sectors. This chapter also attempts to make recommendations for the future studies that have been also included. Despite the existence of numerous techniques developed to recover useful constituents as value-added products from the wastewater sources, following stringent environmental norms. Recent regulatory norms also force every industrial sector to adopt advanced water treatment technologies to meet international standards to treat and dispose of treated water to the environment. Therefore, to fulfill these conditions waste management infrastructure managers, industries and municipal bodies need to develop or adopt sustainable alternatives for the treatment of highly contaminated wastewater. Among all, membrane-based technologies have been proven efficient technologies for wastewater treatment. On the other hand, engineered osmotic process-induced forward osmosis (FO) uses semi-permeable membrane and osmotic gradient to separate water from the contaminants or dissolved solutes is now being widely used for the treatment of wastes. The main driving force in FO for the selective separation of water from contaminated feed is the osmotic pressure gradient that exists between the solution of lower concentration generally called "feed" and high concentration draw solute or draw solution that exists across the semi-permeable membrane as illustrated in Figure 4.1. Mainly FO can be distinguished from reverse osmosis (RO) in which hydraulic pressure is used to drive the separation of water from wastewater. In RO high applied pressure is applied on the feed side to drive only water across the membrane, on the other hand, draw solute and/or draw solution across semi-permeable membrane exerts osmotic pull to extract pure water through osmotic phenomenon. [22, 23]

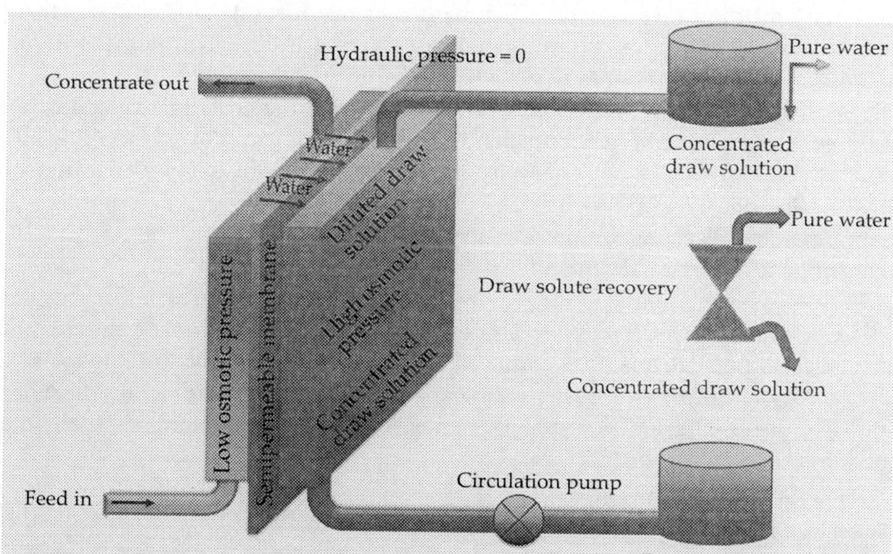

Figure 4.1. Schematic illustration of lab-scale working FO process highlighting all three important components, namely, semipermeable membrane, draw solution and draw solution recovery system and other operational specifications. NaCl, green ionic salts, and magnetic nanocomposite are selected as the draw solutes as low cost, abundant and high osmotic draw agent.

Hence, the energy requirement for the separation of water from the industrial wastewater or contaminants in RO process is relatively considerably higher than that of FO. Osmotic pressure difference in FO process can be utilized to influence a net flow of water from the feed into the draw solution chamber through the semi-permeable membrane without any external pressure source. As explained in the previous section the difference between operating principles of RO and FO, the overall net energy consumption for FO process compared to RO is negligible. However, major energy consumption in the FO process comes from the draw solute or draw solution recovery step in which diluted draw solution is separated from reusable quality water. However, in some cases diluted draw solution can be directly used as a value added product. For example, if the draw solute has mineral quality such as fertilizer, then afterdilution of the draw solution drawing from feed side directly can be used as liquid fertilizer. In such cases, the operational cost of the FO process can be minimized compared to pressure-driven processes like NF and RO. [24] Although a draw solution can be replenished over the course of FO operations that nees to be taken into account. However, if a used drawing solute or solution necessitates recovery step then it is essential to recycle efficiently to make the overall process sustainable. Nonetheless, if the draw solution recycling step is energy intensive, then the FO process would be equal to or even more costlier than that of conventional NF and RO systems. However, it should be emphasized that these estimates are made considering standard feed as seawater and complex industrial wastewaters yet to be seriously studied. However, FO can play an important role in specialized applications containing organics such as pharmaceutical compounds, dyes, trace organic contaminants (TrOCs), inorganic pollutants like heavy metals, biological wastes in which RO may not be effective. In case of high concentrated feed systems it will be difficult if not impossible to treat by standalone RO and in such case it is necessary to adopt pre-treatment and sanitization or post-polishing. This may invariably add up to the cost and use of external chemicals for cleaning or post-treatment process at regular intervals. Such situations also demand addition of anti-scalants thereby making the overall process less sustainable. Nonetheless, wherever cost-intensive NF or RO processes fail, FO is expected to shine. Fortunately, innovations in the field of nanotechnology are enabling complex wastewaters treatment and low fouling propensity due to extensive use of nanocomposite anti-fouling membranes and use of no external pressure that makes FO process suitable as well as a sustainable alternative. [25]

Nevertheless, all preliminary large scale studies from fully operating FO technology for desalination applications prove the greater potentials of the process. The use of easily recoverable draw solute in the FO process for desalination has shown superior technological parameters compared to other pressure-driven membrane technologies. On the other hand, the influence of important physicochemical properties of polymers in membranes also plays an important role on mass transport mechanisms. From the beginning it is expected to develop an anti-fouling or fouling resistance membrane which is expected as the membrane is exposed to high concentration feed and draw solution. Initially, cellulose derivative polymers like cellulose acetate (CA), cellulose triacetate (CTA) and conventional RO membranes made of polyamide (PA) were evaluated for water flux ranging from 3 to 13 in different applications. In all cases, it was expected and later membranes experienced some degree of irreversible flux decline over the course of operation. In addition to the conventional physicochemical properties, it is estimated that charge neutralization, osmotic swelling and deswelling also contribute to infusion of pollutants into the polymer matrix. In this direction, modified versions of CA, CTA and PA membranes have also been adopted in composite forms with improved anti-fouling properties. However, due to the presence of both high concentration, draw solution and feed contaminants still pose a serious threat to the surface properties of membranes. [26, 27]

A significant aspect of novel membranes is to develop anti-fouling surface coated membranes parallel to TFC membranes specifically for the FO process. For such purposes, hydrogels have been widely used to induce hydrophilic layer coating to reduce reverse solute flux (RSF) by 50% and enhance antifouling properties. The removal of micro-organic pollutants like aniline, phenol and nitrobenzene have been successfully carried out using conventional polyamide-TFC RO membranes. Interestingly, these membranes were successfully adopted for FO mode surpassing flux performance that of commercial RO membranes operating under high applied pressure. On the other hand, most of the initial FO experiments were tested using TFC-RO membrane against 2M NaCl draw solution for various feed solutions. Even though there is limited experimental data available on rejection patterns on enhanced rejection and fluxes, novel materials combinations pose challenges. [28]

4.3 Advantages of Forward Osmosis in Water Treatment

Every technology that aims to replace existing methods requires fulfilment of various parameters affecting the performance during water treatment. In addition to this, the method or technology is also evaluated based on their sustainability, energy efficiency and feasibility to integrate with waste mitigation. For example, industries expect a high degree of reusable quality water to adhere to environmental agencies' norms with minimum disposable concentrated waste. Also, modern water treatment technologies are expected to include recovery of value-added materials which would also increase the circular economy and prevent the release of toxic and hazardous compounds into the environment. In very rare cases less harmful sludge or concentrates would be directly released to the environment, however futuristic treatment technologies are expected to integrate steps to address toxic end concentrate. In this direction, FO technology is accepted as complete water treatment technology providing a good semi-permeable membrane, effective draw solute or solution and economical draw solution recovery in place must be socially acceptable for further developments. Considering these parameters FO is now proving a relatively simpler process in fulfilling various criteria among other membrane-based technology platforms. Further section discusses the performance of the FO process in view of the sustainability, economics and value addition criteria which are missing in the existing methods. [29]

As a standalone process, FO technology has proven its ability to be a feasible energy efficient technology for complex red category wastewater dewatering. In the FO process, water is osmotically extracted from the effluents which can be regarded as the primary step involving the treatment of the effluents using such technologies. Nonetheless, initially the FO process was used as a pretreatment technique for the RO process in the desalination process, however, as a complementary step FO showed limited success from the point of draw solution recovery.

Other considerations, standalone FO processes give relatively low water quality in a single stage, draw solution properties decline with dilution, membranes experience reverse salt flux and the lack of dedicated ideal FO membrane limits standalone processes. However, all advanced FO systems are critically considering these factors for the development of the FO water treatment system. Subsequently, mixed matrix membranes (MMM), inorganic salts and innovative engineering have been used for the tuning of the FO process. In membranes, researchers have adopted a series of natural as well as synthetic clays and nanomaterials as fillers in the polymer matrix to induce high surface area adsorptive porosity which enhances water permeation further leading to separation properties. [30] This approach is mainly due to two reasons: first, to counter the inherent hydrophobicity of the polymer matrix and secondly, a hydrophilic filler alters the rigidity of the polymer matrix creating water permeation pathways via the plasticization effect. In this regard, a new class of bio-mimicking macromolecules have been adopted with defined water pathways and semi-permeable barriers. One such class of materials comes from protein family such as aquaporin which has been successfully used as selective layer in both on a flat sheet as well as hollow fibre membrane substrates. Now, the FO membranes with designed precursor formulations can be fabricated via phase inversion technique. These methods can be further divided into thermally induced phase separation (TIPS), precipitation by controlled evaporation or solution casting, vapor phase precipitation, and non-solvent induced phase separation (NIPS). Phase inversion is the most important and successful technique used for the preparation of flat sheet and hollow fibre FO membranes. Phase inversion technique involves mixing polymer desired precursor with solvent and as prepared polymer solution is known as dope solution. Subsequently, the dope solution is cast on either glass plate for free standing membrane or on porous polymer substrate to result in composite membrane. As a cast solution film is immersed in a coagulation bath containing non-solvent which is also miscible with solvent and strictly non-solvent to polymer which results in precipitation of membrane. On the other hand, porous hollow fibers are prepared following the dry-jet wet spinning process. Here, polymer dope solution is extruded through a specially designed spinneret at a volume rate that was fixed by a gear pump. Spinneret is hung on a coagulation bath with a specific gap which provides evaporation time which varies for different dope solution and its composition. [31-33]

Electrospinning is another nanomaterial preparation technique used for the preparation of nanofibrous mats of various configurations and functionality. These nanofibrous mats are packed with different densities to convert them into flat sheet membranes. Different polymer precursors are converted into nanofibrous mats by applying a high electrical field to a polymer dope solution in a syringe. This will result in layer technique (L-b-L) deposition with high packing density OF membranes. This technique is now widely used to prepare prefabricated polyelectrolyte type membranes with opposite charges. With the optimization of dope solution concentration, composition and applied voltage uniform and homogenous ultra-thin FO membrane of specific selective layer and thickness can be collected on the substrate surface. Many recent approaches have utilized several combinations of charged polyelectrolytes in different configurations such as multi-layers morphology with tuneable flux and rejection patterns. Also, commercially successful polyamide thin-film composite (TFC) RO membranes prepared following interfacial polymerization between *m*-phenylenediamine (MPD) and trimesoyl chloride (TMC) monomers are still used for FO applications. However, nanocomposite versions TFC RO membranes having a series of nanomaterials in the interfacial thin layer selective layer matrix yielding better ant-ifouling and improved flux performances. Largely, these membranes have asymmetric structures with controlled rejection and permeability in FO processes. [34, 35]

However, there are still several obstacles in preparing an ideal FO membrane such as finding suitable hydrophilic at the same time inert polymer precursor, anti-biofouling nature and achieving stable permeate flux in long-term operations. Also, interfacial thin layer morphology architecture in thin-film composite membrane design suffers from several restrictions due to some serious issues with compatibility. For instance, porous substrate used for coating thin selective layers is susceptible to hold high concentration solute in the asymmetric pores. This is crucial in controlling mass transportation from feed to the draw solution chamber. In such a situation, drastic decrease in permeate flux as a result of blocked asymmetric pores which causes a gradual increase in mass transfer resistance. In addition to this, deposited solutes in the pore also induce internal concentration polarization (ICP). Further, the ICP also occurs due to the precipitation of saturated solute on both the sides, upon dilution of the concentrated draw solute yields drastic increase in pollutants in feed side selective thin layer on the other hand, stationary draw solute in asymmetric pores gradually precipitate across the selective layer resulting in decrease in the permeate flux of up to ~80%. Therefore, detailed understanding of ICP phenomena, optimizing the selective layer thin film structure for their porosity, thickness, tortuosity, and affinity through their preparation technique is essential. Nonetheless, extensive research is being made to prepare an ideal FO membrane through an appropriate support layer and optimizing phase inversion procedure. Significantly, composition of dope solution and interfacial adherence in nanoscale at phase inversion preparation conditions. The combination of solvent and non-solvent systems also has been paid attention to have better mixing and demixing during the coagulation process that can regulate pore size and their distribution. Further, some efforts are also being made to include green solvent components in these solvent/non-solvent systems to regulate pores in FO membranes. [36, 37]

On the other hand, the concentration of the polymer precursor and nanomaterials-based additives were also widely studied. Thus, by replacing dope solution recipe, altering combination of solvent and non-solvent systems, and controlling membrane preparation process parameters, an ideal FO membrane can be realized. Thus far, several FO membranes have prepared several substrate and selective layer combinations starting from cellulose, cellulose derivatives like cellulose acetate, triacetate, biopolymer, biomimicking biomacromolecules and so on. These combinations have been brought together with different FO membrane preparation techniques following various fabrication parameters as documented in Table 4.1.

On the other hand, there are several innovative and nature inspired materials that have been tried and tested as semi-permeable media to induce effective osmotic barriers such as aquaporin. Traditionally used cellulosic derivative TFC membranes will remain in use for FO applications due to their high range tolerance to pH which is from 2 to 11 and temperature stability ~60°C. These membranes can also exhibit higher water permeability which can tolerate harsh operational conditions even in pressure retarded osmosis (PRO) mode. However, two-stage preparation techniques like phase inversion integrated with interfacial polymerization need to be modified to make the overall process economical. Even though, hollow fibre FO membranes have shown promising potential, however wider use of hollow fibre configuration most of the time is restricted by blocking of hollow channels and fouling proficiency in the narrow inner diameters. This technological barrier limits the effectiveness of FO processes for highly contaminated water treatment like industrial effluents. [38]

Table 4.1. Summary of different types of FO membranes with their respective porous substrate, selective layer component,t and their preparation method.

Substrate	Selective layer	Preparation method
Polyamide thin film composite	Polyethylene terephthalate nanofibres backing layer	Phase inversion
PESU-*co*-sPPSU 11 sulfonated PES polymer	PA TFC	Phase inversion
Nanoporous PES flat sheet	Polyamide active layer	Phase inversion
PSf support layer and polyamide active layer	CTA support layer	Phase inversion
	TFC flat sheet	Interfacial polymerization (IP)

Polyester mesh	Cellulose acetate (CA)	Phase inversion technique
Woven fabric mesh	Cellulose triacetate	Phase inversion technique
Polysulfone (PSf) support layer	Polyamide active layer	Electrospinning
	Polyamide active layer	Interfacial polymerization (IP)
Polyvinylidene fluoride (PVDF) nanofibres support layer		Electrospinning
Polyamide active layer		Interfacial polymerization (IP)
PSF and sulfonated poly(phenylene oxide) support layer		Phase inversion
Polyamide active layer		Interfacial polymerization (IP)
PVA nanofiber support layer		Electrospinning
Polyamide active layer		Chemical crosslinking via acid
PSf support layer PA/silicon dioxide (SiO_2) active layer	Polyamide thin film layer	Phase inversion interfacial polymerization (IP)
Polysulfone–titanium dioxide (PSf/TiO_2) support layer polyamide active layer		Phase inversion interfacial polymerization (IP)
Polyethersulfone (PES) support layer	TFC hollow fiber	Dry-jet wet spinning process with advanced co-extrusion technology
Polyacrylonitrile (PAN) support layer	FO modified flat sheet	Layer-by-layer (L-b-L) assembly
Poly(amide–imide) (PAI) polymer polyamide active layer	Double skinned TFC hollow fibre	Phase inversion, IP chemical
Polysulfone UF support	Aquaporin	Surface coating

4.4 Advances in Draw Solutions and their Recovery and Reuse

In addition to focusing on developing effective FO membranes, a great deal of research and developmental efforts have been made to explore appropriate and effective draw solutes or solutions for FO processes in the recent past. As discussed in the previous section several combinations of draw solute/solution and FO membranes have been tried for several wastewater feed solutions. However, detailed studies have revealed both positive aspects as well as technology limitations. Further, recent studies have used a series of inorganic salts and polymer-based draw solutes for various FO applications. A summary of various types of draw solute/solution, and their recovery methods have been documented in Table 4.2. Largely, 2M NaCl, ammonia-carbon dioxide (NH_3-CO_2), ammonium bicarbonate, EDTA sodium salt, synthetic sea water, $MgSO_4$, Glucose, Na-EDTA, Mg-acetate, Na-acetate, Glycine, Na-formate, and polymeric draw solutes like polyacrylamide sodium salt (PAA-Na) have been widely used as draw solutions. [39-41]

Table 4.2. Summary of type of draw solutes or solutions widely used in various FO processes and their possible recovery method.

Feed Solution	Draw Solute/Solution	Recovery	Recovery method	Value addition
Simulated sludge	EDTA sodium salt	Yes	NF	None
Raw coke oven wastewater	NaCl, $MgSO_4$ and $CaCl_2$	Yes	NF	No
Treated landfill leachate	Ammonium bicarbonate	No	–	–
UF pretreated secondary effluent	NaCl solutions	Yes	Solar power-driven ED	–
Anaerobic digester concentrate	NaCl solution	Yes	RO	–
Secondary and tertiary treated effluents	Synthetic seawater (instant ocean, mentor, OH) solutions	Yes	SWRO	–
MBR effluent, 23 TOrC + 9 inorganic ions + humic acid	Synthetic sea salt	Yes	Seawater RO	–
Municipal wastewater from MBR	Seawater brine from desalination plant	No		Value addition to seawater brine from RO units
Activated sludge	NaCl solution	No		–

Activated sludge from MBR pilot plant at wastewater treatment plant	$MgSO_4 \cdot 7H_2O$, NaCl, $NaNO_3$ and $ZnSO_4 \cdot 7H_2O$, & NaCl	No		–
Raw municipal sewage	NaCl+$MgSO_4$ & NaCl (0.5M)	No		–
Seed sludge from water reclamation plant	NaCl	No		–
Digested sludge from full scale wastewater treatment plant	$MgSO_4$ Glucose Na-EDTA Mg-acetate Na-acetate Glycine Na-formate NaCl	No		–
Synthetic wastewater	NaCl	No		–
Synthetic wastewater	NaCl solution	Yes	Postulated to be done by RO	Water from human metabolized wastewater.
50 Acid orange dye solutions	Polyacrylamide sodium salt (PAA-Na)	Yes	MD	–
Raw sewage	NaCl	Yes	MD	
Synthetic oil contaminated wastewater	NaCl, 2M NaCl	Yes	MD	
Anaerobically digested sludge	$MgCl_2$	Yes	MD	Recovery (P&NH^{4+}) in feed
Real urine samples	M NaCl	Yes	MD	Recovery of water from human metabolic wastewater.
Raw domestic wastewater	NaCl	Yes	RO	–

Contd...

Municipal wastewater	NaCl		–	Phosphorous recovery
Municipal wastewater	Seawater brine from RO	Yes	SWRO	Phosphorous recovery
Mixture of dyes and inorganic salts as simulated textile industry wastewater (3000 ppm TDS)	Different NaCl solutions		–	–
DNA & protein concentration	Deep eutectic solvents	Yes	Freezing	Use of high-quality water
Sugarcane juice concentration	Sea bittern	No	–	Epsom salt recovery
Seawater	SO_2 + seawater SO_2 + alcohol SO_2 + (KNO_3)	Yes	Heat treatment	Water recovery

In general, the draw solutes/solutions listed in the above table can be divided mainly into two categories, namely: (i) commercially feasible compounds, and (ii) and specially designed synthetic compounds. Initial FO process explorations were carried out using commercially available high osmotic pressure inducing compounds as draw solutes. Further, these category substances can be subdivided into four sections based on their physicochemical properties: (i) inorganic salts, (ii) organic salts or polymeric substances, (iii) volatile compounds, and (iv) nutrient compounds. However, based on composition and modifications in the constituents, these categories may overlap and can be adopted as a new class based on their overall performance. Among many inorganic salt combinations, Neff et al., in 1964 first used a mixture of ammonia and carbon dioxide gases as a highly saturated draw solute for seawater desalination applications. [42, 43] The preparation and use of ammonium bicarbonate from ammonia and carbon dioxide has been widely used for industrially important feed separations such as sugar, and drug molecule concentration; however, technological feasibility is only realized for seawater desalination. On the other hand, water recovery and regeneration of the draw solute was achieved by heat treating the diluted draw solution which consumes a considerable amount of energy. Another limitation is relatively low osmotic pressure exerted by NH_4HCO_3; this is also due to its low solubility in water which restricts larger applications of the draw solution. To overcome this issue ammonium salts were additionally mixed such as mixing of NH_4HCO_3 with ammonium hydroxide (NH_4OH) at specific ratio of NH_3 to CO_2 to enhance the draw solute solubility. This resulted in inducing high osmotic pressures to the resulting draw solution thereby resulting in better mass transportation performance. [44]

Further, sulphur dioxide (SO_2) was utilized as a draw solute in seawater to prepare a relatively high osmotic pressure draw solution to be used in FO seawater desalination. Further, diluted draw solution heat treated or air stripped to separate water from SO_2 which was diluted from pure water transported from the feed seawater. Even though use of SO_2 in the draw solution showed relatively good water flux, there is no encouraging rejection pattern recorded. Further, to improve the pollutant rejection while maintaining high water flux, volatile compounds such as SO_2 in different liquid media as draw solutes. In one such attempt volatile solute is dissolved in aliphatic alcohols to result in high osmotic pressure draw solution in the FO process. Interestingly, addition of SO_2 in alcohol drastically reduces the water activity thereby increasing the osmotic pressure, which also helps in the recovery of product water and SO_2 following distillation process. Lately, 2002, SO_2 was combined with potassium nitrate (KNO_3) to prepare novel draw solutions for seawater desalination in the two-step FO process. The saturated KNO_3 solution as a draw solution at one side was prepared against warm seawater desalination in the first FO unit. Once the concentrated KNO_3 draw solution got diluted, the draw solution was again re-concentrated using saturated SO_2 solution in the second FO unit to make the complete process. Further, SO_2 was separated from product water applying heat to produce reusable quality water. Therefore, in this unique process along water both draw solutes KNO_3 and SO_2 were efficiently recycled making it one of the feasible processes. Even though SO_2-based draw solutes showed relatively better FOR performance in various applications in which physicochemical properties of solute components played an important role, however, SO_2 is corrosive, volatile substance, redox agent and unstable in acidic conditions. These physicochemical properties make SO_2-based draw solutes susceptible to readily react with both reducing and oxidation agents. Nevertheless, volatile draw solutes can be easily separated from diluted draw solution to produce quality water either by distillation, heating and dissolving additional volatile gases into water. Further, extended studies are still required to assert the usefulness of volatile draw solutes which is currently limited due to insufficient experimental data. [45, 46]

On the other hand, the second category of draw solutes is mainly composed of nutrient compounds. These new classes of naturally abundant draw solutes were first used for food ingredients concentration (food and wine) and later extended to wastewater treatment. In this direction, low-cost glucose was initially explored extensively as a draw solute for seawater desalination in the FO process. Major advantage with the use of glucose is that the diluted draw solution can be directly used for edible or drinking purposes in special applications like emergency lifeboats. This also solves the problem of draw solute recovery and recycling which make it a cost-effective option. In another attempt pure fructose was used as the draw solute due to its higher osmotic coefficient compared to glucose which exerts higher osmotic gradient against several feed solutions. Further, carbohydrate-based nutrients were developed as nutrient solutes in the FO process in which combinations of glucose/fructose were used as solutes. This combination of glucose/

fructose showed better water drawing capabilities during the FO process compared to pure glucose or fructose solutes. Further, glucose and fructose were regenerated using low operating pressure RO processes. However, this approach requires a two-step hybrid process involving an FO for dewatering and an additional low operating pressure RO process for draw solute recovery. Interestingly, these commodity nutrients have shown great potential as draw solutes in various commercially important feed concentrations and product recovery. [47, 48].

Recently, many researchers and innovators are revisiting the use of nutrients as draw solutes. Now, the use of sucrose as a draw solute is being adopted in the novel FO process having double skinned hollow fibre membranes for wastewater treatment applications. Highly saturated sucrose solution (up to 1M) was found to be highly effective in recovering high quality water from different feed sources. This saturated solution yielded better flux performances even compared to commercially widely adopted 1 M $MgCl_2$ solution in the PRO mode. This is also due to larger molecule size of sucrose compared to $MgCl_2$ retained high osmotic gradient across semi-permeable membrane which resulted in negligible decrease in the overall water flux and maintaining minimum reverse solute (sucrose) flux. Overall, food ingredients which are produced at commodity level can be effectively adopted to serve as effective osmotic agents. These also play an important role in product water recovery, can be directly used or recycled using easy steps. [49]

Inorganic salts: Among many classes of draw solutions inorganic salts have shown reasonably high fluxes performance and can be easily regenerated using either NF or RO processes. RO Earlier mention of using inorganic salts as draw solutes dates backs to 1970, in which researchers have used various acids like manganic acid, valeric acid, glucuronic acid, and different water-soluble inorganic salts such as sodium, calcium, potassium, barium, and cesium salts as draw solutes mainly for FO seawater desalination applications. One of the major advantages using inorganic salts is that draw solutes can be easily recovered by varying solubility of a particular salt through pH or temperature variation. For instance, solubility of draw solutes can easily be changed by varying temperature or their reactions with carbon dioxide (CO_2), sulfuric acid (H_2SO_4) thereby recovery of high-quality product water from precipitated draw solutes. Particularly, aluminum sulfate ($Al_2(SO_4)_3$) as a draw solute and can be used to extract high quality of water from different feed sources. Further, treatment with calcium hydroxide ($Ca(OH)_2$) would yield precipitates of calcium sulfate ($CaSO_4$) and aluminum hydroxide ($Al(OH)_3$) were segregated from the product water. Subsequently, H_2SO_4 and CO_2 were added to neutralize the excessive $Ca(OH)_2$ entered the product water portion. However, the main limitation of these approaches comes from their draw solute recovery complexities which involve several non-friendly additional chemicals in the recycle step. These chemicals vary from bases like ($Ca(OH)_2$ and $Al(OH)_3$) to strong acids like (H_2SO_4) which are highly corrosive. Initial studies have also attempted to use hygroscopic $MgCl_2$ as a draw solute in fundamental studies of FO process, however in these studies regeneration and recycle steps have been omitted. [50]

Further, inorganic salts-based draw solutes can be categorized into three types based on their feasibility of regeneration and their methods, namely: (i) first type draw can be categorized in which thermal recovery methods are feasible. For instance, ammonium bicarbonate (NH_4HCO_3), used in desalination of seawater, can be easily regenerated by applying heat (~60°C). The high flux produced drawing water from seawater feed can be regenerated by heating a dilute draw solution around 60°C at which NH_4HCO_3 salt decomposes into carbon dioxide gases and ammonia. Once the draw solute decomposes into its original components, simultaneously product water is separated. Further, to make the overall process sustainable, the recovered components can be re-dissolved back into water. In many cases waste heat from industrial processes can be redirected to decompose NH_4HCO_3 in a diluted draw solution which can minimize stress on energy resources. [51]

However, some of them have shown limitations as they are used in saturated forms. For instance, draw solute scaling results on membrane surface and their impact on osmotic pressure and transportation phenomenon in membranes have been thoroughly studied on a FO membrane. Further, membrane scalants result as a result of precipitation of commonly used draw solutes like NaCl, calcium carbonate ($CaCO_3$) and ammonia-carbon dioxide (NH_3-CO_2). Even though these scaling phenomena may have insignificant changes in flux performances for NaCl draw solution, when NH_3-CO_2 was used as draw solution drastic flux decline was recorded due to prolonged scaling phenomenon. Interestingly, these scaling effects are being studied in detail using advanced analytical tools like FESEM and AFM on a microscopic level. These tools provid detailed information on scaling phenomena on selective layers of FO membrane. Therefore, through these images it is noted that the $CaCO_3$ scale pattern on selective layers is formed due to the reverse diffusion of carbonate species from the draw solution chamber. Nevertheless, membrane cleaning agents like EDTA have been constantly used to maintain flux performance to avoid draw solute scaling when high concentrate draw solute like NH_3-CO_2 is used as a draw solution. [52]

The second category inorganic salt-based draw solutes can be those compounds which act as direct value-added products upon extracting water from waste sources such as fertilizers. In this direction, several inorganic salts have been studied that are generally used as one or main component of fertilizers in the agriculture sector. These inorganic salts can be directly used in FO process for desalination processes and diluted draw solution post-FO can be directly used for irrigation, in such cases separation of product water from the draw solution may not be required. In general, most of the draw solute recovery process post-FO is regarded as the energy-intensive step in seawater desalination. Therefore, in the absence of regeneration step, FO process will be highly energy efficient compared to all other membrane-based separation processes. However, there are still some limitations associated with use of fertilizer-like inorganic salts for draw solution purposes, for instance most fertilizers possess acidic properties when dissolved or diluted in water. Thus, an acidic environment creates detrimental conditions to FO-membranes like cellulose acetate or triacetate which drastically reduces the shelf-life of a system. In acidic conditions, FO membranes containing ester groups readily undergo hydrolysis resulting in inconsistent FO process performance. Another major limitation with fertilizers like draw solutes is their limited or partially dissolvability in water. The partial dissociation of low solubility salts in aqueous media provides limited osmotic pressures to extract water from feed water through an FO process. Therefore, demand for highly soluble fertilizer type draw solutes can fulfil the high osmotic coefficient solution, however, saturated, or high concentration draw solutes cause significant concentration polarization in FO membrane which leads to overall decline in both water permeability and product water recovery. To overcome such limitations, recently researchers have attempted to use a combination of fertilizers as draw solutes, but balancing bulk osmotic pressure, permeability, recovery, and reuse requires further studies. Further, all other inorganic solutes excluded in these two categories can be categorized as third type of draw solutes due to the involvement of pressure-driven processes in their recycling and regeneration step. Thus, to mitigate internal concentration polarization due to higher concentration solutes in FO membranes must be addressed by adopting characteristic polymers that can provide high permeability and enhanced hydrophilicity having small structure parameters. On the other hand, novel draw solutes must have an enhanced Van't Hoff factor, high diffusion coefficient, low viscosity and relatively moderate ion or molecule sizes.

Therefore, a perfect balance between osmotic pressure, product permeability and reverse flux should be the main selection criteria for optimal draw solutes. However, draw solutes such as $MgCl_2$, sucrose, MNPs, NH_4HCO_3, organic salts, polyelectrolytes, and seawater bittern, deep eutectic solvents (DESs) and RO brine have demonstrated great potentials for some specific applications of industrial importance.

4.5 Advances in FO-based Desalination

Among the different membrane-based water treatment and desalination techniques in practice, FO has shown promising performance compared to all other pressure driven membrane-based technologies. The FO process is experiencing revolutionary changes in terms of their design and engineering of the working units considering various energy consumption parameters and environmental aspects when in use. Future desalination application of FO processes should integrate sustainable and green aspects of origin draw solutions, recovery of both product water and draw solute. Therefore, desalination applications of FO process are feasible when cost-effective, green, and renewable draw solutes/solutions are utilized for dewatering. [53]

Nevertheless, developing an ideal combination of FO membrane draw solute or solution for seawater desalination to achieve potable quality water may still take a deeper research and innovative approaches. Desalination of seawater using the FO process is a complex process which requires three main considerations.

(i) Firstly, to be an efficient draw solution in the FO seawater desalination process, it should possess an osmotic pressure considerably higher than seawater (27 atm), nontoxic, sustainable, and low cost.

(ii) Secondly, separation of product water and draw solute requires energy efficient, low cost and affordable steps.

(iii) Finally, the process requires perm-selective FO membranes with less to no reverse draw solute flux in order to satisfy the drinking water standard without complex post-treatments.

Recent studies have highlighted that the major drawback associated with FO membrane fouling is directly associated with the reverse draw solute flux. Fouling due to reverse solute flux is also directly related to affinity of membrane with that of draw solute, relatively higher diffusion rate of draw solute compared to that of feed and interactions with feed ionic species.

Therefore, to nullify the overall negative effects, designing FO membranes with much salt rejection and higher flux as well as drawing solutions having lower reverse are studied. In this direction, focus is increasingly made to develop multivalent ionic species-based solutions to meet requirements of low diffusion coefficient criteria. However, it is still expected that an extreme concentration of multivalent ionic species estimated interaction between the FO membrane and potential feed species upon reverse diffusion in the long run may yield overall decrease in FO process. This is an important factor as the currently used FO membranes have either dense selective structure on porous support or thin-film composite asymmetric configuration on porous support as shown in Figure 4.2. Nevertheless, this effect may be less prevalent as the

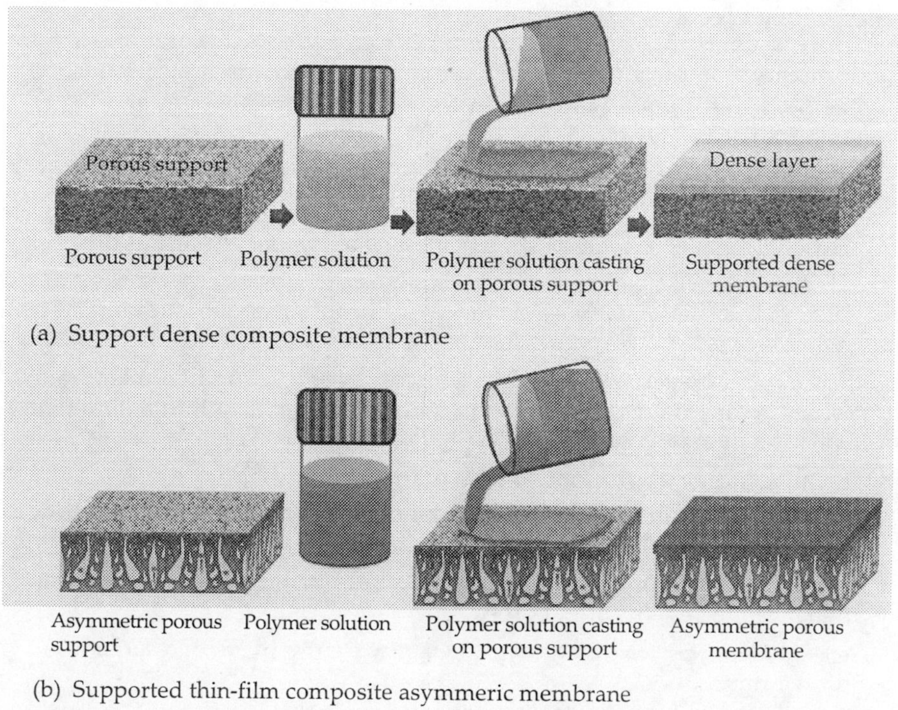

Figure 4.2. Schematics of two important FO membranes used as semi-permeable barrier to establish selective transportation of water from feed to draw solution side.

multivalent ions experience resistance to diffusion due their large ion sizes. Similarly, large molecular weight draw solutes both in ionic or polymeric in nature possess bulky or expanded configurations under fully dissociated forms. For instance, sodium salt of polyacrylic acid (PAA-Na) polyelectrolytes with relatively low molecular weight exert high osmotic pressures however, polymeric draw solutes undergo insignificant reverse fluxes due to their expanded bulky structures. Therefore, to be a successful osmotic agent for seawater desalination application, a draw solute should possess a bulky molecular structure with low diffusion coefficients and minimum ICP effects. This would also help in recovery of draw solute in less energy intensive processes like ultrafiltration (UF). Further, overall combination of process systems can be integrated as point-of-use OF units to work utilizing solar-powered pumps which may potentially make FO futuristic alternative process. Nevertheless, FO technology is yet to mature, however once the limitations are overcome it is expected to help in reducing the both capital as well as operational cost of FO in producing drinking water from seawater. At present several industrial scale seawater desalination FO plants have been commissioned in millions of litre/day capacities. According to recent estimates and business reports, several efforts have been made to multiply the capacity for which high-capacity FO treatment plants are under construction. On the other hand, the currently used pressure driven RO process which is popularly used for seawater desalination is cost intensive and suffers from various process limitations. Similarly, dwindling freshwater resources are also forcing researchers to devise new efficient techniques to recycle wastewater or extract potable quality water from seawater. Therefore, in such cases new, innovative FO systems would play a vital role in the seawater desalination and water treatment industry effluents.

In addition, the major challenges of FO process to be a feasible for desalination applications are:

(i) requires detailed understanding osmotic phenomena in various types of membrane materials,
(ii) lack of deep understanding on factors like fouling and biofouling in FO membranes,
(iii) developing target specific high flux and superhydrophilic membrane which is capable of retaining high salt rejection and a nullifying internal concentration polarization (ICP) phenomena,
(iv) the presence of appropriate draw solutions,
(v) recovery of high quality product potable water from diluted draw solution, and
(vi) flexibility to adopt in combination with the existing processes that can enhance the benefits of FO process.

These challenges necessitate dedicated research to overcome some or all listed limitations.

On the other hand, research is also focused on preparing better FO membranes including membrane material, functionalization to suit the feed and draw solution, large-scale fabrication feasibility and adopting novel methods to enhance membrane property. The first commercially available FO membrane was developed by hydration technology Inc (HTI). This cellulose acetate-based membrane on polyester support mesh has been widely used as one of the best options. Later came polyamide TFC RO membrane adopted for FO desalination applications. However, several second generation TFC FO membranes (Oasys, HTI, Porifera) are readily available in the membrane market. Desalination FO membranes have shown great improvements in their flux performances over the years from the water flux of commercial FO membranes has been improved significantly from about 10 $L.m^2.h^{-1}$ to ~40 $L.m^2.h^{-1}$. Also, these membranes have also been successfully adopted in hollow fibre membrane modes. [54]

On the other hand, biomimetic FO membranes like aquaporin in thin-film format have shown considerable potential in the FO desalination applications. Interestingly, in addition hollow fibre FO membranes have shown higher competitive flux compared to flat-sheet membrane counterparts and hollow fibre configuration offers greater packing density in FO membrane module. Lately, biomimetic membranes have been incorporated with biomaterials like aquaporin which are also known as natural water channels in biological systems with molecular gates. These biomacromolecule based water channels provide remarkable selectivity and exceptional permeability toward water molecules for desalination applications. Several studies using biomimetic aquaporin membranes showed an average water flux of ~16 $L.m^2.h^{-1}$ and high selectivity for NaCl as high as 7000 against sucrose draw solution. However, biodegradation of aquaporin is the major concern in the long-run and advanced studies should include the wider testing parameters for FO desalination such as pH effect, operational temperature and possible chemical cleaning that may damage biopolymer components. On the contrary aquaporin membranes are susceptible for biofouling which is less understood in FO operations. This is important due to the presence of high natural organic matter (NOM) present in natural water sources and seawater tends to interfere in membrane performance. Largely, the NOM composed mainly of biopolymers and protein-like substances form a thin layer on the FO membrane which adversely impacts water flux.

The support layer of FO membrane also plays a vital role in overall FOR seawater desalination. The polyester or polypropylene support layer used in FO TFC membrane when exposed comes in contact with transparent exopolymeric particles (TEP) matter of seawater. These microscopic clusters are difficult to detect in the absence of stain due to their transparent physicochemical properties. Therefore, it is important to identify TEPs and develop appropriate analytical, monitoring, and regular removal techniques in place for making FO desalination processes commercially feasible. Nonetheless, membranes for PRO configuration are still scarce therefore limit their application in seawater desalination. Even though at lab-scale conditions, power density of pressure retarded osmosis (PRO) membrane has reached higher than 20 W m^2, however there are no high performing and commercially feasible PRO membrane available in the FO market today.

4.6 Advances in Wastewater Treatment

Recent years have seen numerous study reports in the FO process being developed for wastewater treatment applications. The FO process in addition to treatment of hypersaline wastewater desalination and municipal wastewater, has shown great potential in many industrial wastewater treatment applications such as shale wastewater reclamation, mining industry wastewater treatment as standalone and in hybridization with other membrane processes for multiple resource recovery , and value-added product concentrations. Recent trends also show potentials of FO processes in specialized applications such as biological component concentration and biomacromolecules like protein and DNA enrichment. However, FO is also facing tough competition with the existing conventional separation techniques as well as other membrane processes. [55]

It is generally observed that standalone FO processes have several limitations in facing high concentrated feed solutions like industrial wastewater. Therefore, FO as a standalone technique is ineffective and therefore generally integrated with hybrid processes like membrane distillation (MD) process for the municipal wastewater treatment. In this configuration, FO integrated with MD process has shown a stable water flux in a continuous operation achieving up to 80% water recovery rate. During this time, in the hybrid FO-MD process, FO step showed a moderate to high rejection for organic contaminants while MD step rejected all residue contaminants to achieve a near complete purification of feed water. Similarly, innovative approaches have been adopted to reusable quality clean water recovered from secondary clarified effluent using advanced photovoltaic powered forward osmosis in combination with electrodialysis (FO-ED) process. This new configuration of hybrid process showed enhanced total organic carbon (TOC) removal efficiency from feed wastewater resulting in high quality reusable water. [56, 57]

Membrane bioreactor (MBR) is another component of hybrid processes recently being used widely for biological contaminants wastewater. In one such attempt, MBR-FO hybrid process has been used to separate biologically-activated sludge from wastewater. Further, FO membrane process is most used for recovering clarified water from biological wastewater. Generally, the biological wastewater treatment process is complex, requires many steps and is highly energy intensive. However, integration of FO-MBR could reduce the energy consumption substantially, even in comparison with conventional standalone MBR. Therefore, novel FO-MBR which is also known as osmotic MBR (OsMBR) has shown greater potential in treating highly contaminated biological wastewater at low-cost operations compared to pressure-driven membrane processes like ultrafiltration and microfiltration. Another major advantage with the MBR-FO process is that fouling in hybrid systems can easily be controlled by simple air scouring in the MBR step which retains longer sustainable flux and reliable contaminants removal pattern. Therefore, hybrid MBR-FO can be a reliable, lower fouling propensity and sustainable alternative for wastewater treatment. [58-60]

In another combination of process FO and RO techniques have been integrated to achieve dewatering and recovery of value-added products from anaerobic digester concentrate. In this process, the FO process is designed to dewater nutrient-rich anaerobic digestercon centrate in the first step using saturated NaCl draw solution. Further, diluted draw solution subjected to RO process for recovering the fresh clean water from diluted draw solution. Meanwhile, concentrated nutrients and organic compounds in FO steps are either post-treated to get value added products or sent to safe disposal. Similarly, EDTA sodium salt was used as a draw solute in the FO process for dewatering activated sludge. The use of EDTA sodium salt draw solution achieved enhanced dewatering while simultaneously concentrating nutrients in the sludge. For the draw solution (EDTA based salt) regeneration was achieved using the NF recycle process and nutrients from concentrated sludge were extracted for recovering value added products. Interestingly, these hybrid processes have shown greater future potential of FO process in combination with other conventional treatment steps to drastically reduce disposal sludge volume at lower cost. This is one of the main criteria for future technologies to minimize final disposal of solid waste that helps in easier transportation and handling. [61]

Industrial wastewater treatment, extraction of high-quality reusable water and reclamation of value-added products and their reuse in various sectors is the need of the hour. Industrial wastewater mainly comprises process residues, impurities and leftover both hazardous and non-hazardous contaminants. Nevertheless, more than 90% industrial wastewater is reusable water; however, there needs to be a suitable treatment technique in place to realize maximum dewatering of industrial wastewater. Another major part of polluted water treatment and reclamation is finding suitable treatment. In this regard, the FO-based treatment system has been adopted in various forms and configurations. In one such attempt, quality water is recycled from textile wastewater from the carpet mills industry which is characterized by high concentrations of organic dye used during dyeing processes.

In another FO process application removal of heavy metals from industrial wastewater of various sources. In this standalone process, several contaminant characteristics as well as process parameters like hydrodynamic conditions, temperature, organic fouling, feed and draw solution properties have been considered in detail to effectively devise suitable FO configuration. Among many process optimization conditions, tailor-made FO membrane plays a vital role. However, in this study, TFC RO membrane showed complete rejection for various heavy meals like Cu, Zn, Pb, Cd and so on from wastewater of industrial origin. Recently, many of the FO process applications for oil and gas (O&G) industrial wastewater treatment have been successfully used for the treatment and reclamation of product water. [62]

In another study, oil contaminated wastewater is treated to extract quality water using the FO process. Here, the bench scale FO process was designed for the separation of water from emulsified oil/water. [63] FO process feasibility was tested against the simulated feed having as high as 2,00,000 ppm oily contamination with the help of tailor-made TFC FO membrane. This process yielded a moderate flux of ~11 L $m^{-2}.h^{-1}$. Further, this bench scale was modified to scale up and tested against real drilling wastewater from O&G processing. Further, RO step in FO-RO hybrid operations helped in reclaiming product water (HTI, 2011) and it is estimated to recover up 2,42,000 gallons per day from drilling wastewater. This enormous amount upon reclamation consequently drastically reduces the stress on freshwater resources. Similarly, several studies are being conducted to develop a feasible combination of membrane materials, fabrication technique, appropriate module and combination of draw solutes to induce optimal osmotic pressure for wastewater treatment. Outcome of many researchers attempts clearly reveal that the FO process can yield high percent of quality water recovery from waste sources. Therefore, advanced designed FO processes can easily concentrate feed of as high as ~10,000 ppm solution and achieve high rate of water recovery economically even outperforming commercially popular RO processes. Nevertheless, due to the complex nature of industrial effluent that are characterized with both organic and inorganic substances, traces of these compounds are still expected to remain in the produced water. Further, to make product water safe for reuse, either pretreatment or posttreatment process may be adopted.

FO process for specialized applications: Recent years have seen increased efforts to impart sustainable aspects to separation and purification technologies by design and development of greener materials. In this direction, imparting sustainability into separation technologies requires use of green materials and their simple preparation protocols and using such materials for the treatment of polluted wastewaters and other speciality applications. In this direction, naturally occurring relatively greener substances in the form of solvents have been used as a draw solution. In one such approach, sea bittern is a leftover brine highly concentrated salt mixture from salt farms after production of common salt (NaCl). This leftover sea bittern was used for the concentration of sucrose solution from sugarcane juice. Generally, sugarcane juice is concentrated in sugar industries using a multi-effect evaporator which consumes enormous energy. However, use of leftover waste sea bittern is used as a draw solution against sugarcane juice across polyamide FO membrane. With this, high quality water from sugarcane juice is extracted from the feed side to achieve nearly 4-fold sucrose concentration. This study has shown greater potential of using waste natural resources as a source of energy in the form of draw solution in FO applications.

In another approach, deep eutectic solvents have been used as a draw solution to enrich biomacromolecules of industrial importance. Deep eutectic solvents (DESs) are a new class of relatively greener solvents widely used for a variety of applications including separation and purification. DESs are made of large, non-symmetric ions which possess a low lattice energy and hence they acquire low melting points. Interestingly, several other colligative properties of DES such as high osmotic coefficient, melting point depression, high viscosity and high boiling point make them suitable candidates as efficient osmotic agents in FO applications. DESs are prepared by the complexation of a quaternary ammonium salt, a hydrogen bond acceptor (HBA) with a metal salt or hydrogen bond donor (HBD). This complexation occurs with a charge delocalization through hydrogen bonding between a halide ion in HBA and the hydrogen-donor moiety which drastically decreases the melting

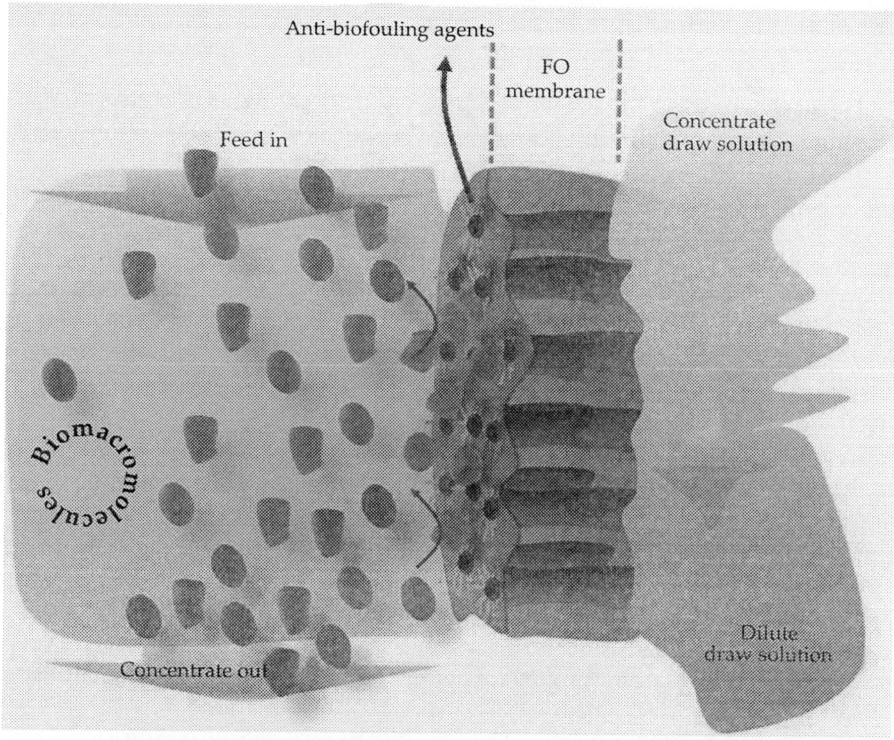

Figure 4.3. FO process in which concentration of biomacromolecules is achieved using deep eutectic solvents as draw solution.

point of the resulting mixture of individual components. In one such attempt, a combination of choline chloride (CC) and ethylene glycol (EG) or glycerol (Gly) were brought together in 1:2 ratio to convert these HBA and HBA into highly viscous draw solution. In particular, CC-EG and CC-Gly were used for the concentration of dilute DNA and bovine protein feed solutions in the FO process. Interestingly, most of the macromolecules are sensitive to external stimuli like temperature, pH and applied pressure. [55] So many conventional biomacromolecule concentration and separation either use one of these external stimuli. However, use of greener solvents like DESs in FO creates sustainable and safe concentration and separation media for these biomaterials as shown in Figure 4.3. In this process, sensitive bio-polymers were efficiently concentrated to 3-7-folds via such a green sustainable approach. Interestingly, upon dilution of the draw solution, it was subjected to freezing which separates active DES from water components using their freezing point depression properties. In these both attempts, researchers have used natural but misplaced resources as energy sources for industrially important applications. These draw solutions have shown great potential for the water treatment as well as speciality applications. Therefore, owing to the green and environmentally benign nature of sea bittern and DEs for sucrose and macromolecules concentration, respectively can be a better alternative for synthetic origin draw solutes or solutions.

4.7 Advances in Industrially Important Water Reclamation Applications

As discussed in previous sections, FO has been used for desalination, waste water treatment as well as speciality applications where conventional separation and purification processes fall short. It is also largely undertaken that the FO process is relatively energy saving and can be potentially used as an alternative to conventional desalination and water treatment processes such as NF and RO. For such applications FO process has been generally used in two different approaches, namely, direct in which standalone process is utilized, and secondly indirect approaches where FO process has been used in combination with other conventional processes in a hybrid system. The former method so far has been successfully applied for desalinating seawater using a draw solution of higher ionic strength solution, which is generally recycled and reused. The second approach comprises the use of highly saline seawater as the draw solution against contaminated water or industrial wastewater as the FS. This leads to the partial dilution of seawater which can be further treated using low pressure reverse osmosis (LPRO) or nanofiltration step as part of an FO/LPRO hybrid process. Such approaches have been widely applied to treat industrial wastewater. In one such case, municipal wastewater was used as an FS, which yields highly concentrated sludge that can be converted further into a bio-energy via an anaerobic process. This overall approach makes wastewater treatment more cost-effective and provides an opportunity to recover value added products. Indirect FO desalination of wastewater treatment processe uses a high salinity water like seawater or RO reject or concentrated brine as a natural DS. Further to recover dilated draw solutes, nanofiltration or low operating pressure reverse osmosis (LPRO) processes can be effective. It is therefore estimated that recovered water from diluted draw solution following the FO-LPRO hybrid process is more economical compared to standard reverse osmosis. Further, FO has utilized to treat various types of industrial wastewater such as municipal or domestic wastewater, oil contaminated wastewater, leather industry wastewater, automobile industry wastewater, dairy and food processing industry wastewater, textile industry wastewater and several manufacturing units produce a variety of synthetic wastewater. [64]

Lately, the FO process is gaining wider recognition as an effective alternative to RO, NF and UF processes for wastewater treatment. As discussed briefly in the previous section, FO-based processes can be further subdivided into three groups, namely; FO standalone, FO-based hybrid processes, and integrated FO processes. Both in FO standalone and FO-based hybrid process configuration, the FO membrane is primarily used to recover freshwater by concentrating pollutants on feed side using suitable semi-permeable FO membrane. On the other hand, in both FO-hybrid process and the integrated FO systems, the multi-step approaches are followed for effective as well as economical industrial wastewater treatment. In particular, integrated FO membrane process gradually replaces conventional pressure driven membrane processes in the bioreactor, such as the FO membrane in membrane bioreactor (FO-MBR). In all the cases, hydrophilic and tailor-made functional membrane is applied for the concentration of wastewater. Therefore, FO in all three forms and configurations have been extensively employed in wastewater treatment and draw solution reuse, resource recovery, seawater desalination, and industrial process or manufacturing purification and separation. Among all, the FO process has shown promising results compared to pressure-driven processes where the treatment of wastewater is limited and, in some cases, failed to yield desired results. Further, FO-based processes have shown several advantages in comparison to the conventional wastewater treatment techniques. [65]

On the other hand, even though FO standalone systems have shown limited success for desalination, FO-based hybrid processes have been advantageous over conventional systems. Hybrid desalination systems using emerging FO processes and combined with traditional processes like membrane distillation, electrodialysis (ED), reverse osmosis, and nanofiltration could remarkably reduce the energy consumption in the desalination process and enhances the potentials of combined processes for real field applications. However, in most of the hybrid processes, FO system has been used as a pretreatment step, while ED, NF or RO have been used for either water recovery or draw solution regeneration step. [66]

The main advantage with the FO-hybrid process is that the performance of the individual process in different steps can be improved when they are in combination taking benefit of the distinctive strengths of the individual processes. For this purpose, generally the FO process is frequently combined with an MD process. For instance, the FO-MD hybrid process was employed for raw municipal wastewater with water recovery near 80% from feed. This hybrid process interestingly achieved high removal efficiency (91–98%) for even trace organic contaminants (TrOCs). Similarly, another study used FO-MD hybrid system was used for oily wastewater treatment. The findings specified that >90% feed water could be recovered readily even removing traces of oil and NaCl species. Further, a vapor pressure driving FO-MD hybrid process was designed to treat the high saline hazardous waste to the level of landfill leachate. In addition to this, total nitrogen (TN) and total organic carbon (TOC) rejection rates are as high as 98% while rejection rate of ionic species is ~96%. Further, the hybrid process was highly effective for the removal of NH^{4+}-N, and heavy metal ions from contaminants in feed sources. Similarly, several recent studies have shown high removal efficiencies for various other industrial wastewater such as textile industry, dairy wastewater, and food processing wastewater treatment. [67, 68]

Traditionally, several desalination processes have been developed and applied for desalination of seawater as well as ground to produce potable quality water. Among several existing processes, RO has been the most consistent and reliable in producing instant high quality product water which also offers several benefits due to its high salt rejection rates, high water recovery, and considered s relatively greener technology. However, despite the several above-mentioned advantages, the RO process possesses several shortcomings, such as high energy consumption and severe fouling propensity which pose greater challenges to researchers and technologists. Interestingly, a new combination of hybrid FO-RO has been uniquely designed to treat wastewater feeds. Wastewater treatment is increasingly gaining importance in the FO segment as the other competitive treatments fail to achieve required standards. Initially, hybrid combinations of FO and RO have been used for seawater desalination to produce potable quality water. Traditionally, the RO process has been widely used for seawater desalination, however, several limitations associated with RO membranes such as susceptibility to fouling and increased intensive process operations have forced researchers and technologists to adopt hybrid processes such as FO-RO. Therefore, recent studies have proven hybrid FO and RO processes as economic alternatives for cost-intensive standalone RO for desalination as well as wastewater treatment applications. Generally, in this configuration, the first stage is designed to have a FO process in which quality water is osmotically drawn from seawater feed into a draw solution chamber. Further, in the second stage diluted draw solution is recovered from the RO process as the fresh product water.

Several other studies have listed the comparative analysis of hybrid FO-RO process over standalone RO process for desalination applications. These studies clearly point out four major benefits over standalone RO desalination, namely:

(i) lower energy consumption,
(ii) multi-barrier safety of drinking water,
(iii) beneficial reuse of recovered high quality water, and
(iv) reduction membrane fouling.

Similar studies have also been conducted to compare the advantages of hybrid FO-RO system over standalone RO process for both seawater desalination and wastewater treatment. The study demonstrated that the hybrid FO-RO process can be economical depending on the sources of seawater and type of organic content and type and concentration of the draw solution. Interestingly, hybrid processes have consumed a total power consumption in a FO-RO process much less than that of RO process and standalone FO process as well. Also, studies have estimated the total power consumption in FO-RO hybrid processes to only 2–4% of the total power consumption in earlier cases. Therefore, even in hybrid processes, the most of the power consumption was realized in the RO regeneration step where high hydraulic pressure is applied for recovery processes. In another study, the FO membrane process was used as a pretreatment for a hybrid FO-RO desalination process. The optimal parameters such as water recovery, water flux and final draw solution of this FO pre-treatment process were estimated by modeling and were experimentally validated by using real brackish water. In a further study, the FO-RO hybrid process for wastewater from coal-fired power plants using concentrated seawater from UF was investigated as DS. Results presented that the total energy consumption of the FO-RO system was merely 15% compared to typical seawater desalination using the RO process.

Further, novel combinations of FO-NF hybrid systems were designed to replace the relatively cost-intensive FO-RO process. This approach has shown several advantages even over the FO-RO process on several fronts not only energy consumption but also overall process economics and environmental potential in designing large-scale FO-based hybrid systems. However, there are only a few reports available on the use of the FO-NF process in the literature. For instance, pilot-scale FO in combination with NF and distillation processes for the treatment of seawater as well as wastewater effluents. Further, a pilot-scale FO-NF hybrid process in a closed loop was developed for the treatment of tannery industrial wastewater at a rate of 52–55 L/m^{2}h with COD rejections as high as 98.5%, 98.2% sulfate and 97.2% chlorides were achieved in this approach. In addition, a hybrid FO–NF process was also designed for brackish water desalination which showed promising results such as utilization of lower hydraulic pressure, less flux decline and inducing low potential fouling. In another study, a hybrid FO-NF process, two-step NF process used as posttreatment for desalinating seawater.

Interestingly, another configuration of hybrid FO-ED processes was designed as an alternative to both FO-RO and FO-NF processes. Electrodialysis is an ion-exchange membrane-based separation process in which ion species were allowed to transport across ion-selective membranes known as cation-exchange membrane (CEM) and anion-exchange membrane (AEM) under an electric field. A FO-ED hybrid process was examined using diammonium phosphate (DAP), as a draw solution to achieve dewatering of wastewater through the mitigation of salinity build-up from the feed side. Electrodialysis in this configuration was able to considerably recover the ~96 % reverse-fluxed DAP under applied potential of 3.0 V daily operation. Forward osmosis process was tested upstream to the ED-RO hybrid process for an access to draw solution with higher electrical conductivity in the FO-ED-RO hybrid process. In another study, the FO-ED-RO hybrid system was projected to produce good quality water from secondary-effluent produced in the first step. Therefore, results showed that the recovered water from this hybrid system contains a low concentration of carbonate, cations, and total organic carbon (TOC) derived from the feed water. [69]

On the other hand, pressure retarded osmosis (PRO) which is a specially configured FO process where osmotic pressure can be converted into hydraulic pressure for energy generation. Further, this can be engaged as an energy recovery process in an assimilated process, such as RO-PRO. However, lack of suitable FO and PRO membranes is still a major constraint and limiting factor in large scale adaptation of this hybrid process for potential application.

4.8 Advances in Industrial Waste Process Applications

Forward osmosis (FO) is the membrane process that has been now widely used as one of the energy efficient techniques to treat industrial wastewater along with leachate. Repeated studies have established that FO has shown great potentials of the process to be a sustainable technique in recovering high quality water from both seawater as well as industrial wastewater. The FO process is advantageous over other membrane techniques as it is being run at nearly zero or minimum hydraulic pressure. Use of an efficient osmotic agent in the FO process makes overall water recovery from wastewater a high efficiency by rejecting contaminants in the feed chamber. The use of low operating pressure also helps in maintaining low membrane fouling proficiency compared to pressure driven membrane processes. Therefore, due to high concentration gradients between feed wastewater and high concentration of draw solution which possess high osmotic pressure naturally induce transportation of pure water from feed side to draw solution. In general, several inorganic ionic species originating from salt precursors such as NaCl, $MgCl_2$, $CaCl_2$, etc., have been widely used as osmotic agents for wastewater treatment applications. Recent studies have also used RO concentrated brine, seawater, and synthetic seawater composition as active draw solutions for the treatment of industrial wastewater.

These FO experiments have demonstrated the ability of FO membranes to reject pollutants such as chemical oxygen demand (COD), contaminants like heavy metals, organic pollutants, or nutrients such as phosphate, nitrogen so on from wastewater, especially. Further, a submerged hollow fiber configuration membrane module which makes it possible to adapt this technique as an efficient primary clarifier that can be directly employed for partial demineralization of wastewater. The recent findings have shown that advances in the design and development of FO membranes lead to the high rejection of pollutants like organics, heavy metals and so on present in the wastewater up to the efficiency ~99%. However, in some cases, biopolymeric contamination such as biomacromolecules have shown biofilm formation which leads to biofouling in the long run on the membrane surface. In addition to standalone FO process, several hybrid FO-process have been adopted for the treatment of wastewater. In one such attempt, osmotic membrane bioreactor (OMBR) has been adopted as derivative of FO-process for municipal wastewater treatment.

Further studies are underway should be conducted to improve the FO process efficiency by

(i) developing hydrophilic and thin barrier FO membranes,
(ii) deeper understanding of osmotic phenomenon,
(iii) attaining sustainable the process under long-term operation, and
(iv) improving the FO working set-up design to address the various technical niches associated industrial wastewater handling.

In addition to this, to attain the reusable quality water recovery parameters, special attention is given to transport or retaining of low molecular weight minerals like nitrogen and carbon compounds from the feed side solution to the draw solution chamber, which may in long term act as substrate for bacteria to grow. This would lead to the formation of a gel-layer thereby causing biofouling. Therefore, advanced techniques are focusing on adopting new numerical modeling that can be used to predict and prevent biofouling in FO membranes. However, with all limitations and technological difficulties, FO can still play an important role as a connection to integrate upstream as well as downstream processes to reduce the energy dependency and consumption of the overall process, thus can be an effective alternative to dewatering industrial wastewaters.

In one such approach, the FO integration with membrane bioreactors (MBRs) is an attractive way that has generated interest in the industrial wastewater treatment sector. Traditionally, MBRs employ other waste management technologies such as MF or UF processes that are energy effective. However, FO assisted membrane bioreactors (FO-MBRs) have shown great potential to treat highly contaminated wastewater feed by optimizing the important operational parameters. Here, the commercially available CTA-FO membrane in comparison with TFC RO-membrane was used for activated sludge treatment in the OMBR process. In this process, 0.5 M NaCl relatively concentrated brine used as draw solution which produced a flux of 6.2 L/m^2.h at 20°C. This study at laboratory scales has been designed to explore the effects of various types of draw solutes on treatment of polluted effluents. For instance, the study revealed the effectiveness of monovalent ($NaNO_3$, and NaCl) and bivalent ionic draw solutes ($MgSO_4$ and $ZnSO_4$) for recovery of high quality water activated sludge. Overall, the monovalent ionic salts-based draw solutions showed better flux performance than the bivalent draw solution. Alternatively, another study designed the innovative submerged FO process in combination with the MBR system for water extraction from the simulated wastewater feed using commercial CTA FO membranes which resulted in extracting potable quality water with pure water flux of ~18-12 LMH. However, over a period of 30 days, an average flux of the hybrid system dropped to 9 LMH with a relatively low reverse transport of solutes from the DS side. Similarly, a pilot scale FO-MBR study achieved stable osmotic flux of 3 L/m^2 of dewater from severely contaminated mixed liquor tested over 73 days of operation with minimum biofouling. Even in case of severe and accelerated test conditions, possible foulants were removed effectively using a continuous air plug at the feed side of the pilot FO-MBR system for the domestic sewage treatment having high concentrations of bio-wastes. In this condition a constant flux of 3 and 7.2 L/m^2.h was accomplished using NaCl and $MgSO_4$ with different osmotic pressure coefficient draw solutions over 150 hours of operation, respectively. Similarly, a study by Lay et al. [70] showed the importance of air injection to control and mitigate fouling mitigation. This study also revealed that the absence of air injection drastically reduced the flux by ~30% resulting in low water recovery. This study also concluded that the air injection or air scouring provides safe and uninterrupted interfacial passage to water to transport from feed to draw solution chamber.

Some studies also investigated the effect of process design, configuration and operational parameters on overall performance of the FO-based process. For example, use of hollow fiber membrane modules in submerged conformation was used to recover water and other value-added products in the FO-MBR hybrid process. In this approach, two configurations were examined including: (a) active layer facing feed solution (AL-FS), and (b) active layer facing draw solution (AL-DS). Further in this study, the FS containing mixture of contaminants such as mono- and divalent ions, bacteria, organic contaminants and extracellular polymeric materials exposed to membrane or passed through the porous substrate create an interfacial contact with the selective layer. Using the AL-DS configuration, 45% flux decrease was observed compared to that of AL-FS configuration however, flux decline was negligible due to membrane fouling via pore blockage from bacterial clusters and extracellular polymeric debris. More significantly, the study reveals that the FO process in AL-FS dynamic configuration mode saves energy and proved to be highly productive.

Interestingly, Huang et al. used two different modes of FO-MBR having mesophilic FO-MBR and room temperature FO-MBR under similar process operating conditions. Both these approaches showed high productivity for the removal of NH^{4+}-N, PO_4^{3-}-P, and total organic carbon (TOC) from wastewater sources. Mesophilic FO-MBR revealed a lower flux compared to ambient-FO-MBR. [71] Similarly, another study achieved 96% removal of PO_4^{3-}-P, and NH^{4+}-N as well as 100% for TOC, using 1 M NaCl draw solution in FO-MBR. In addition to this, the phosphorus was occasionally collected from concentrated feed portions up to 890–990 mg/L in the feed by adjusting the pH using Na-Hydrogen phosphite hydrate solution. Also, Qiu et al. used MF-FO-MBR hybrid process for the treatment of real municipal wastewater using a biopolymer-based FO membrane that showed improved removal of nitrogen and organic matter. FO-based processes have also been used to efficiently remove heavy metals and persistent emerging pollutants. Aftab et al. have designed FO-MBR with CA membrane for the removal of heavy metals from the sludge supernatant. Heavy metals such as Cr and Pb in 2 to 10 mg/L concentration in sludge have been effectively removed in this mode. In several studies use of both CTA and polyamide TFC have been used as best performing FO membranes however, these membranes have shown several limitations and did not satisfy sustainability considerations for large-scale FO-MBRs process development. However, to improve the surface properties of conventional membrane several have been followed on both CTA and TFC-RO to modify the surface (Li et al.) using thin layer coatings of polydopamine (PD) and grafting of poly(ethylene glycol) (PEG) and (PD-g-PEG). This has to some extent improved the stability of FO membranes during their use in a submerged FO-MBR configuration. These approaches have reduced the bio-material, organic pollutant adhesion and bio-cake formation on the FO-membrane surface. Further, Wang et al. used polysaccharides (β-D-glucopyranose) as a biofouling agent in feed solution to evaluate anti-fouling properties of both commercial CTA and polyamide–TFC FO-membranes. Study revealed that the polysaccharides and microorganism build-up were ~77% of the total bio-volume on CTA FO membrane, while polysaccharides depositions were the only dominant bio-foulants on TFC FO-membrane. [72]

Futuristic approaches focusing on turning a hybrid combination of FO with other existing treatment technologies to satisfy the sustainability criteria and to enhance wastewater treatment efficiency thereby enhance the quality of the treated water for reuse. From the economics point of view, due to the relatively moderate OPEX, most of the FO processes have been integrated as a pretreatment step. An overview of FO integrated hybrid processes shows greater potential of new approaches in wastewater treatment. In one such system, the FO-MBR system reported for the treatment of wastewaters with special emphasis on direct use of diluted draw solution which may address the real field capabilities of FO-based hybrid processes. However, FO-MBR in the long run system has proven limited practicality.

Lately, FO has been used as a pretreatment step to supplement the pressure drive processes. For example, FO can be effectively used as an economical alternative for other pretreatment processes attached to RO wastewater treatment units. FO has proven more effective than other pretreatment techniques like MF process which suffers from various limitations involving drastic flux decline due to exposure to high concentration feed solutions like BOD, COD and organic matters in the wastewater. In the absence of effective pretreatment step high concentrated pollutants can create serious problems to the RO membranes. Therefore, FO-RO combinations have been considered good candidates to enhance the quality of product water recovery from various industrial wastewaters.

In one such attempt, Qin et al. designed indigenous hybrid FO-RO process in which FO has been used as a pretreatment for the RO process. This combined FO-RO process operated in synergistic association with RO results in unsteady water flux and retaining of the pollutants from the nutrient-rich digested biomass. In this arrangement, the FO step retained nearly 70% of the solutes and final RO step continuously; the final recovery in RO resulted in complete retention of pollutants in feed side in both lab scale as well as large continuous processes. Similarly, Holloway et al. followed modeling approaches to design a pilot scale FO-RO process which resulted in ~70% of water recovery from industrial wastewater, and Qin et al. reclaimed good quality water from secondary municipal effluent in a RO process. This was first assisted by the FO step in which a considerable amount of turbidity (~15 to 3 NTU) and salinity was achieved. Further, Boo et al. designed FO to investigate the effect of organic foulants in seawater desalination as well as wastewater feed on FO membrane deterioration. In this approach water treatment parameters were analyzed using synthetic seawater with 30 g/L as the draw solution against wastewater effluent feed characterized by high concentrations of inorganic, humic acid and TOCs. Further, elevated concentrations of industrial wastewater containing more than 20 different types of different contaminants were treated with the help of commercial RO spiral wound membrane testing unit up to 40 days long run. This approach resulted in nearly 99% rejection of most of the pollutants thereby resulting in high-quality reusable water reclamation. Investigators further combined RO and FO processes in two different arrangements and configurations as shown in Figure 4.4 to evaluate the process efficiency and high quality water recovery from various industrial as well as simulated feed mixtures.

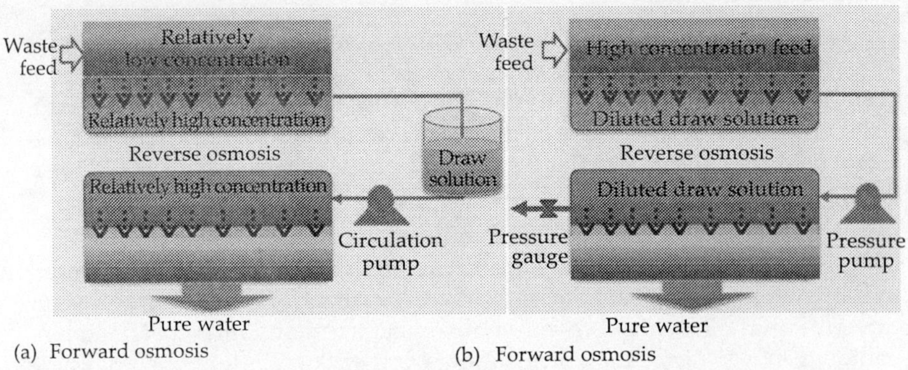

Figure 4.4. Schematic representation FO-RO hybrid process showing two different configurations, namely: (a) closed loop draw solution in which FO recovers water from a contaminated feed and in second step RO is used to re-concentrate simultaneously product water and reusable DS recovery, and (b) an open loop draw solution in which FO step is used to recover water from an contaminated high concentration wastewater feed and diluted draw solution was further used as feed in RO to recover purified water.

In one such attempt, Hau et al. explored the perspective of EDTA Na-salt as a novel draw solution in a hybrid FO-RO dewatering process from high concentration nutrient-containing sludge. For this purpose, simulated feed mixture having 2 g/L Na-alginate, 2 g/L BSA, and 2 g/L natural organic matter, which was collected from a river stream along with 7 different salt mixtures with concentrations ranging from 0.6 to 2.0 mM. Further, comparative studies were also carried out using different draw solutions on raw seawater and RO brine water which resulted in obtaining recovery of pure water with flux of ~7 and 9.5 $L.m^{-2}.h^{-1}$, respectively. Interestingly, this study also evaluated the hybrid process parameters to understand the fouling phenomena of organic pollutants on the FO-membrane surface. With a large set of data analysis, researchers have proposed suitable membrane chemical method and procedure of cleaning the foulants deposition by controlling physic-chemical methods as well as regulating operating parameters. These combinations of data analysis and development of cleaning procedures concluded interesting observations to follow in further operations, namely:

(i) importance of varying the feed solution and draw solution cross-flow velocity,
(ii) introducing the constant turbulence continuously or intermittently through feed channel spacers, and
(iii) operating the FO-process in a pulsed flow manner that leads to reliable process output. [73]

As deliberated before, in large-scale wastewater treatment applications, the economic considerations are amongst the most significant criteria. It also applies mainly in membrane-based water treatment technology applications due to the involvement of continuous energy consumption. In this direction, replacing energy intensive processes with economical methods and approaches is also gaining more significance. For instance, the development of nanofiltration with RO along with FO have been adopted for several hybrid combinations for wastewater treatment applications. Therefore, NF is gaining significant momentum in replacing RO in recent times for industrial wastewater as well as ground water desalination applications. Likewise, Phuntsho et al. designed a hybrid pilot scale FO-NF process using ammonia/ sulphate-based fertilizer type inorganic salt systems as the draw solutions for extracting high quality water from groundwater. In another study, inorganic contaminants were removed from the coal mining wastes using the same set of draw solutes. Interestingly, the hybrid NF-FO process provided the essential rejection for SO_4^{2+} and NH_4^+ and the low rejection for monovalent feed species (Na^+, Cl^-) was observed by FO, while the commercially successful CTA-FO membrane suffered serious reverse solute flux as well as cake formation in a long run resulting in a rapid decline in flux. Further, Kumar and Pal used FO-NF hybrid process for treating the coke-oven wastewater in a long run which was reliably operated and optimized for their flux and contaminant rejection. During this hybrid process also evaluated effects of operating parameters such as FO process cross-flow rate pressure, pH of the FS and draw solutes through which researchers achieved nearly 96% removal of contaminants like phenols, cyanide, COD and NH^{+4}-N from the real coke-oven using 1.5 M NaCl across FO membrane as the draw solution. [74]

Generally, conventional membrane distillation (MD) uses heat and vapor pressure gradient to drive the separation and purification of contaminated groundwater as well as industrial wastewaters. MD process works using applied thermal heat or utilizing waste heat in which only vapor molecules transmit through the hydrophobic porous membrane. Interestingly, vapor pressure gradient across the hydrophobic membrane drives the transfer and separation of product water. This process configuration has been proven to be highly effective in treating biological contaminated wastewater having high metabolic and biological characteristics. Water from wastewater feed is heated to form vapor and allowed to pass through hydrophobic membranes in MD process and other side of the membrane vapors are condensed to produce fresh water.

This process has been adopted in combination with the FO step by Martinetti et al. to recover water from high concentration brines via the treatment of metabolic wastewaters. This process used FO with direct osmosis (DFO or DO) in combination with modified MD known as membrane osmotic distillation (MOD). This process step requires the use of temperature and concentration gradient as the driving forces. Interestingly, the FO-assisted MD hybrid process exhibited remarkable ~25 times more flux and over 99% rejection of urea with a slight increase in temperature (3-5°C). Realizing the importance of this study Ge et al. designed vacuum assistance FO-MD in which poly(acrylic acid) sodium (PAA-Na) salt was used as the draw solutes to obtain improved dewatering of sewer mining wastewater effluent. This arrangement was also further used to recover high-quality water from the RO retentate brine having total dissolved solids (TDS) ranging from 7500 to 17,500 mg/L. Conversely, vacuum assisted MD did not accomplish enough water recovery as expected and recovery was only limited ~81% due to precipitated salt in feed chamber onto the membrane surface which caused steady decline in the permeate flux. Meanwhile, FO step produced ~90% water recovery in FO-MD hybrid process itself, making it one of the potential arrangements for efficient water recovery from waste sources. The effectiveness of PAA-Na towards dewatering wastewater in the FO step showed a regeneration capacity of 0.48 g m/L at 66°C in MD stage making it one of sustainable hybrid methods for wastewater treatment.

To ascertain the importance of recovery resources, high quality product water and draw solute/solution in FO-based processes, it is important to assess processes for their sustainability aspects. In this direction, Song and Xie et al. used the FO-MD hybrid process to extract the valuable resource like phosphorus from the digested sludge waste. [75] This robust hybrid FO-MD process was tested in lab-scale continuous mode using raw sewage feed which resulted in >80% recovery of high-quality product water. Further modifications in MD-step have also been tested for their performance which showed enhanced TOrCs concentration in draw solution leading to high recovery rate of water. Alternatively, researchers proposed adding granular activated carbon for the selective adsorption or including UV oxidation step to prevent the organic contaminant accumulation in the testing unit. Further, Zhang et al. used the modified FO-MD process for possible sustainable recovery of water from the oil contaminated wastewater in which both minerals like phosphorous in the form of struvite (in the form of $MgNH_4PO_4 \cdot 6H_2O$) and high-quality reusable water were successfully mined. For this purpose, researchers designed MD process steps to the recovery of the draw solution ($MgCl_2$). Interestingly, dewatering of feed solution led to the significant concentration of acetic acid in the feed chamber, which was utilized as an additive to produce crude oil in the MD process step. Therefore, this important study demonstrated the potentials of not only recovering high quality water from oil-contaminated wastewater, but also recovery of value-added products like organic additives via FO-MD hybrid systems. [76]

In addition to recovering high value resources, hybrid FO-MD process, hybrid some also inherently possess some minor drawbacks such as low rejection pattern for complex minerals. For instance, hybrid FO-ME is ineffective towards the rejection of total nitrogen (TN), TOC, and ammonium nitrogen ($NH^{4(+)}$-N) from the human urine contaminated wastewater sources. Further, Nguyen et al. employed the highly charged phosphate (P)-based as the draw solute to result in permeate fluxes as high as 12 LMH in FO-MD process for concentrating high-nutrient waste sludge. In this process, the efficiency of the draw solution and high rate of product water from the diluted Na_3PO_4 were accomplished using 4-different types of hydrophobic polytetrafluoroethylene (PTFE) membranes in MD operations. Similarly, Chen et al. achieved a high rate of water reclamation from the shale-gas drilling flow-back fluid (SGDF) using a hybrid FO-vacuum-MD system. This hybrid FO-based system produced ~90% water recovery from SGDF feed in which FO was used as a pretreatment step to retain most of the contaminants.

This step greatly reduced the possibilities of clogging or membrane fouling in subsequent MD steps. [77]

In another approach, electrodialysis (ED) process has been used in combination with FO process for wastewater treatment applications. The ED process is a technique that uses ion exchange membranes (cation-exchange membrane, CEM and anion-exchange membrane, AEM) to facilitate transportation of ionic species from the concentrated feed chamber (feed) to dilute chamber under the influence of applied electric potential difference. Hybrid combination of FO-ED as shown in Figure 4.5 can be an effective process for industrial wastewater as well as brackish water treatment. In one such attempt, high quality reusable water was reclaimed from the secondary wastewater treatment plant effluent by concentrating carbonates, TOC, and cationic pollutants. Further recent studies have demonstrated the effective recovery of draw solute from the ED process via a sustainable approach utilizing naturally rich solar energy. An effective way of producing high quality product water from secondary waste treated sources in an FO-ED process integrated with solar energy supply which makes the overall hybrid process a cost effective and a sustainable alternative. Similarly, FO process integrated with electrochemical oxidation (FO-ECO) was designed to effectively concentrate the trace pharmaceutical contaminants in the bio-waste effluents. Here, use of FO-ECO process achieved recovery of trace antibiotics with over 98% rejection at a current density of J=1 mA/cm^2 and treated water attained safety standard for direct disposal norms. [78]

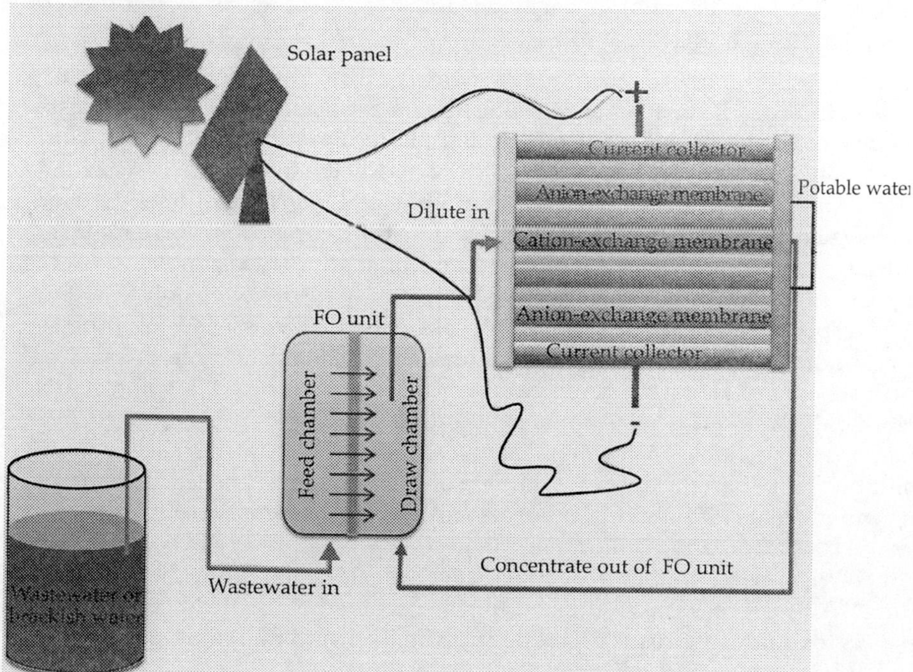

Figure 4.5. Schematic presentation of recovery of potable quality water from wastewater feed solution in hybrid FO-ED process.

Recently, researchers have adopted an innovative FO process in combination with microbial fuel cell (FO-MFC) process for potential recovery treatment of wastewaters. [79] Microbial fuel cell (MFC) is an evolving technology used for the generation of clean energy following sustainable future criteria. In this approach, innovatively designed hybrid FO-MFC process in two different modes have been adopted, namely: (i) in this mode recycling of water from the waste sources is achieved for direct reuse purpose, and (ii) in this second mode, direct seawater desalination is achieved in MFC mode known as microbial desalination cell (MDC). In this hybrid process the performance of FO-MFC was closely evaluated using both NaCl and seawater as a catholyte or draw solution. More significantly FO-MFC produces efficient electricity than standalone MFC in batch experiments, which may be ascribed to enhanced proton-transport phenomenon with appropriate water flux in FO-membrane. Similarly, an air-cathode assisted FO-MFC hybrid process improved the energy recovery where the use of a suitable FO-membrane with minimum internal resistance to flow resulted in higher power production up to 43 W/m^3. On the other hand, the high removal efficiency of carbon-based compounds like acetate derivatives up ~90% resulted in attaining stable flux of 4.1 LMH in the FO-MFC process. Further, MDC process was also adopted for wastewater treatment following an energy-efficient design in which organic waste treated in an MDC process was subjected to water recovery in a FO mode. This resulted in reducing the wastewater volume by dewater up to the concentration of 64% thereby enhancing the conductivity of the source by twice. Nevertheless, MDC technology at present is in its initial developmental stage and researchers are rigorously engaging in adopting further studies to understand practical applications.

Largely, MF and UF membranes work well in low operating pressure ranges resulting in relative high-water fluxes extracted from various waste sources. Now the MF-assisted membrane photo-bioreactor (MF-MPBR) and FO-assisted photo-bioreactor (FOMPBR) have been innovatively designed by a group of researchers for the treatment of *Chlorella vulgar* in tertiary wastewater treatment step. [80] Through FO-MPBR researchers have removal and recovery of nutrients like nitrogen and phosphorus from waste sources up to 86-99% and 100%, respectively. This was only restricted to only 97% and 46%, respectively in absence of FO process making FO-assisted the process appropriate for the treatment of microalgae-based wastewater. In similar approach Qiu et al. used a combination of microfiltration and FO bioreactors (MF-FOMBR) to recover nutrients including PO_4^{3-}, Mg^{2+}, Ca^{2+}, etc., up to 70% from the municipal wastewater using concentrated seawater as the draw solution. Employing seawater as draw solution not only ensued in nutrient enrichment up to 97% for PO_4^{3-}-P, but also resulted in the removal of TOC (>90%) and NH^{4+}-N. Similarly, Holloway et al. used FO-MBR process in combination with UF has been applied to extract the reusable water from the activated sludge. This process is also utilized to concentrate the dissolved constituents such as total phosphorus (TP), total nitrogen (TN), and inorganic salts and so on. For this, researchers designed a FO process integrated with aerobic and UF hybrid systems for the sludge treatment which achieved simultaneous water and nutrient recovery. Thus, application of FO-based hybrid systems has been highly efficient and effective in water treatment, pollutants removal from industrial wastewater sources, helps in recovery of value-added products like nutrients and in some cases produces energy which can be further used for running treatment systems on site. [81]

4.9 Future of Forward Osmosis (FO) in Water Treatment

The earlier sections have evaluated use of FO-based membrane separation process for the treatment of water and various wastewaters. Efficiency of FO-based processes, both standalone as well as hybrid processes, can be evaluated based on their sustainability criteria involving in the selection of the most appropriate material adoption and choice of technology combination to deal with the highly polluted wastewater systems. An overview on the recent studies reveals that FO processes, especially in the form of FO-based hybrid processes having FO step as a pre-treatment step can be effectively realized to treat various types of contaminated resources, wastewater streams as well as recovery of value-added products. Several other studies have also indicated the technological advantages of FO-based hybrid system in reducing as shown in Figure 4.6, namely:

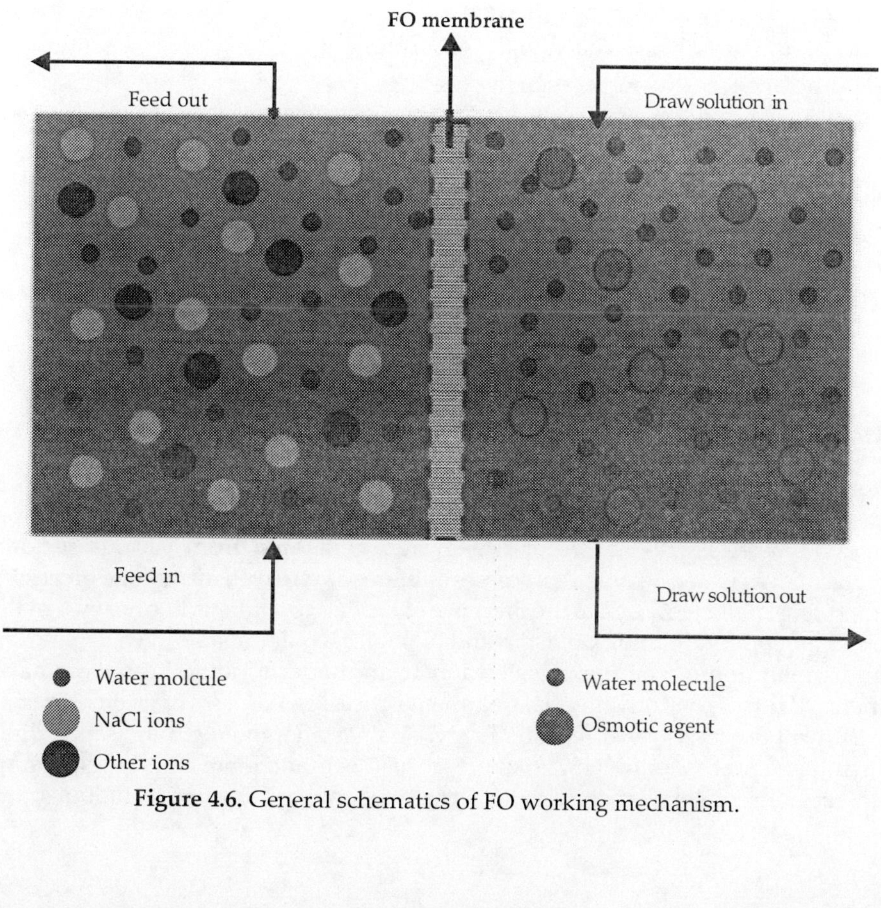

Figure 4.6. General schematics of FO working mechanism.

(i) reduced CAPEX and OPEX,

(ii) attaining highly concentrated sludge which can minimize the risk associated with disposal and management,

(iii) FO process is advantageous over other pressure-driven processes as osmotically driven processes pose less risk of membrane fouling and scaling,

(iv) selection of appropriate FO membrane reduces the maintenance cost,

(v) FO-based processes can be easily adopted where other membrane-based separation process fail to address the severely contaminated waste sources,

(vi) FO-based processes can be designed to suit the end use applications,

(vii) it is proven in various studies that use of FO-based steps drastically reduces the energy consumption leaving low energy footprints,

(viii) along with aforementioned advantages, FO-based processes are relatively clean which makes them future alternative processes for both water treatment applications as well as recovery of value-added products from various waste sources.

Benefits and advantages of FO in comparison to other conventional desalination and wastewater treatment

- Significant energy savings (up to 50%)
- Minimizes discharges back to the environment
- Eliminating scaling and reducing fouling
- Membrane self-cleaning and clean In Place (CIP) systems
- Cheaper and longer lifetime membranes, valves and pumps
- Lower levels of boron without additional treatment
- Reduced chemical consumption
- Design Scalability and retrofit is possible
- Reduced carbon footprint
- Reduced OPEX and CAPEX
- More sustainable and cost effective alternative to traditional systems

4.10. Conclusion

Forward osmosis technology is emerging as an alternative to pressure-driven membrane processes. Increasing complexities in the water contamination and difficulties with nature of wastewater generated from various sectors, conventional water treatment processes and pressure-driven membrane processes are facing challenges. Due to the presence of modern pollutants like heavy metals, emerging pollutants, microplastics and biological contaminants in water sources and generation of complex wastewater from modern industrial sectors have increased stress on conventional treatment processes. Also, several conventional treatment processes have been ineffective against emerging pollutants including membrane processes. In such situations, researchers and innovators are exploring advanced treatment processes such as nanotechnology assisted techniques and forward osmosis. In particular, forward osmosis process is proving effective where all other processes prove ineffective. This chapter has given detailed account of potentials of forward osmosis processes for addressing various types of water treatment applications.

References

1. Angelo Basile and Kamran Ghasemzadeh, Current Trends and Future, Developments on (Bio-) Membranes Silica Membranes: Preparation, Modelling, Application, and Commercialization, ISBN 978-0-444-63866-3, 2017, Elsevier.
2. UN Water. World Water Day-Factsheet. United Nations; New York, NY, USA: 2019.
3. Sakhel, S.R., Geissen, SU. & Vogelpohl, A. Virtual industrial water usage and wastewater generation in the Middle East and North Africa 2011–2015. Euro-Mediterr J Environ Integr 2, 7 (2017). https://doi.org/10.1007/s41207-017-0018-9.
4. Thirugnanasambandham Karchiyappan, Rama Rao Karri, Mohammad Hadi Dehghani2022, Industrial Wastewater Treatment, Emerging Technologies for Sustainability, Springer Cham, Hardcover ISBN978-3-030-98201-0, eBook ISBN978-3-030-98202-7.
5. Şahin, D. Forward Osmosis Membrane Technology in Wastewater Treatment. In: Ince, M., Ince, O. K., editors. Osmotically Driven Membrane Processes [Internet]. London: IntechOpen; 2021. Available from: https://www.intechopen.com/chapters/76491 doi: 10.5772/intechopen.97483.
6. Ashesh Mahto, Kanakaraj Aruchamy, Ramavatar Meena, Mohammadreza Kamali, Sanna Kotrappanavar Nataraj, Tejraj M. Aminabhavi,Forward osmosis for industrial effluents treatment – sustainability considerations, Separation and Purification Technology, Volume 254, 2021, 117568, ISSN 1383-5866.
7. Qingwu Long, Yongmei Jia, Jinping Li , Jiawei Yang, Fangmei Liu, Jian Zheng and Biao Yu, Review Recent Advance on Draw Solutes Development in Forward Osmosis, Processes 2018, 6, 165.
8. Muhammad Amjad, Jabbar Gardy, Ali Hassanpour, Dongsheng Wen, Novel draw solution for forward osmosis based solar desalination, Applied Energy, Volume 230, 2018, Pages 220-231, ISSN 0306-2619.
9. Gordon T. Gray, Jeffrey R. McCutcheon, Menachem Elimelech, Internal concentration polarization in forward osmosis: role of membrane orientation, Desalination, Volume 197, Issues 1–3, 2006, Pages 1-8, ISSN 0011-9164.
10. W. Suwaileh, N. Pathak, H. Shon, N. Hilal, Forward osmosis membranes and processes: A comprehensive review of research trends and future outlook, Desalination 485 (2020) 114455.
11. P.S. Goh, A.F. Ismail, B.C. Ng, M.S.J.W. Abdullah, Recent progress of forward osmosis membranes formulation and design for wastewater treatment, Water 11, (2019) 2043.

12. M. Elimelech, W.A. Phillip, The future of seawater desalination: energy, tech[1]nology, and the environment, Science 333 (2011) 712–717.
13. X. Song, Z. Liu, D.D. Sun, Energy recovery from concentrated seawater brine by thin-film nanofiber composite pressure retarded osmosis membranes with high power density, Energy Environ. Sci. 6 (2013) 1199–1210.
14. S. Zhao, L. Zou, C.Y. Tang, D. Mulcahy, Recent developments in forward osmosis: Opportunities and challenges, J. Membr. Sci. 396 (2012) 1–21.
15. B.D. Coday, N. Almaraz, T.Y. Cath, Forward osmosis desalination of oil and gas wastewater: Impacts of membrane selection and operating conditions on process performance, J. Membr. Sci. 488 (2015) 40–55.
16. F. Zhang, K.S. Brastad, Z. He, Integrating Forward Osmosis into Microbial Fuel Cells for Wastewater Treatment, Water Extraction and Bioelectricity Generation, Environ. Sci. Technol. 45 (2011) 6690–6696.
17. A.J. Ansari, F.I. Hai, W.E. Price, J.E. Drewes, L.D. Nghiem, Forward osmosis as a platform for resource recovery from municipal wastewater - A critical assessment of the literature, J. Membr. Sci. 529 (2017) 195–206.
18. D. Mondal, A. Mahto, P. Veerababu, J. Bhatt, K. Prasad, S.K. Nataraj, Deep eutectic solvents as a new class of draw agent to enrich low abundance DNA and proteins using forward osmosis, RSC Adv. 5 (2015) 89539–89544.
19. M. Kamali, D.P. Suhas, M.E. Costa, I. Capela, T.M. Aminabhavi, Sustainability considerations in membrane-based technologies for industrial effluents treatment, Chem. Eng. J. 368 (2019) 474–494.
20. M. Kamali, M.E. Costa, T.M. Aminabhavi, I. Capela, Sustainability of treatment technologies for industrial biowastes effluents, Chem. Eng. J. 370 (2019) 1511–1521.
21. J.H. van't Hoff, A suggestion looking to the extension into space of the structural formulas at present used in chemistry, and a note upon the relation between the optical activity and the chemical constitution of organic compounds, Archives Neerlandaises des Sciences Exactes et Naturelles 9 (1874) 445–454.
22. J.S. Guest, S.J. Skerlos, J.L. Barnard, M.B. Beck, G.T. Daigger, H. Hilger, S.J. Jackson, K. Karvazy, L. Kelly, L. Macpherson, J.R. Mihelcic, A. Pramanik, L. Raskin, M.C.M. Van Loosdrecht, D. Yeh, N.G. Love, A New Planning and Design Paradigm to Achieve Sustainable Resource Recovery from Wastewater, Environ. Sci. Technol. 43 (2009) 6126–6130.
23. D. Puyol, D.J. Batstone, T. Hülsen, S. Astals, M. Peces, J.O. Krömer, Resource recovery from wastewater by biological technologies: opportunities, challenges, and prospects, Front. Microbiol. 7 (2017) 2106.
24. Al Bazedi, G., Soliman, N., Sewilam, H. Novel organic draw solution in forward osmosis process for fertigation: performance evaluation and flux prediction. Environ Sci Pollut Res (2022). https://doi.org/10.1007/s11356-022-20674-4.
25. Giagnorio M, Ricceri F, Tagliabue M, Zaninetta L, Tiraferri A. Hybrid Forward Osmosis⁻Nanofiltration for Wastewater Reuse: System Design. Membranes (Basel). 2019 May 6;9(5):61. doi: 10.3390/membranes9050061. PMID: 31064140; PMCID: PMC6572449.

26. Islam M.S., Sultana S., McCutcheon J.R., Rahaman M.S. Treatment of Fracking Wastewaters Via Forward Osmosis: Evaluation of Suitable Organic Draw Solutions. Desalination. 2019; 452: 149–158.
27. Yip N.Y., Tiraferri A., Phillip W.A., Schiffman J.D., Elimelech M. High Performance Thin-Film Composite Forward Osmosis Membrane. Environ. Sci. Technol. 2010; 44: 3812–3818.
28. Q.C., Ling M.M., Chung T.S. Draw Solutions for Forward Osmosis Processes: Developments, Challenges, and Prospects for the Future. J. Membr. Sci. 2013; 442: 225–237.
29. Xie M., Nghiem L.D., Price W.E., Elimelech M. Toward Resource Recovery from Wastewater: Extraction of Phosphorus from Digested Sludge Using a Hybrid Forward Osmosis-Membrane Distillation Process. Environ. Sci. Technol. Lett. 2014; 1: 191–195.
30. Goyal, P.; Sundarrajan, S.; Ramakrishna, S. A Review on Mixed Matrix Membranes for Solvent Dehydration and Recovery Process. Membranes 2021, 11, 441. https://doi.org/10.3390/membranes11060441.
31. Yip N.Y., Tiraferri A., Phillip W.A., Schiffman J.D., Elimelech M. High Performance Thin-Film Composite Forward Osmosis Membrane. Environ. Sci. Technol. 2010; 44: 3812–3818.
32. Tiraferri A., Yip N.Y., Phillip W.A., Schiffman J.D., Elimelech M. Relating Performance of Thin-Film Composite Forward Osmosis Membranes to Support Layer Formation and Structure. J. Membr. Sci. 2011; 367: 340–352.
33. Tiraferri A., Kang Y., Giannelis E.P., Elimelech M. Highly Hydrophilic Thin-Film Composite Forward Osmosis Membranes Functionalized with Surface-Tailored Nanoparticles. ACS Appl. Mater. Interfaces. 2012;4:5044–5053.
34. Wanlin Wu, Ling Yu, Liangzhong Li, Zhenfeng Li, Jiatong Kang, Shasha Pu, Da Chen, Ruixue Ma, Kun An, Guoqiang Liu, Yang Yu, Electrospun nanofiber based forward osmosis membrane using graphene oxide as substrate modifier for enhanced water flux and rejection performance, Desalination, Volume 518, 2021, 115283, ISSN 0011-9164.
35. M. Obaid, Hend Omar Mohamed, Ahmed S. Yasin, Olfat A. Fadali Khalil Abdelrazek Khalil, Taewoo Kim and Nasser A. M. Barakat, A novel strategy for enhancing the electrospun PVDF support layer of thin-film composite forward osmosis membranes, RSC Adv., 2016,6, 102762-102772.
36. Liu, Q., Li, J., Zhou, Z. et al. Hydrophilic Mineral Coating of Membrane Substrate for Reducing Internal Concentration Polarization (ICP) in Forward Osmosis. Sci Rep 6, 19593 (2016). https://doi.org/10.1038/srep19593.
37. Bhinder, Amrit, Shabani, Simin, Sadrzadeh, Mohtada. "Effect of Internal and External Concentration Polarizations on the Performance of Forward Osmosis Process" In Osmotically Driven Membrane Processes: Approach, Development and Current Status, edited by Hongbo Du, Audie Thompson, Xinying Wang. London: IntechOpen, 2017. 10.5772/intechopen.71343.
38. Zhenyu Li, Rodrigo Valladares Linares, Szilárd Bucs, Luca Fortunato, Claus Hélix-Nielsen, Johannes S. Vrouwenvelder, Noreddine Ghaffour,

TorOve Leiknes, Gary Amy, Aquaporin based biomimetic membrane in forward osmosis: Chemical cleaning resistance and practical operation, Desalination, Volume 420, 2017, Pages 208-215, ISSN 0011-9164.

39. Chung, T.-S. et al. Emerging forward osmosis (FO) technologies and challenges ahead for clean water and clean energy applications. Curr. Opin. Chem. Eng. 1, 246–257 (2012).
40. Mi, B. & Elimelech, M. Organic fouling of forward osmosis membranes: Fouling reversibility and cleaning without chemical reagents. J. Membrane Sci. 348, 337–345 (2010).
41. Elimelech, M. & Phillip, W. A. The future of seawater desalination: energy, technology and the environment. Science 333, 712–717 (2011).
42. Review Recent Advance on Draw Solutes Development in Forward Osmosis Qingwu Long, Yongmei Jia, Jinping Li , Jiawei Yang , Fangmei Liu , Jian Zheng and Biao Yu, Processes 2018, 6, 165.
43. Andrea Achilli, Tzahi Y. Cath, Amy E. Childress, Selection of inorganic-based draw solutions for forward osmosis applications, Journal of Membrane Science, Volume 364, Issues 1–2, 2010, Pages 233-241, ISSN 0376-7388.
44. Yufeng Cai, Xiao 'Matthew' Hu, A critical review on draw solutes development for forward osmosis, Desalination, Volume 391, 2016, Pages 16-29, ISSN 0011-9164.
45. McCutcheon J.R., McGinnis R.L., Elimelech M. A Novel Ammonia-Carbon Dioxide Forward (Direct) Osmosis Desalination Process. Desalination. 2005; 174: 1–11.
46. Elimelech, M. & Phillip, W. A. The future of seawater desalination: energy, technology and the environment. Science 333, 712–717 (2011).
47. N.T. Hau, S.-S. Chen, N.C. Nguyen, K.Z. Huang, H.H. Ngo, W. Guo, Exploration of EDTA sodium salt as novel draw solution in forward osmosis process for dewa[1]tering of high nutrient sludge, J. Membr. Sci. 455 (2014) 305–311.
48. H. Yuan, I.M. Abu-Reesh, Z. He, Mathematical modeling assisted investigation of forward osmosis as pretreatment for microbial desalination cells to achieve con[1]tinuous water desalination and wastewater treatment, J. Membr. Sci. 502 (2016) 116–123.
49. Giagnorio M., Ricceri F., Tiraferri A. Desalination of Brackish Groundwater and Reuse of Wastewater by Forward Osmosis Coupled with Nanofiltration for Draw Solution Recovery. Water Res. 2019; 153: 134–143.
50. Achilli A., Cath T.Y., Childress A.E. Selection of Inorganic-Based Draw Solutions for Forward Osmosis Applications. J. Membr. Sci. 2010; 364: 233–241.
51. Mohan Qin and Zhen He, Self-Supplied Ammonium Bicarbonate Draw Solute for Achieving Wastewater Treatment and Recovery in a Microbial Electrolysis Cell-Forward Osmosis-Coupled System, Environmental Science & Technology Letters 2014, 1(10), 437-441, DOI: 10.1021/ez500280c.

52. N.T. Hau, S.-S. Chen, N.C. Nguyen, K.Z. Huang, H.H. Ngo, W. Guo, Exploration of EDTA sodium salt as novel draw solution in forward osmosis process for dewa[1]tering of high nutrient sludge, J. Membr. Sci. 455 (2014) 305–311.

53. Chung, T.-S., Zhang, S., Wang, K. Y., Su, J. & Ling, M. M. Forward osmosis processes: Yesterday, today and tomorrow. Desalination 287, 78–81 (2012).

54. HTI starts producing thin-film composite FO membrane, Membrane Technology, Volume 2012, Issue 7, 2012, Pages 5-6, ISSN 0958-2118.

55. D. Mondal, A. Mahto, P. Veerababu, J. Bhatt, K. Prasad, S.K. Nataraj, Deep eutectic solvents as a new class of draw agent to enrich low abundance DNA and proteins using forward osmosis, RSC Adv. 5 (2015) 89539–89544.

56. Y. Zhang, L. Pinoy, B. Meesschaert, B. Van der Bruggen, A natural driven membrane process for brackish and wastewater treatment: photovoltaic powered ED and FO hybrid system, Environ. Sci. Technol. 47 (2013) 10548–10555.

57. P. Liu, H. Zhang, Y. Feng, C. Shen, F. Yang, Integrating electrochemical oxidation into forward osmosis process for removal of trace antibiotics in wastewater, J. Hazard. Mater. 296 (2015) 248–255.

58. A. Achilli, T.Y. Cath, E.A. Marchand, A.E. Childress, The forward osmosis membrane bioreactor: a low fouling alternative to MBR processes, Desalination 239 (2009) 10–21.

59. E.R. Cornelissen, D. Harmsen, K.F. de Korte, C.J. Ruiken, J.J. Qin, Membrane fouling and process performance of forward osmosis membranes on activated sludge, J. Membr. Sci. 319 (2008) 158–168.

60. J.-J. Qin, K.A. Kekre, M.H. Oo, G. Tao, C.L. Lay, C.H. Lew, E.R. Cornelissen, C.J. Ruiken, Preliminary study of osmotic membrane bioreactor: effects of draw solution on water flux and air scouring on fouling, Water Sci. Technol. 62 (2010) 1353–1360.

61. Nguyen Thi Hau, Shiao-Shing Chen, Nguyen Cong Nguyen, Kuo Zue Huang, Huu Hao Ngo, Wenshan Guo, Exploration of EDTA sodium salt as novel draw solution in forward osmosis process for dewatering of high nutrient sludge, Journal of Membrane Science, Volume 455, 2014, Pages 305-311, ISSN 0376-7388.

62. Kerri L. Hickenbottom, Nathan T. Hancock, Nathan R. Hutchings, Eric W. Appleton, Edward G. Beaudry, Pei Xu, Tzahi Y. Cath, Forward osmosis treatment of drilling mud and fracturing wastewater from oil and gas operations, Desalination, Volume 312, 2013, Pages 60-66, ISSN 0011-9164.

63. PHH Duong, TS Chung, Application of thin film composite membranes with forward osmosis technology for the separation of emulsified oil–water, Journal of Membrane Science, 2014, 452, 117-126.

64. Mudhar A. Al-Obaidi, Chakib Kara-Zaïtri, Iqbal M. Mujtaba, Modeling the Performance of Low Pressure Reverse Osmosis Membrane System for N-nitrosamine Rejection, Editor(s): Anton Friedl, Jiří J. Klemeš, Stefan Radl, Petar S. Varbanov, Thomas Wallek, Computer Aided Chemical Engineering, Elsevier, Volume 43, 2018, Pages 19-24, ISSN 1570-7946, ISBN 9780444642356.

65. A. Achilli, T.Y. Cath, E.A. Marchand, A.E. Childress, The forward osmosis membrane bioreactor: a low fouling alternative to MBR processes, Desalination 239 (2009) 10–21.
66. A. Bernardes, M. Rodrigues, J.Z. Ferreira, Electrodialysis and Water Reuse, Springer, 2016, https://doi.org/10.1007/978-3-642-40249-4
67. Q. Liu, C. Liu, L. Zhao, W. Ma, H. Liu, J. Ma, Integrated forward osmosis-membrane distillation process for human urine treatment, Water Res. 91 (2016) 45–54.
68. N.C. Nguyen, H.T. Nguyen, S.-T. Ho, S.-S. Chen, H.H. Ngo, W. Guo, S.S. Ray, H.- T. Hsu, Exploring high charge of phosphate as new draw solute in a forward osmosis–membrane distillation hybrid system for concentrating high-nutrient sludge, Sci. Total Environ. 557–558 (2016) 44–50.
69. R. K. McGovern, S. M. Zubair & J. H. Lienhard V (2014) Hybrid electrodialysis reverse osmosis system design and its optimization for treatment of highly saline brines, IDA Journal of Desalination and Water Reuse, 6:1, 15-23, DOI: 10.1179/2051645214Y.0000000016.
70. Winson C.L. Lay, Jinsong Zhang, Chuyang Tang, Rong Wang, Yu Liu, Anthony G. Fane, Factors affecting flux performance of forward osmosis systems, Journal of Membrane Science, Volumes 394–395, 2012, Pages 151-168, ISSN 0376-7388.
71. L.-Y. Huang, D.-J. Lee, J.-Y. Lai, Forward osmosis membrane bioreactor for A. Mahto, et al. Separation and Purification Technology 254 (2021) 117568 19 wastewater treatment with phosphorus recovery, Bioresour. Technol. 198 (2015) 418–423.
72. Jialing Song, Mengying Yan, Jingling Ye, Shengyang Zheng, Liang Ying Ee, Zhiwei Wang, Jun Li, Manhong Huang, Research progress in external field intensification of forward osmosis process for water treatment: A critical review, Water Research, Volume 222, 2022, 118943, ISSN 0043-1354.
73. Hau, N. T. et al. Exploration of EDTA sodium salt as novel draw solution in forward osmosis process for dewatering of high nutrient sludge. J. Memb. Sci. 455, 305–311 (2014).
74. Panpan Guan & Duo Wang (2016) The improvement of CTA forward osmosis membrane performance by hydrophilic modification on interface between support layer and non-woven fabric, Desalination and Water Treatment, 57:57, 27505-27518.
75. Hongwei Song, Fang Xie, Weiwei Chen & Jinrong Liu (2018) FO/MD hybrid system for real dairy wastewater recycling, Environmental Technology, 39:18, 2411-2421.
76. Sui Zhang, Peng Wang, Xiuzhu Fu, Tai-Shung Chung, Sustainable water recovery from oily wastewater via forward osmosis-membrane distillation (FO-MD), Water Research, Volume 52, 2014, Pages 112-121, ISSN 0043-1354.

77. Gang Chen, Zhouwei Wang, Long D. Nghiem, Xue-Mei Li, Ming Xie, Baolong Zhao, Mengxi Zhang, Jianfeng Song, Tao He, Treatment of shale gas drilling flowback fluids (SGDFs) by forward osmosis: Membrane fouling and mitigation, Desalination, Volume 366, 2015, Pages 113-120, ISSN 0011-9164.
78. Pengxiao Liu, Hanmin Zhang, Yujie Feng, Chao Shen, Fenglin Yang, Integrating electrochemical oxidation into forward osmosis process for removal of trace antibiotics in wastewater, Journal of Hazardous Materials, Volume 296, 2015, Pages 248-255, ISSN 0304-3894.
79. Fei Zhang, Kristen S. Brastad, and Zhen He, Integrating Forward Osmosis into Microbial Fuel Cells for Wastewater Treatment, Water Extraction and Bioelectricity Generation, Environmental Science & Technology 2011 45 (15), 6690-6696.
80. P. Praveen, J.Y.P. Heng, K.-C. Loh, Tertiary wastewater treatment in membrane photobioreactor using microalgae: Comparison of forward osmosis & microfiltra[1]tion, Bioresour. Technol. 222 (2016) 448–457.
81. G. Qiu, S. Zhang, D.S. Srinivasa Raghavan, S. Das, Y.-P. Ting, The potential of hybrid forward osmosis membrane bioreactor (FOMBR) processes in achieving high throughput treatment of municipal wastewater with enhanced phosphorus recovery, Water Res. 105 (2016) 370–382.

CHAPTER

5 Advances in Functional Polymer Materials for Water Treatment Applications

— ABSTRACT —

The functional materials, particularly functional nanomaterials, are a versatile class of advanced materials recently gaining importance in water and wastewater treatment. The advanced class of functional materials involving polymers, nanomaterials and other smart materials are used in various process applications in separation, intermediate purification, and contaminant removal from water and wastewater sources. Conventionally, resins type functional materials with polymeric and inorganic active site are used for decontamination of water having fluoride, chromium, and arsenic pollutants. However, recent innovations in finding new classes of functional materials reveals that polymer-based active smart materials possess greater potential for their water remediation and purification applications. This chapter presents an overview of recent innovations in advanced material developments for water treatment applications.

5.1 Introduction

Polymers is a class of natural or synthetic substances made of small to very large molecules called macromolecules, which are large repeating chemical units called monomers. The term polymer has its origin in Greek meaning 'many parts'. Polymers are everywhere and thousands of polymers can be in different forms and types. Polymers can be either natural which can be traced even in the living organisms or synthetic origin known as man-made materials. Polymers are naturally found in animals, plants in nature which are termed *natural polymers* and man-made polymeric systems are known as *synthetic polymers*. For instance, the strand of our DNA is a biopolymer or largely known as biomacromolecule which is naturally occurring in living beings. Other types of polymers are man-made such as polypropylene which is widely used as multi-purpose plastic. Various polymers possess properties with several unique physical, chemical and physicochemical properties due to which polymers have been widely used in everyday life. [1]

Among many polymers, the monomer molecules are arranged in various architectures which may be all alike, or polymers may further contain two, three, or more types of compounds or molecular structures in their backbones. Generally, polymers are termed with at least 100 monomer molecules connected to each other with stable interaction. Less than that of monomer integration is called oligomers. These polymers with stable molecular structure and specialized properties make a product which has certain unique physical, chemical and physicochemical properties such as mechanical properties, elasticity, high tensile strength, or ductility that can form fibers. Therefore, polymers are composed of smaller and simpler molecules often in many thousands of monomer units. The monomers interact with stable covalent chemical bonds between the monomers via polymerization process, and crystallization in which large number of molecules come together under the influence of weak intermolecular forces. [2, 3]

Polymers are broadly classified into 4 categories as shown in Figure 5.1, namely:

(i) Class of polymers based on their origin, these are subdivided into three categories, namely, natural polymers, synthetic polymers and semi-synthetic polymers. Natural polymers have their origin in nature and are found or extracted from plants and animals. For instance, proteins, cellulose, starch and rubber. These are largely biodegradable in nature. Synthetic polymers are man-made polymers. Thousands of plastic products that are commonly used in day-to-day life are made up of synthetic polymers. Synthetic polymers also have been widely used in various industries, housing, textile industry, automobile industry, packaging industry and so on. Some of the widely produced and applied synthetic polymers are polyethylene, nylon-6, 6, polyethers, polyesters, PET, etc. Semi-synthetic polymers are derived from naturally occurring or extracted polymers and further subjected to chemical modification such as cellulose nitrate, cellulose triacetate, cellulose acetate.

(ii) Class of polymers based on their polymerization technique, which is further divided into condensation polymers and addition polymers.

(iii) Polymers category, based on their structure, are subdivided into linear polymers, branched polymers, crosslinking polymer,s and speciality polymers.

(iv) Polymers with varying molecular forces have their properties been classified into elastomers, fibers, thermoplastic and thermosetting polymers.

Generally, many polymers are prepared following mainly two polymerization techniques, namely:

(i) Condensation polymerization and

(ii) Addition polymerization.

In condensation polymerization, monomers react with each other to form larger molecular structures with repeating monomer units while releasing by products such as water or methanol molecules. Therefore, every polymer process formation in this process is accompanied by the formation of a molecule of some simple compound, generally water. In addition to polymerization, polymerization of monomers is carried out in the presence of catalysts, which in certain cases is carried over structural details that have important effects on the properties of the polymer. Therefore, additional polymerization follows with the reaction of a monomer to form a polymer without the formation of any byproducts. More significantly, additional polymerizations are generally carried out in the presence of catalysts, which in most of the cases exert control over structural growth and induce specific signature properties that have important effects on the polymer end use. Interestingly, the type of polymerization mechanism followed depends on the type of individual or multiple functional groups attached to the precursors or reactants. [4]

5.2. Classification of Polymer Systems (Fig. 5.1)

On the other hand, based on structure and their architecture, polymers are further classified into (a) linear polymers that contain long and straight chains that fall in this category involving polymers such as polyvinyl chloride (PVC) that is largely used for preparing pipes, electric cables and so on; (b) branched-chain polymers are formed when linear chain polymers are further modified via grafting the side chain to the linear backbone of a polymer such as low-density polythene. (c) Cross-linked polymers are composed of bifunctional and trifunctional monomers in their backbones which can be easily altered to stabilize via crosslinking of polymer chains. This inter-polymer chain crosslinking is carried out to stabilize a polymer via a stronger covalent bond. Chitosan gels, Bakelite and melamine are examples of this category. However, the most popularly known polymers are segregated into three main groups: thermoplastics, thermosets, and elastomers. The thermoplastics are further divided into two categories, namely, crystalline and amorphous polymers.

On the other hand, advanced polymeric materials represent large either, functionalized, crosslinked, blended, and composite forms of polymers which can act as an ion exchange intermediate, or media that can extract or interact with foreign ionic species, particularly in water treatment cases, pollutants. Advanced polymeric active materials have been lately attracting immense interest due to their futuristic potential in water treatment applications. Active functional polymeric materials have unique properties of high adsorption capacity, interactive linkings, reversible and interchangeable possibilities between a solute in contaminated water and wastewaters. Polymer materials have been traditionally used to eliminate hazardous, harmful contaminants or ionic contamination like hardness in water in general, replacing exchangeable sites in polymers with contaminants to be eliminated from effluents. Therefore, advanced polymeric materials to be used in water treatment applications essentially consists of two main components: (i) the matrix material or or activated polymer, which is generally the low-to-medium molecular weight and high activity, and (ii) the functionalized either with blending, composite with active foreign and compatible substances which is relatively highly active and regenerative in nature. [5, 6]

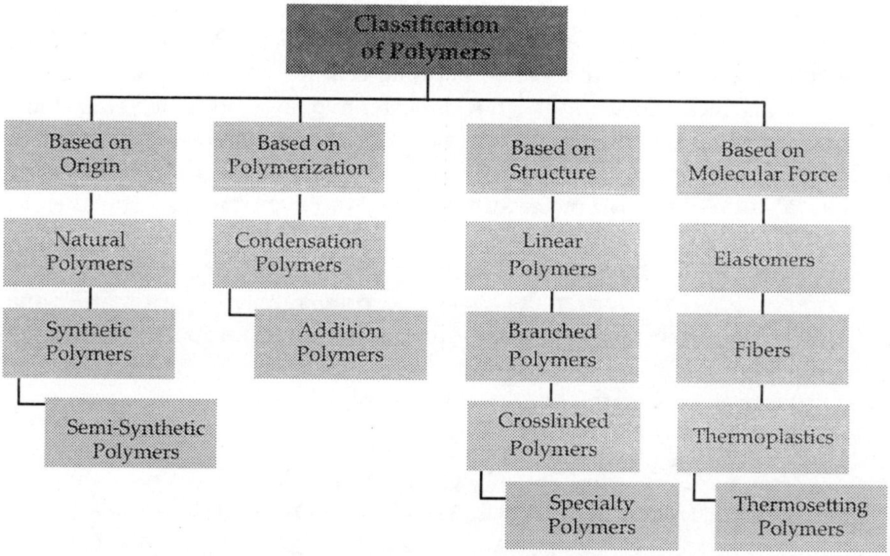

Figure 5.1. Detailed polymer classification details based on their origin, structure, preparation techniques and properties.

In the past, several polymer-based active materials have been developed with suitable stabilizing agents or crosslinking agents. Largely published literature suggests that most viable polymeric materials have been used in water treatment applications in the form of ion-exchange resins. These conventional polymeric materials were initially used for selective separation of hardness causing solutes such as magnesium, calcium, and potassium. This is conventionally known as ground or surface water softening by removing hardness from water sources. Before synthetic polymers came into practice, natural polymers as well as inorganic substances like natural clays have been widely used for water treatment applications. In addition to natural clays like bentonite, montmorillonite and so on have been widely used on the other hand, first reported synthetic inorganic active materials have been aluminosilicates which are well known as zeolites. [7]

Functional polymers are tailor-made macromolecules that have unique properties or applications. The properties of polymeric materials are often evaluated by the presence of chemical functional groups that are dissimilar to those of the backbone chains. For instance, chemical functionality like polar or ionic groups on hydrocarbon backbones or hydrophobic groups on polar polymer chains. Such chemical heterogeneity generated due to chemical modification on the polymer chains can lead to increased reactivity, phase separation, or association. In addition to inducing chemical alteration, the functionalization process also induces other physicochemical changes to parent polymers such as self-assemblies or supramolecular structures as added incentive as shown in Figure 5.2. When a polymer is functionalized it alters the polymer assemblies triggered by chemical, physical and physicochemical stimuli which turns macromolecules into "smart" polymer materials. Most of the polymers that can be functionalized are based on simple linear backbone architecture. These linear backbone polymers can be telechelic having chain-end, in-chain, block copolymer, graft polymer and so on in which side chains are altered to be active via functionalization process. However, recent advances in polymer innovation have specialized functionalization or property tuning on special topologies or architecture routes are encouraged to suit the end application. These polymers include specialized 3-dimensional polymers architecture such as hyperbranched polymers, stars, or dendrimers or tree-like structures. This chapter discusses the preparation route of functional polymers through either direct free radical, anionic, cationic, and coordination polymerization, surface as well as the bulk modification of polymer backbones to be used for water treatment applications. [8]

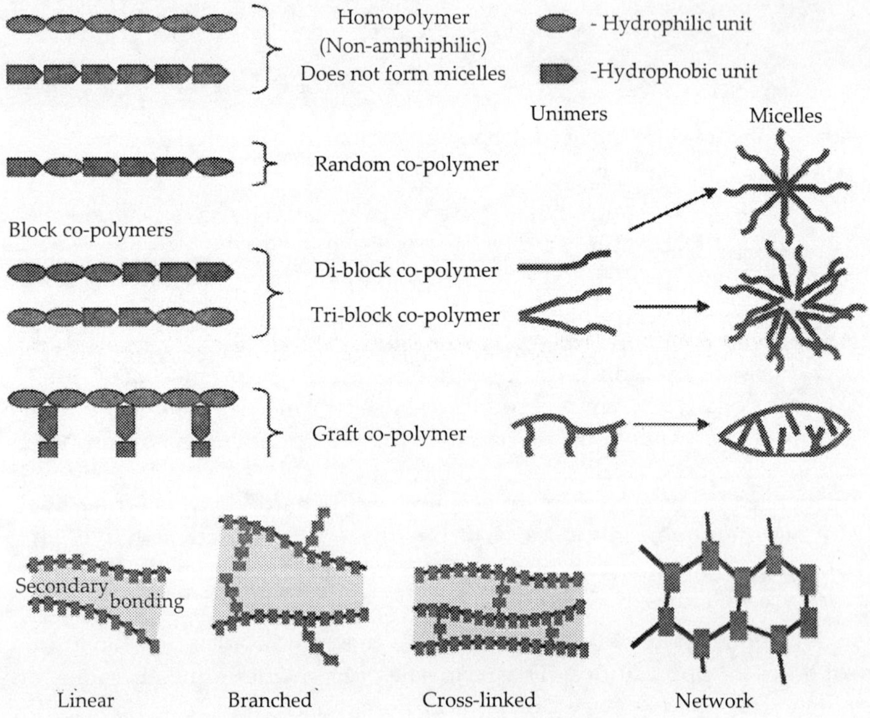

Figure 5.2. Schematic representation of polymers, polymer structures, methods, and routes of polymer functionalization.

Among several other methods, anionic polymerization perhaps is popularly used as a controlled method for the direct functionalization via living anionic polymerization. This method of polymerization proceeds with the absence of chain transfer and termination process leading to polymers with predictable molecular weights (MWs), and narrow molecular weight dispersion (MWD). This particularly defines the architecture, a high degree of functional purity and activity of polymer. This method is used to prepare linear or star polymers with MWDs less than 1.1 and functional purity over 95%. Interestingly, both chain-end and in-chain functional polymers can be prepared via anionic methods. For instance, termination of "living" lithio-macromolecules with electrophilic reagents which result in a wide variety of telechelic polymers. [9]

On the other hand, cationic polymerization is a relatively inferior method to induce functionality via this route. The main limitation comes from cationic polymerization to induce well-defined functionality to polymers due to the existence of inter-or intramolecular transfer or rearrangement. Nonetheless, advanced research activities in this field have revealed that several monomers have exhibited potentias of sufficient control over initiation step, propagation and termination to allow the preparation of advanced functional polymers. [10]

Linear polymers composed of chain-like molecules are generally viscous liquids or long chain macromolecular solids having different degrees of crystallinity. Crosslinked polymers on the other hand have a molecular network structure that can either be turned into a thermosetting that can form under the influence of heat but do not melt or soften upon further reheating or do not dissolve in solvents. On the other hand, thermoplastic polymers undergo deformation and dissolution under heating conditions and suitable solvents, respectively. Both linear and crosslinked polymers can be prepared either by addition or condensation polymerization techniques. [11, 12]

Therefore, due to the advances in polymer chemistry, functional polymers have gained great interest as key drivers in the development of several high-end technologies. The main requirement in selecting polymer is its potential of having chemical reactive or otherwise functionalizability for further specific modifications. In polymer chemistry, a functional group or functionally active site is a substituent or moiety in a polymer backbone that causes the polymer's characteristic chemical reactivity. The same set of functional groups on the polymer backbone are used for the functionalization that can undergo the chemical transformation regardless of the rest of the polymer composition. These active sites also enable systematic prediction of chemical reactivity and behavior of polymers and the design of functional polymer preparation. Further, the reactivity of a functional group on polymer back can also be altered through neighboring functional groups. Some of the important functional groups include amino, alcohol, ether, halides, thio, hydroxyl, methyl, carboxyl, carbonyl, phosphate, sulfhydryl, sulfonyl groups and so on as shown in Figure 5.3.

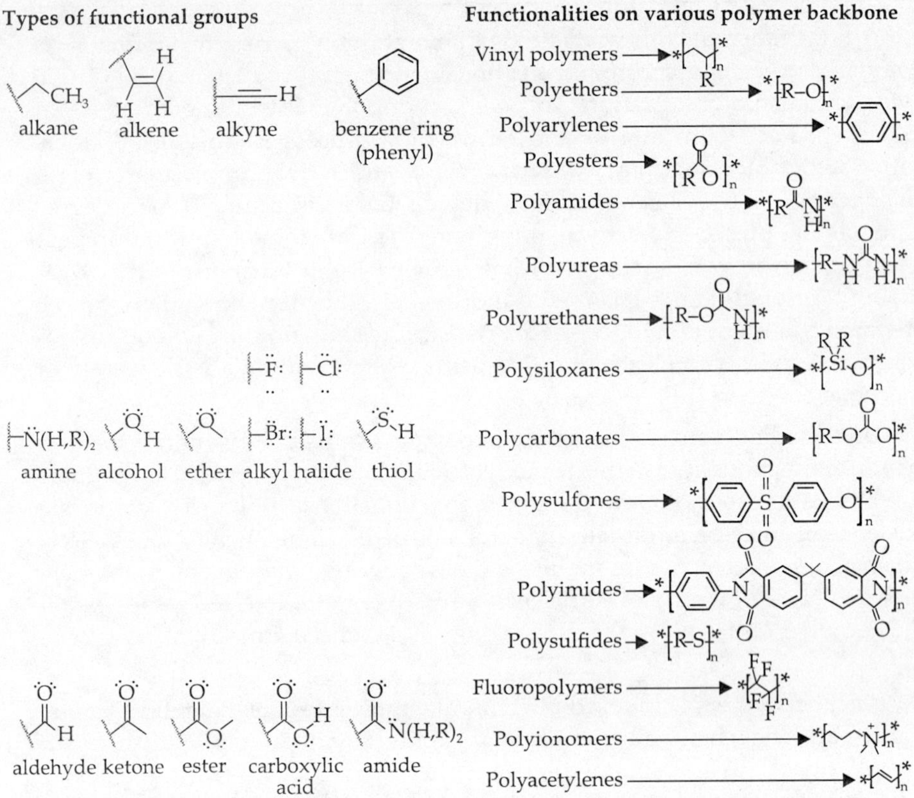

Figure 5.3. List of functional groups generally found in polymer backbone and another list showing examples of series of polymers carrying different functional groups used in various applications such as water remediation.

5.3. Functional Groups in Polymers

As it is clear that a functional group is an atom or active site in a polymer backbone with distinctive chemical characteristics regardless of the other in the molecule. The atoms in a functional group are linked through various interactions like covalent bonds with each other and to the rest of the molecule. For instance, the repeating units of polymers, functional groups attach to their non-polar core of the carbon backbone and thus add chemical characteristics to polymer backbone or carbon chains. Functional groups can be either charged, neutral, for example, in carboxylate salts (–COO–) it is negatively charged, which turns the molecule into a polyatomic ion or a complex ion. Further, the functional groups binding to a central atom in a polymer coordination site are called ligands. Solvation and complexation in a product preparation or application result due to specific interactions of functional groups. As it is commonly known in organic chemistry systems, "like dissolves like", similarly, a polymeric system is either solvated or complexation via a shared or mutually well interacting functional group which results in solubility. For instance, sugar dissolves in water due to both shared strong hydroxyls and hydroxyl functional groups (–OH) interacting with each other. In addition to this, when functional groups are more electronegative than atoms they are attached to, the functional groups turn more polar, and on the other hand, nonpolar molecules containing these functional groups turn polar which undergo soluble in some aqueous environment. [13, 14]

In principle, chemical modification of polymers can be achieved in bulk on polymer backbone, solution and/or on surfaces of polymers using surface activity. However, considering the macromolecular nature of the polymer, the reactivity of a functional group instilled to a polymer backbone or surface may or may not be the same as it is attached to a small organic molecule. For instance, the reactivity of a group can be controlled or influenced via physicochemical interactions that are influenced by various parameters such as: (a) the type of adjacent active groups, (b) the type of the polymer backbone and their conformation, (c) the crystallinity and morphology of the backbone, (d) miscibility or solubility of reactant, and (e) catalyst phase or reactive media. In addition to this, functionalization reactions on polymers pose the added complexities of polymer crosslinking or polymer backbone/chain cleavage. These polymer solution reactions induce diverse chemical modifications such as hydrogenated, cyclized, isomerized, epoxidized, and carbenes or enophiles. Most of these transformations enhance the glass transmission temperatures (T_g) of the polymers with a subsequent change in the physical, chemical, physicochemical, and mechanical properties of the materials.

Among the many sets of polymers which were used for various environmental applications are polyolefins. The polyolefins are amongst the largest group of thermoplastics that are widely used as commercial end products. Polyolefins class mainly involves alkene backbones with the general formula C_nH_{2n}, such as propylene, ethylene, isoprenes, butenes, and pentenes, copolymers and modified versions thereof. These polyolefins when used in blending or composites incorporate extraordinary chemical, physical and mechanical properties, which directly help in processability and recyclability of polymer during and post-application at low cost. However, absence or less reactive groups on polymer backbone limit some of their end applications, particularly where water treatment applications, adhesion, paintability, dyeability, printability, and so on for which further compatibility or hybridization is required. Consequently, processes like free-radical grafting are carried out to introduce the functionality to polyolefins, for example, reaction with maleic anhydride.

One the other hand, amine functionalized polymers have been widely adopted for water treatment applications particularly in the field of membrane-based separation processes. Amine functionalized polymers can be obtained via derivatization of carboxylic anhydride functionalized polyolefin and other branched polymers with amines. Polymeric materials with amino groups have shown great potential due to their chemical versatility to interact with pollutants used in wastewater treatment. Although most of the functionalities in polymers are tailor-made to suit the end applications via appropriate modification. Such protocols are carefully designed to control molecular weight, degree of functionality and their distribution in the polymer. For all these reasons, surface modification, functionalization and realizing new applications using such polymers has now become the main area of research. Therefore, surface functionalization in polymers induces novel properties based on nature and type of modifications without affecting the characteristics of the bulk phase. For instance, surface alteration changes the interfacial properties of polymers, such as biocompatibility, adhesion, permeability, printability, wettability, weatherability, affinity towards specific foreign elements, friction and so on. Further, polar group thiols lead to changes in wettability, as evidenced by contact angle measurements which are useful for water treatment applications.

Furthermore, the surface modification can be achieved to induce additional or added functionality via various post-treatment like heat treatment. On the other hand, polyolefins can be further surface modified following either oxidation, sulfonation, chlorination, etc., With all these functionalized polymers imparting new chemical, physicochemical, biophysical, or optoelectronic properties that can be directly applied as product materials. These applications include organic catalysis such as supported catalysts, medicine as red blood cell substitutes, conducting polymers, magnetic polymers in optoelectronics and polymers for nonlinear optics, functionalized biopolymers in biomedical applications, polymer paints and varnishes, photographic materials, building materials as well as lube and fuel additives. [15-17]

Nonetheless, polymer supported photocatalysts designed for water treatment applications have several advantages due to the ease of preparing supported- catalyst, their stability during applications and ease of recovery. Indeed, immobilization of active nanomaterials act as active functional sites or adsorptive catalysts onto a polymer matrix support is a unique strategy for avoiding antagonistic or undesired reactions between them, and to produce secondary pollutants. For instance, polymer supported nanocatalysts have been used as the Lewis acid catalysts in the carbocationic polymerization of polyalkenes like polyisobutylene to be further used for adsorptive removal of pollutants in contaminated water. After the application, supported catalysts like functional polyisobutylene can be regenerated by simple filtration. Similarly, several nanosized and microsized metal functionalized polymers have shown great potential for a wide range of applications including catalysis, controlled-release medicaments, biocides, electronic devices, energy, and environmental applications. In water treatment applications, photosensitive polymers can be suitably tuned to achieve high surface area and surface activity to be directly used for water treatment applications. [18-20]

5.4. Conventional Ion-Exchange Resins

Traditionally, inorganic, naturally abundant and low cost materials have been used for water treatment applications. Historically, natural clays extracted from volcanic ashes like Bentonite, Montmorillonite, Greens and so on have been refined and used for water treatment applications. On the other hand, aluminosilicate-based synthetic zeolites have been successfully converted into an ion-exchange material as an alternative to natural clays. Among many natural clay materials, Greensand was first used as ion-exchange media for exchanging ionic contaminants in various contaminated water, however, soon proved to be ineffective. Nevertheless, Greensand has been noted to be a stable ion-exchange media during its application in water treatment compared to zeolitic material at large scale. [21-27]

However, zeolites, microporous, 3D crystalline solids of aluminum silicate with small openings of defined size which allow small foreign molecules or ionic species to pass through them easily. This possibly occurs via intercalation and deintercalation through an ion-exchange process. However, larger molecules are restricted to pass through the intercalation channel. Therefore, such channels are sometimes called molecular sieves (Figure 5.4). During the water treatment process, zeolite exchanges specific ions (pollutants) from the contaminated water being treated for another exchangeable ion that is in the resin such as chlorine is one component of softening salt, with sodium being the other. Active zeolitic resin exchanges Na^+ for calcium and magnesium in the ion-exchange media. In zeolite softening, water containing scale forming

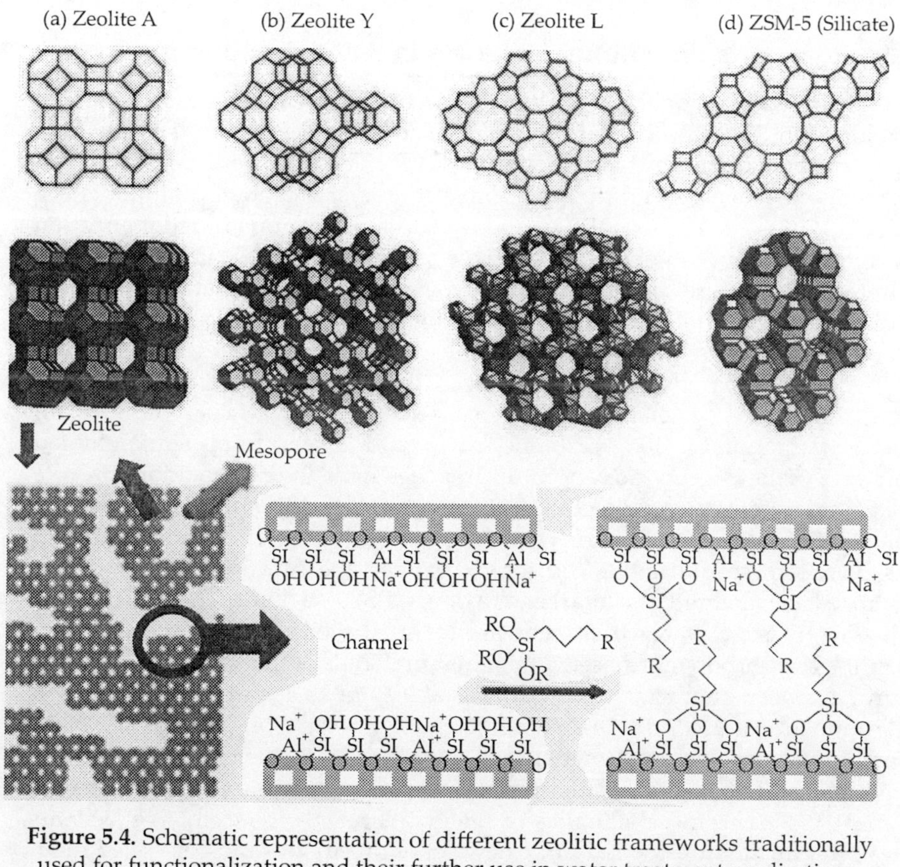

Figure 5.4. Schematic representation of different zeolitic frameworks traditionally used for functionalization and their further use in water treatment applications.

ions, such as Ca^{+2} and Mg^{+2}, passes through a resin bed containing strong acid cation (SAC) resin in the form of sodium. Therefore, in the zeolitic resin, the hardness causing ions such as Ca^{+2}, Mg^{+2} and so on are exchanged with sodium through the diffusion mechanism in which the sodium diffuses into the bulk contaminated water to reach pollutants. Nevertheless, the ion-exchange capacity of any resin or media plays an important role in retaining pollutants from contaminated water. Therefore, ion-exchange capacity (IEC) is generally used to characterize and calculate performance of an ion-exchange material. Thus, ion-exchange capacity is defined as the number of exchangeable ions a unit quantity of resin or media which removes from a solution or contaminated water. [28, 29]

Further, another set of naturally occurring resources like coal has been sulfonated to create cation-exchange media which is commercially known as carbonaceous zeolite and further used for hydrogen cycle operation in which reduction hardness in contaminated sources has been achieved. Further, advanced innovations carried out condensation reaction products between polyamines and formaldehyde for the preparation of anion-exchange resin. These advanced anion-exchange resins are further used for the removal of all dissolved ionic species in wastewater or contaminated water sources.

5.5. Polymeric Functional Materials in Water Treatment

Polymer based functional materials in the form of ion-exchange resins were first developed in the early 1940s following the copolymerization of styrene long chains crosslinked with divinylbenzene (DVB). The polymers functionalized with ion-exchangeable groups for both cationic and anionic pollutants are effectively instilled through various processes. Upon functionalization polymer backbone attains highly stable ion-exchange media for different pollutants. Therefore, polymer based ion-exchange water treatment systems have largely proven to be stable and effective treatment media having higher ion-exchange capacities than inorganic counterparts. The polystyrene-divinylbenzene based (PS-DVB) has been the first block copolymer system turned into anion-exchange resin popularly used for efficient removal of anionic contaminant like carbonic acids and silicic from contaminated sources. Further, recent advanced approaches have led to the adoption of a series of polymers as an effective water treatment media through innovative functionalization processes for removing various types of contaminants . In this direction, several advanced ion-exchange type functionalized polymer systems have entered the market, however PS-DVB ion-exchange systems still govern large water treatment arenas in the commercial space. Major advantage with polymer-based functional materials is that polymers in resin form have been characterized by their stability in various harsh application environments and conditions. Nonetheless, for a very long period the basic constituents of polymeric functional systems have not changed; however, the polymeric systems have greatly improved in many forms and configurations to meet the emerging challenges of water treatment. Ion-exchange systems have been further improved with a new set of precursors and combined them as effective treatment media such as polymer blends, polymer composites, functional polymers and microporous systems using nanotechnology routes to suit water treatment applications. [30-32]

In advanced approaches, researchers have prepared a series of polymer-based functional systems with highly porous both with microporosity and macroporous architecture to obtain highly permeable end products. This facilitates the use of such advanced polymeric systems either in the form of packed beds or membranes. Likewise, the polymeric resins with macroporous structure allow the selection of suitable chemical and physical parameters to effectively remediate contaminated water. However, conventional polymeric resin attains gel-like physical characteristics upon crosslinking using suitable chemical reagents that simultaneously induce an exchangeable active site. However, conventional resins are inefficient against severely contaminated wastewater or contaminated water. This is due to the main reason that the conventional resins undergo structure, chemical modification, and mechanical deformation under severe applied conditions. Therefore, recent advanced studies focusing on developing chemically stable and mechanically robust systems with high ion-exchange capacity.

Therefore, polymer-based advanced macroreticular functional materials feature distinct pores, surface characteristics and morphological parameters within a highly surface-active polymer network structure. The highly surface functionalized network structure is expected to possess rigid and mechanically robust systems. These combined and synergetic surfaces as well as bulk properties provide greater stability against thermal degradation and oxidizing agents. Also, more significantly macroreticular anion-resins exhibit much desired anti-fouling properties to water treatment systems when applied to treat organic and biological contaminated wastewater.

Nevertheless, scope of advanced water treatment systems has been expanded to accommodate a series of polymeric systems to convert them into pollutant capturing media such as acrylic polymers in addition to conventionally PS-DVB systems. In addition to possessing high ion-exchange capacities, acrylic polymers are also known to acquire affinity towards blending materials and offer ease of composite formation thereby acting as antifouling materials. These added properties play an important role when used in contaminated water, even in harsh environments. These modifications can also be made inducing nanomaterial fillers against organic contaminated water treatment applications. On the other hand, a stable backbone ionizable functional moieties/groups play an important role in pollutant capturing or ion-exchange property and capacity. These functional groups or moieties comprise both positively charged cation elements and negatively charged anion elements in which one of the ionic species is mobile or exchangeable. For instance, the sulfonyl group functionalized PS-DVB (sulfonic-PS-DVB) has changeable ionic sites consisting of immobile anionic (SO_3^-) radicals and mobile cationic pollutants can be chromium, calcium (Ca^{2+}) and sodium cations (Na^+) as shown in Figure 5.5. When these types of ion-exchangeable polymeric systems are added to contaminated raw water, pollutant ions

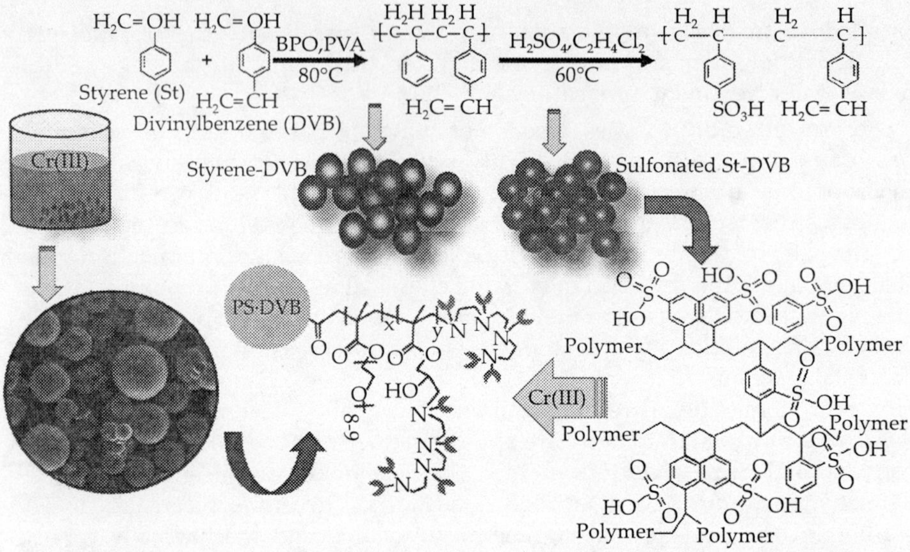

Figure 5.5. Sulfonic polystyrene-divinylbenzene (PS-DVB) functional system showing immobile sulfonyl group and mobile cations useful for cationic pollutant removal such as chromium from contaminated water.

diffuse into the polymeric matrix and facilitate exchange for the mobile moiety of the functional group and pollutant ions are displaced from the functional matrix diffuse back into the water solution when pH is altered. [33-40]

5.6. Polymer Blends and Complexes in Water Treatment

Advances in polymer technologies have led to the discovery of novel material properties in polymers which can offer several useful characteristics, namely, remarkable flexibility, chemical stability due to their large molecular weights, high mechanical strength, and high to ultra-high surface area when prepared in nano-size architecture. Among all properties, polymers can host many of these characteristics of at least some of them simultaneously. Therefore, polymers are excellent hosts for several functional groups of different organic and inorganic origin. Thus, polymer-based blends, composites and nanomaterials can be used as effective water treatment systems prepared via various methods with target properties. further, the polymer-polymer blends, polymer composites and polymer-nanomaterial induced composites offer the possibility of tuning the adsorptive capacities against various hazardous pollutants through crosslinking, blending, and surface functionalization. Another major advantage following these strategies of blending, compositing is their simplicity in preparation, tuning both physical, chemical and physicochemical properties to suit the applications. Interestingly, composite fillers or functionality moieties in these complexes also induce high chemical stability when used in harsh conditions. However, major disadvantage remain with these systems currently is the high cost of production. [41, 42]

5.6.1. Polymer-based adsorbents for dye removal

In the recent past, many polymers have been combined in various compositions and formats to prepare highly functionalized water remediation media for organic pollutants contaminated water. In one such attempt, sodium alginate/ chitosan-based polymer/polymer blend systems were prepared for the purpose of dyes removal from wastewater. Further, Elkady (et al.) prepared a block copolymer with the combination of styrene and acrylonitrile through a solution polymerization process. As obtained, copolymer is turned into a nanofiber mat using an electrospinning technique with high surface and targeted functionality for dye contaminated water treatment. The styrene and acrylonitrile copolymer with additional carboxylic acid groups functionality in nanofiber mat was effectively used for improved dyes absorptivity. Performance of polymer-based adsorbents were confirmed using various analytical techniques. Scanning electron microscope (SEM) and Fourier transform infrared spectroscopy (FTIR) have been successfully used to record the changes in the nanofiber after functionalization process, and further similar analytical tools were used to confirm the dye adsorption on the polymeric adsorbent. All analytical results indicated the presence of a uniformly shaped carboxylated nanofibers and these were compared with non-functionalized nanofibers. However, post-functionalization induced morphological changes after adsorption of dye molecules on the surface of polymeric nanofibers from wastewater. Among many pollutants, copolymer-based adsorbent showed 67.11 mg/g adsorption capacity for basic violet dye in less than 30 minutes. Particularly, the adsorption capacity of the basic violet dye reached maximum adsorption capacity within 30 minutes on surface modified copolymer nanofiber. [43]

However, the plain polymer systems showed relatively low adsorption capacity; this is due to the relatively low surface area and the lack of appropriate surface porosity in the polymeric system. Therefore, to increase the adsorption capacity in polymeric systems, it is important to either functionalize or make composites with the help of foreign systems either nanoparticles or additives. Hence, it is important to tune the surface zeta or charge of polymers via introduction of charged functional groups to create an excessive surface-active site on polymers. This can also be achieved by tuning morphological changes of the polymer nanofiber by tuning dope solution composition, electrospinning parameter and other process parameters thereby altering the physicochemical structure of end adsorbate.

In addition to this, it is also important to consider economic and environmental aspects such as selection of sustainable precursors and adopt simple and easily scalable process conditions. In this direction, a low-cost, multifunctional, and sustainable sourced straw-based adsorbent was prepared using imbibing ternary carboxylic acid and acrylamide units on straw-based powders. This study therefore uses abundant amino and carboxyl groups added to the composite structure in the final adsorptive powder to provide a high adsorption capacity for different organic pollutants dyes from wastewater. Specifically, straw-based functional adsorbent was highly effective in retaining dyes like MB and MO reaching 120.84 and 3053.48 mg/g, respectively. This sustainable adsorbent material showed maximum adsorption capacity for MB and MO dyes within 3 and 54 times than those of unmodified straw, respectively. [45]

On the other hand, recent studies focus on the use of biogenic polymers with suitable modification in various applications including energy and environmental sectors. In one such study, bio-based cyclodextrin was used as the main precursor for the preparation of sustainable adsorbent for the removal of organic contaminants like dyes from industrial wastewaters. Generally, biopolymers show greater antifouling properties when used for wastewater treatment that contains biological pollutants as co-pollutants. Therefore, biopolymer-based adsorbents have drawn great attention as a new generation of wastewater remediation media due to their extraordinary physicochemical properties and high adsorption capacity. Further studies have also attempted to improve the adoption capacity adopting advanced preparation techniques which helped in tuning structural properties of cyclodextrin-based composites. Some of the observations in these studies have shown greater potential of biopolymer-based functional adsorbents in the practical industrial application compared to native cyclodextrin. Also, cyclodextrin was crosslinked with polydopamine to produce an ecofriendly and sustainable composite for the removal of different dyes with high adsorption capacity. These adsorbents have been further characterized for their structural feature before and after pollutant adsorption. Further, β-cyclodextrin was immobilized in starch which showed similar adsorption behavior for organic dye removal from wastewater. These combinations of biopolymers resulted in irregular surface characteristics with rich NH_2, COOH, and OH functional groups of the β-cyclodextrin/starch helping blend polymers to attain enhanced adsorption capacity. These functional groups in blend polymers played significant roles in the selective dye capture and removal. On the other hand, comparative studies have shown the inferior performance of both individual polymers β-cyclodextrin and starch compared to blend adsorbents. Interestingly, polymer blends with dissimilar polymer systems have provided improved pollutant diffusion pathways to reach functional groups in the blend polymer. More significantly, NH_2, COOH, and OH functional groups in the blend system interact with pollutant moieties in wastewater via improved interparticle diffusion and the electrostatic interaction which resulted in high adsorption capacities. In a similar study, cyclodextrin grafted amino hyperbranched polymer was successfully prepared as an efficient blend adsorbent for the removal of dye molecules from industrial wastewater. These graft copolymers have shown remarkable adsorption capacities for dye removal due to the presence of functional groups. [44]

In the same direction, agro-wastes were used to extract cellulose and further modified to obtain functional active material for their potential application in water applications. In this study, biomass extracted cellulose was grafted using the monomers of acrylic acid and 2-acrylamido-2-methylpropane sulfonic acid in the presence of a crosslinker to produce highly efficient polymer composites for dye removal from wastewater resources. Upon modification of cellulose was characterized using SEM tool to analyze the morphological changes in the end product which showed smooth and long fiber-like morphology in unmodified cellulose, on the other hand, in case of the crosslinked graft copolymers showed flaky and highly thick threads. These modifications indicated the complete surface functionalization of cellulose. Further, graft modified cellulose polymer composite was tested for their adsorption behavior against both anionic and cationic pollutants. These functional polymer composites showed higher adsorption capacity for cationic-malachite green and crystal violet and anionic Congo red and CR) dyes, respectively from aqueous contaminated sources. Further, this study evaluated the effect of varying process conditions like concentration variation, pH effects and presence of co-pollutant on the overall adsorption capacity of modified cellulose adsorbents. Detailed studies showed that the cationic dyes were efficiently removed at pH 7.0 within 90 minutes, while the anionic dyes showed maximum adsorption capacity at pH of 2.2 in 8 hours of adsorption experiment. Interestingly, adsorption isothermal data fitted using different equations which showed a well fitted Langmuir isotherm and the pseudo-second-order model. Thus, cellulose-based grafted copolymers showed great potential for the removal of both anionic and cationic pollutants from industrial wastewater.

In another set of studies, natural cellulose was also functionalized using hyper branched polyethylenimine for selective adsorptive removal and separation of various dye pollutants from industrial wastewater. Particularly, polyethylenimine was covalently grafted to the backbone of cellulose polymer molecules via a Schiff-Base formation in which the CHO groups on oxidized cellulose linked with NH_2 groups of polyethylenimine resulting in efficient polymer composite. With this, adsorption behavior adsorbent was investigated for dye removal efficiencies using CR and basic yellow dyes in contaminated water. In these cases, a maximum adsorption capacity of 2100 mg/g and 1860 mg/g was achieved for Congo red and basic yellow dyes, respectively. The study also extended to remove further sets of dye contaminants like reactive red and brilliant blue from pollutant sources, however, the absorptive performance for eosin and bright yellow dyes were recorded to be relatively very low. It is important to note that the hyperbranched structure of the grafted copolymer composite worked like a brush, which improved the interparticle diffusion of contaminated dye molecules into the copolymer composite. However, these adsorption mechanisms were highly dependent on the size of the pollutant molecule size and their concentrations. Further improving the performance of polymer composites, the diethylenetriamine was incorporated in polyacrylonitrile following electrospinning technique which resulted in highly functional composite adsorbent for pollutant dyes. The adsorbent morphologies were then analyzed in detail using the SEM tool. Surface area of post-modified composite nanofiber systems were altered and became rougher compared to the pristine polymer nanofibers. The roughness in the modified nanofiber composite was attributed to the presence of incorporated diethylenetriamine units. In addition to physical changes, the chemical modification induces additional surface functionality in the composite structure that directly influences the removal efficiency of pollutants like dyes. [45-48]

In recent times, polymeric functional materials have been converted to attain high surface area while retaining high ion-exchange capacities for selective pollutants. Therefore, the polymer with ion-exchange sites even in the fibers forms was first developed by Economy (et al.) decades ago. Polymeric fibrous ion-exchange materials possess several advantages over the traditional ion-exchange resins, both bulk and inorganic resins. In addition to this, polymeric systems in fibrous morphology also provide added advantages such as improved spinnability of precursor, control over the fiber size and diameter, improved control over the morphology and functionality in the end product and ease of preparing various composite forms in fiber form make them one of the most preferred candidates for water treatment applications. Further, polymer functional systems can be fabricated in the form of fabrics or felts in which the contact efficiency of treatment media can be greatly improved. Further, the fibrous morphologies in the polymer system can be prepared as self-standing filters which can eliminate the requirement of presently used support such as screens, canisters, etc. Thus, in this direction the new concept of designing various anionic and cationic exchange materials both of inorganic and polymeric origins. One such attempt has been the use of glass fiber and polymer substrates with specific ion-exchange sites on their backbone and active sites.

Further, polymer active materials were hybridized with carbon precursors to surface modify to achieve greater adsorption capacity for hazardous pollutants. In one such attempt, activated carbon was coated on a glass fiber substrate to induce high surface area adsorbent. Earlier to this, carbon-based adsorbents have been widely used for treating issues of environmental concern, in particular ultrahigh surface area activated carbon powder and now activated carbon fibers (ACFs) following tailor-made structures' pore sizes and chemistries on both surface as well as pore perimeters. These strategies now have been improved to interchange precursor combinations to achieve stable and possessing extra surface functionality in their bulk structure as well as surface sites. This study also reveals the usefulness of simple and effective strategy in obtaining cost-efficient procedure for the production of activated carbon coated on glass fiber substrates. The carbon coated glass fibers active materials have been then effectively used for the removal of trace organic contaminants like toluene, benzene, xylene, ethylbenzene and so on from contaminated sources even below the part per billion (ppb) level which is greater than that of pristine granular activated carbon. In addition to this, the carbon coated glass fibers have shown a 7-fold enhancement in adsorption and desorption capacities for different pollutants. This in turn nullifies the disadvantage of glass core which is low surface and poor porosity characteristics, thereby decreasing the overall adsorption capacity of end product. Similarly, various ion-exchange active materials coated on inorganic substrate like glass fibers demonstrate several advantages over the conventional ion-exchange resins. Therefore, most of these strategies provide simplistic preparation or synthesis procedures, efficient post-modification potential and benefits against contaminants. [49]

Other benefits using polymer, carbon and functionalized adsorbents include the ability to easily fabricate them into papers, felts, or fabrics that provide improved filter media via enhanced water contact efficiency, physical, chemical and mechanical robustness with effective interactive adsorption and regeneration. Further, these research outcomes outline that advanced functional systems can be extended to remove toxic metal ions in excessive concentration like lead, arsenic, mercury, copper, as well as fluoride and carcinogen bromate from both groundwater as well as wastewater sources.

5.6.2. Carbon-based functional materials

Recent studies in the field of carbon-based materials using innovative and advanced approaches have led to the exploration of new sets of materials in both energy and environmental applications. Also, advances in nanoscience and nanotechnology protocols have provided new insight into the surface functionalities of both polymers and carbon in their individual and hybrid form have shown great potential in water treatment applications. Both these classes of materials designed in simple device form have proven ecofriendly materials from preparation to end use. Further, simple functionalization processes in preparing make them economically viable for water and wastewater treatment applications. Thus, recent efforts have been made to prepare advanced and more sustainable and easy to scale-up materials which has led to the use of polymer and carbon-based nanomaterials, consisting of various low-dimensional structures of carbaneous materials such as carbon, activated carbon, graphene, graphene oxide, carbon nanotubes, activated carbon, hydrothermal carbon and so on.

Nonetheless, carbon-based materials possessing exceptionally large surface area with oleophilic and hydrophobic nature have shown to exhibit varying morphological patterns and structural, porosity and surface characteristics. Therefore, carbonaceous materials have attracted tremendous attention in the area of research for their exceptional advantages arising due to their exceptional physical, chemical and physicochemical characteristics such as surface area, functionality, easily accessible active sites, chemical stability, large area of contact, electrical and heat conductivity, greater ability to direct, flexibility, advanced optical characteristics and robust mechanical characteristics which make them suitable candidates for water treatment applications.

Another major advantage with carbon-based nanomaterials is that carbaneous materials offer vast tunable structures and surface properties that are key to environmental applications particularly in water treatment and remediation. Most importantly, many of the carbon-based adsorbents can be directly converted into an end use device and potentially modeled as point-of-use devices both in the form of cartridges, filters, or membranes. [50, 51]

5.6.3. Graphene/Graphene oxide-based nanocomposites

Of all the well-designed two-dimensional (2D) materials-based membranes comprising either graphene (G), graphene oxide (GO) and its derivatives have shown great potential as nanomaterial-based filters for water treatment. Graphens are made up of a 2D grid of densely packed carbon atoms impermeable to all other atoms and molecules have become nano-components on the rise for the fabrication of membranes with adjustable physicochemical characteristics. On the other hand, graphene/GO-based nanocomposites in the form of membranes have shown exceptional molecular separation characteristics for various pollutants from wastewater. Among many nanomaterials, carbaneous 2-D materials have exceptional pollutant retention properties. Nanocomposite membranes fabricated using 2D nanomaterials have been considered next-generation separation membranes due to their high mechanical strength simultaneously retaining high flexibility, electrical and thermal properties, free-standing nature of composite membrane have ease of adaptability in selective separation and remediation of pollutants from wastewater. Nevertheless, advances in nanotechnology devices and scale-up activities in grapheneous materials have led to the exploration of new 2D materials as separation media for various water source remediation. These materials have been explored both in pristine as well as functionalized form for water treatment applications as shown in Figure 5.6. The freestanding nature of GO membrane with nano-dimensions have exhibited an excellent mechanical stability compared to conventional filter media in selective filtration applications. Interestingly, the selectivity of graphene induced nanocomposite membrane can be easily adjusted by tuning thickness of the membrane, modifying chemical functionality of composite constituents. Among many advantages, sheet-like physical structures of graphene nanomaterials can be assembled to attain tunable porosity ranging from reverse osmosis to ultrafiltration membranes to selectively allow water, ionic species or biomacromolecules depending on end use requirement. These arrangements can create preferable water channels through their self-assembled alignments in which inter-sheet voids are regarded as pores. Similarly, graphene nanosheets have also been dispersed in polymer membrane matrix to create additional porosity and preferable wettability. Through these strategies, properties such as dispersibility, self-assembly, functionalization,

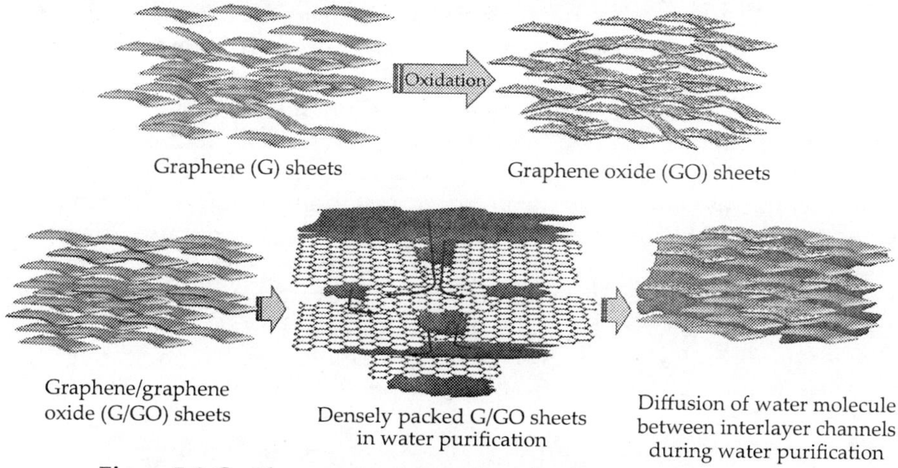

Figure 5.6. Graphene and graphene-based polymer nanocomposite membranes used in water treatment applications.

and compatible nature of overall membrane constituents can be altered to suit the nature of wastewater. Therefore, advanced research activities have been adopted to fabricate free standing membranes by compositing polymer with suspended graphene and their derivatives. Also, innovative approaches have also led to the exploration of these novel membranes for futuristic applications other than water remediation. [52, 53]

Interestingly, following advanced approaches, solubility of grapheneous 2D nanomaterials have been improved to achieve high levels of self-assembly properties and to retain graphene and its derivatives from clumping together. Further, several functionalization approaches on grapheneous materials have been adopted by immobilizing various nanomaterials like metal oxides, layered double hydroxides, oxyhydroxides and so on which act as intercalating or insertion systems in between nanosheets further enhancing pollutant retention properties. Nonetheless, different graphene-based hybrid materials with specific functions have been prepared via self-assembly owing to less energy consumption in its manufacture following drop casting of 2D nanocomposite solution or spin coating dope solution method, vacuum filtration technique, layer-by-layer deposition, or direct polymer composite solution casting into membranes. Further advanced approaches have also adopted inducing additional functionality such as carboxyl and hydroxyl during or after filter preparation thereby providing functional active sites and hydrophilic properties to nanocomposite filters. Therefore, advanced approaches have resulted in multifunctional metal nanocomposite induced filters having synergistic interaction between GO and metal oxide conceptualized to yield exceptional physicochemical properties. In this direction, group of researchers (Chao Xu et al.) created a novel and high-quality output graphene oxide/TiO_2 nanocomposites with high specific surface area (489.2 m^2/g), which is more than that of both bare GO sheets (<10 m^2/g) and TiO_2 (118.9 m^2/g). These GO-

TiO_2 composite stable dispersions in water were further converted into efficient membranes via assembled structure format by simple vacuum filtration. Functionalization of graphene sheets with TiO_2 resulted in broadening of interlayer distance of the graphene nanosheets, which led to the formation of 2D nanosheet assemblies in film form with unique pores and channels, making them suitable filtration membranes for the abatement of organic dye traces from contaminated wastewater sources. With this strategy, a mean pore size of 3.5 nm was achieved in the prepared membrane to effectively remove nearly 100% of methyl orange pollutant from contaminated water. Further studies also showed an increase in concentration of nanocomposite solutions found to be disadvantageous in tuning appropriate porosity in the end filter. Similarly, in many cases, GO/metal oxide (MO) dispersions in water have been vacuum filtered on various stable substrates like filter paper, polycarbonate filter support and porous alumina to act as advanced functional membranes.

5.7. Polymer-based Functional Nanocomposite Devices

Effective and affordable water treatment technologies are the need of the hour. On the other hand, increased water scarcity forces researchers, innovators and technologists to adopt advanced methodologies for remediation of contaminated water sources. In this direction, many filter media have been tried and tested to address the water-related problems and as an effective way to eliminate contaminants from wastewater with improved adsorption capacity, better pollutant retention properties, enhanced product flux from hybrid membranes or adsorption-based modules are evolving as all-in-one solution for water purification. Even though membrane technology is in practice for the last four decades for the treatment of conventional contaminated sources, emerging complications with persistent pollutants and deteriorated wastewaters demand new classes of membranes. This is due to the involvement of conventional polymers that have low sustainability to retain high flux and rejection patterns. Also, limitations associated with membrane fabrication is also restricting use of membrane precursor materials that are not cost-effective, environmentally hazardous, lack flexibility for a variety of contaminants and limited water permeability. Further, ever-increasing environmental pollution has been the base for exploring potential natural origin materials such as biopolymers that can substitute the conventional synthetic polymer in membrane preparation.

Thus, fabrication of cost-effective, sustainable, and highly efficient nanocomposite filters, and membranes have gained more interest for futuristic applications. Specifically, nanomaterial-based active materials have been tested in the form of filter devices for the removal of hazardous contaminants such as heavy metals, fluoride, and other emerging pollutants. These attempts have been made keeping affordability and sustainability in mind to reduce the dependency on cost-intensive technologies. Also, bio-origin resources have shown great potential in this direction and researchers are rapidly adopting novel protocols to adopt versatile bio-original materials in combination with nanomaterials to be converted as effective water purifiers. Additionally, bio-gen materials possess added properties due to their easy availability, cost-effectiveness, oxidation resistance, high modulus, recycling prospect, renewable nature, and more importantly easy processability, making these precursors as materials of future choice in separation and purification technologies sectors. [54, 55]

On the other hand, biopolymers extracted from various biomaterial sources are considered non-hazardous, renewable, decomposable, and biocompatible to adapt for different environmental applications including water treatment applications. The biopolymers are generally ecofriendly and abundant materials with a high degree of functionalizable active sites that can be tuned to prepare highly advanced and robust filters for a series of emerging pollutants. Among many such attempts hydrothermal, solvothermal and ionothermal functional carbon have been prepared using biopolymer and inorganic active precursors. Cellulose, starch, and polysaccharide-based active hydrothermal carbon was prepared in combination with metal oxide, hydroxide and oxyhydroxide active sites via ionothermal processes to apply them in emerging pollutant treatment such as pharmaceutical ingredients, heavy metals, and biological active nanomaterials even in trace concentrations from contaminated water.

5.8. Cellulose-based Nanocomposites

Cellulose is a complex but naturally abundant carbohydrate that occurs in many resources and forms. Cellulose is found in the environment in abundance and widely distributed across ecosystems. Due to their great affinity towards water, cellulose-based natural resources can be easily adopted to selectively separate high-quality water from contaminated sources utilizing their abundant functionality such as hydroxyl and further tuned to attain additional functionalities like carboxy and metal oxy/hydroxide groups. Further, cellulose-based materials can be easily converted into end-user devices in the form of filter beads, filter disks or cakes, films, and resins for the removal of persistent pollutants, organics, heavy metals and hazardous pharmaceutical ingredients present in various sources of water.

Nonetheless, conventional cellulose-based adsorbents suffer from low absorptive capacity towards high concentration feed solutions. However, the limited adsorption ability of cellulose observed against organic molecules like dyes in wastewater are keenly studied to overcome the drawbacks in the naturally abundant precursors. Thus, various advanced approaches have been adopted to prepare composites of cellulose that could have potential as a high-performance and affordable adsorbent for wastewater treatment applications. In general, the high-hydroxyl functionality of cellulose-based materials allows easy and facile functionalization. Shu-Ya Tian (et al.) developed a simple technique to prepare vacuum deposited filter beds to create reusable graphene oxide functionalized cellulose composite films or membranes (CGCMs) that were directly used as sieves to eliminate organic pollutants from contaminated water sources of varying concentrations. The study revealed that the innovative CGCM filters can effectively clean wastewater with high flux performance while simultaneously retaining the pollutant species on their surfaces. These vacuum deposited filter systems showed adsorption capacity of 86.4 mg/g for Rhodamine B while retaining regenerative properties which indicate that they have high ability to remove pollutants on multiple cycles making them the main contender for future water treatment media. Several extended studies using cellulose-based materials reveal their potential in wastewater treatment applications as well.

In another study, Bo Li (et al.) fabricated Fe-doped cellulose (MCFBs) with FeOx nanomaterials in 2-3 mm in diameter that focused on the use of ecofriendly precursors and preparation route to prepare sustainable adsorbents for heavy metal and dye removal from contaminated sources. In this study, MCFBs prepared in bead form showed an adsorptive capacity of 1186.8 mg/g for methylene blue and 151.8 mg/g for rhodamine B from wastewater due to their characteristic surface morphologies and pore structures with carboxyl functional groups. In addition to this, specific chemical composition of the MCFBs also possess enhanced heat stability. Thus, biopolymer-based nanocomposite water filters are easy to prepare, assemble and use as end-use filter as green adsorbents and the study further reveals that processes like hydrothermal, solvothermal and ionothermal routes can be efficient methods for the development of innovative porous beads. Use of iron oxide-based nanomaterials can be tuned to be magnetic along with high adsorption capacity to regenerate post-use of the nanocomposite for water treatment. Further, study on 3D cellulose-based nanomaterials that are mechanically robust with higher porosity and large surface area have been reported by team of researchers (Zhang et al.) wherein a highly efficient adsorbent was prepared for the adsorptive removal of phenol from contaminated water. In this study, cyclodextrin functionalized cellulose (CD/C-CNF) nanofibrils composed of aerogel were prepared and applied for pollutant adsorption applications. The use of CD/C-CNFs findings suggested a high adsorption potential of 148 μmol/g for trace organic pollutants on CD/C-CNFs aerogel. Further, the CD/C-CNFs aerogel established remarkable reusability, ecology-friendly material composition as promising adsorbent for the elimination of phenol pollutants in water sources indicating that it has the potential to be used in large scale applications as well as use in industrial wastewater treatment applications where other conventional materials fail to treat under unfavorable circumstances. [56, 57]

Lately, recent advancements in nanotechnology research have shown graphene oxide, nanocellulose and their composite derivatives or composite have lately attracted a lot of attention in the field of water treatment due to their unique adsorption abilities, mechano-chemical properties, ability to coordinate with pollutant metal ions, and surface charge. In this direction, Zhu (et al.) have prepared bio-hybrids following self-assembly method following vacuum filtration of graphene oxide sheets (GO) along with the dispersion of 2,2,6,6-tetramethylpiperidine-1-oxyl radical mediated oxidation nanofibers (TO/CNFs). The TO/CNFs with unique dispersion, hydrolytic stability and recyclability make them a suitable precursor for preparing a highly stable filter. These biohybrids-filters have demonstrated promising adsorption capacity for metal-based contaminants like Cu(II) in water via their ionic crosslinking between the negatively charged biohybrid and the adsorbed Cu(II). The findings demonstrate that nano-GO-hybrids and TO/CNF can be prepared as a new generation of nanomaterial-based membranes or filters with cooperative features and characteristics such as flexibility, high performance, hydrolytic stability, and mechanical robustness.

5.9. Chitosan-based Nanocomposites

After cellulose, chitosan (CS) has been the second most abundant polysaccharide found in nature with useful characteristics such as non-toxicity, biocompatibility, and biodegradability. Chitosan-based materials have also shown incredible attraction for heavy metal ions due to their unique chelating ability owing to their characteristic functionalities like $-NH_2$, -OH chitosan. In one such study, Sankar (et al.) prepared a chitosan-based all-in-one potable water purifier which does not require external stimuli or pressure like electric powered pumps to operate. In this strategy, advanced water purifiers used metal oxyhydroxide functionalized chitosan nanocomposite materials for cost-efficient, sustainable, and effective water filtration. These filters have also been further functionalized with nano-sized-silver which constantly monitor the microbial contamination to yield safe water at the end. These highly active sustainable nanocomposite materials were assembled to fabricate a cost-effective water purifier that provides a family for $2.5 per year safe and healthy drinking water by removing both conventional pollutants and biological contaminants which can prevent water-borne disease. Similar attempts have been made towards affordable production of arsenic-free potable water using a new set of biopolymer-based nanocomposite filters. In this study reported by the same research group prepared a biopolymeric caged metastable 2-Line ferrihydrite (CM-2LF) at ambient temperature for hazardous metal-based pollutants. As prepared material was used for the removal of both As (III) and As(V) with exceptional adsorption capacities thereby making CM-2LF composite a low-cost and ecofriendly water purification system. It is estimated that the CM-2LF -based systems can be effectively established as future water filters.

On the other hand, advanced version of aerogels modified with nanomaterial-active sites have been converted into filter devices for metal ion removal which received a lot of interest for future. Moreover, when compared to series of conventional adsorbents, aerogels have exhibited certain unique characteristics like relatively large however adjustable surface characteristics, tunable pore size and volumes and their light weight make them vital to adsorb various contaminants. In a recent study, Nataraj and his team have prepared a unique biopolymer-based aerogel filter functionalized with nanosized Fe-Al composites for robust water purification application as a cost-effective, sustainable, and easy to regenerate water purifiers. This study uniquely assembles sustainable materials stocks to prepare, easy to use and flexible to adopt to condition filters. In this study, newly prepared functionalized aerogel filters in directly usable membranes forms showed a removal efficiency of ~89% emerging pollutants, 95% of dye, ~99% rejection for F^- and ~90% rejection for As (V). Overall, a novel approach in adopting aerogel-based highly porous material with high adsorptive capacity for a variety of pollutants has proven its potential for removing fluoride and arsenic with remarkable reusability. These advanced materials have also shown unique point-of-use applications. Thus, ecofriendly and a low-cost aerogel membrane filters have also shown remarkable promise in addressing real-world wastewater and is being developed as a possible alternative for a series of pollutants as "universal water filter." Therefore, this study proved for the first time that functionalized aerogel filter membranes have been utilized for the purification of organic pollutants like dyes in water, persistent contaminants like hormones, heavy metals, surfactants, pharmaceuticals, and oil/water contaminants. Interestingly, all the purification experiments for a series of contaminants through aerogel membranes were conducted under the driving force of gravity. Nonetheless, Yun Liao (et al.) reported similar study outcomes on a new 3D aerogel prepared using polydopamine-functionalized attapulgite/chitosan (PD-AP-CS) systems. These PD-AP-CS aerogel membranes showed high retention property for uranium (VI) in water. Detailed study report revealed that the most of uranium adsorption occurred within 40 minutes of their exposure to each other at pH 5.0 with a maximum adsorption capacity of 175.1 mg/g. More significantly, all aerogel membranes have been recycled and reused over several cycles maintaining over 75% retention properties at the end of five cycles.

All these studies reveal that biopolymer-based functional nanomaterials have shown considerable benefits in water purification and remediation the potential to replace the conventional synthetic polymer-based (PSf, PES, PVDF, polyesters, polypropylene, and polyacrylic acid) membranes soon. Along with technological advantages, several economic benefits have also been recorded as the advanced filters provide remarkable permeability of product water and salt rejection for almost all contaminants without the use of applied pressure. Similar attempts are made using biopolymer-based systems in adsorption systems in converting lab-scale filters to real world cartridges and filter beds. [58, 59]

Nevertheless, these biopolymer-based functional materials also suffer from some disadvantages such as weak mechanical strength which limit their large-scale end-use applications. Nonetheless, natural plasticizer additives are being adopted and stable composite with synergetic contribution are being adopted as ways to circumvent these disadvantages.

5.10. Polyelectrolyte Complexes

Recent advances in polymer science and chemistry have led to the understanding of innovative approaches that can be adopted to convert macromolecular systems to be an efficient pollutant retention media. The polymer or set of polymers in blending or composite form that contain a net positive or negative charge at near neutral pH are known as polyelectrolytes. Polyelectrolytes are generally soluble in water; however these are stabilized using suitable agents depending on the end use. In general, solubility of polyelectrolytes is driven by the electrostatic interactions between the charged monomer and water. For instance, biomacromolecules like protein, DNA, biopolymers like chitosan, carragenan, cellulose polymers and their derivatives can be directly used or converted into polyelectrolytes.

Recent years have witnessed the great interest in a set of polymers that responds to some external stimuli or applied changes such as pH, temperature, electric or magnetic field. Interestingly, the driving force like stimuli that influences the transitions in polyelectrolytes toward neutralization overall group charges are mainly caused by either a pH shift or the addition of an oppositely charged ingredient like polymer or additives. This influences the changes in hydrogen bonding in polyelectrolytes with an increase in pH, temperature or ionic strength and collapse of hydrogels or complexes or polymer interpenetrating networks. Generally, many polyelectrolytes fulfill requirements of biocompatibility and affordability and these can be adapted to meet the various parameters like carrier substances and components for active substances. Therefore, polyelectrolyte complexes (PECs) are the association of polymer/polymers interactions formed between oppositely charged polymers/polymer or polymer/charged organic system, or polymer/pollutant ion. These unique architectures can be effectively used to retain pollutants from contaminated water through their electrostatic interaction between oppositely charged polyions and in case of water treatment applications interactions will be polymer/pollutant/polymer. Largely, PECs preparation techniques avoid the use of chemical crosslinking agents or stabilizing agents to avoid breaking of complexes. This strategy is also followed to reduce the possible toxicity, decrease the surface activity and other undesirable effects of the reagents. In recent advances, PECs have been used for eliminating ionic pollutants like nitrates, perchlorate, phosphates, arsenic in the form of oxo-anions, nickel, and vanadium. In some cases, PECs also have been used for the removal of known ionic organic pollutant like natural organic matter (NOM). Even though known NOMs like fulvic acid and humic have been identified for their fouling propensity on polymeric matrices. Recent advances in the preparation of PECs have led to the production of stable water filtration media. [60]

5.11. Conclusion

In summary, functional materials are and will continue to be in demand due to their multi-functional activities that are useful for various modern applications. The functional materials prepared using either organic or inorganic origin precursors can be easily converted into realistic end use. On the other hand, functional materials having functional groups can be uniquely used to interact with potential pollutants or contaminants in wide range of concentrations and effective unlike their non-functional counterparts. Nevertheless, today most functional materials including polymers, nanomaterials or other ceramic-based precursors can be tailor-made for attaining specific properties to suit task specific applications. However, functional materials having suitable active sites are being utilized keeping commercial aspects as driving factor more and more toward higher value or performance analogs. Consequently, there will be increased expectation on potentials of penetrating functional materials for water treatment applicationsint commodity markets. However, penetration of novel and advanced materials and technologies in real-world applications is still to reach considerable share. This is due the socio-economic reasons associated with the direct use of new materials and technologies replacing conventionally practiced means. Also, one of the major factors in sluggish growth of novel water treatment materials like functional materials is due to their relative complexity in use and cost of their production.

Preparation of functional materials is often difficult due to their chemical antagonisms or phase incompatibilities. This chapter provided detailed discussion on the general methods designed, and followed for the preparation of functional materials targeting water treatment applications. However, functional materials follow multi-step scale-up preparation and functionalization protocols which make them cost-intensive systems. Keeping this in view, this chapter has provided detailed review of scope and limitations of functionalization processes on various materials involving polymers, ceramic materials, biopolymers, carbaneous, grapheneous and advanced 2D materials via cationic, anionic, and organometallic methods. In all cases, to induce optimal functionalities on the backbone of substrates, careful selection of precursor combination and preparation conditions plays a vital role. For example, in situ polymerization and functionalization of macromolecular systems via anionic polymerization requires suitable solvent, appropriate initiator, and temperature to yield optimal end-product. Here, the sensitivity of anionic species toward moisture and oxygen remains an inappropriate factor inherent in anionic polymerization. On the other hand, several novel macromolecules like block copolymers can be functionalized via various in situ process like "living" anionic polymerization. However, issues associated anionic polymerization is highly reactive monomers may hinder uniform functionality in end product.

On the other hand, natural clay minerals that are abundantly available in nature in the lithosphere can be functionalized through simple functionalization process. These natural clays micro- and nanomaterials possess unique physicochemical properties which are very sensitive to the counter ions due to their structural and physical arrangements. Therefore, naturally abundant clays materials can be easily functionalized and applied for specific ion-exchange applications upon exposure to external stimuli. Further, change to the structural properties of natural clay minerals like montmorillonite (MMT) can be achieved with targeted environmental impacts as clay materials can be forced to intake pollutants from contaminated water sources via their swelling properties.

Interestingly, ion exchange process in nanoporous clay depends on their unique microporous structures and relatively high surface area which plays an integral role in attracting pollutants from contaminated water. In

MMT, a common clay mineral in sediments, soils, and muds, can be further utilized for selective insertion and desertion of pollutants which can be used for discretely removal of pollutants through exchange selectivity. In natural clay materials, the ion-exchange processes are believed to be occurring via a process called dynamic delamination and restacking in clay layers, which creates a continuous chain exchange process to take place between the hydration state of the exchanging cation by altering composition of the clay inter-layer. Interestingly, clay particles with different hydration efficiencies are distinct phases with unique binding selectivities. Surprisingly, equilibrium of ion-exchange process is achieved through thermal fluctuations in contaminant sources in which hydration state leads to the exchange ionic species between individual MMT layers. Nevertheless, ion-exchange process for low concentration wastewater or contaminated water can be very effective. Several studies also revealed that clays in water treatment have shown several drawbacks. Among many, it is difficult to choose appropriate reagent for functionalization stage. In addition to this, selective affinity, and ion-exchange mechanisms of pollutant species from contaminated sources by clays and functionalized clays are yet to be clearly understood.

References

1. Stoyko Fakirov, Fundamentals of Polymer Science for Engineers, first published: 23 August 2017, ISBN:9783527341313, ISBN:9783527802180, DOI:10.1002/9783527802180, 2017 Wiley-VCH Verlag GmbH & Co. KGaA.
2. F. Zhao, Y. Shi, L. Pan and G. Yu, Multifunctional nanostructured conductive polymer gels: synthesis, properties, and applications, Acc. Chem. Res., 2017, 50 , 1734-1743 CrossRef CAS PubMed.
3. W. Zhang , B. Aguila and S. Ma , Potential applications of functional porous organic polymer materials, J. Mater. Chem. *A*, 2017, 5 , 8795-8824.
4. Kumar, A., & Gupta, R.K. (2018). Fundamentals of Polymer Engineering (3rd ed.). CRC Press. https://doi.org/10.1201/9780429398506.
5. Gardner, A., Duprez, W., Stauffer, S., Ungu, D.A.K., Clauson-Kaas, F. (2019). Introduction to Biological Macromolecules. In: Labster Virtual Lab Experiments: Basic Biochemistry. Springer Spektrum, Berlin, Heidelberg. https://doi.org/10.1007/978-3-662-58499-6_2.
6. Petr Munk ,Tejraj M. Aminabhavi, Introduction to Macromolecular Science, Wiley–Blackwell; 2nd edition (14 March 2002), ISBN-10 : 0471417165, ISBN-13 : 978-0471417163.
7. Polymer Composites: Volume 1, First Edition. Edited by Sabu Thomas, Kuruvilla Joseph, Sant Kumar Malhotra, Koichi Goda, and Meyyarappallil Sadasivan Sreekala 2012 Wiley-VCH Verlag GmbH & Co. KGaA. Published 2012 by Wiley-VCH Verlag GmbH & Co. KGaA.
8. María Rosa Aguilar and Julio San RománSmart Polymers and their Applications, 2014, ISBN 978-0-85709-695-1, Woodhead Publishing Limited.
9. Martin Wills, Ernest W. Colvin, 2.10 - Alkylboron and-silicon Compounds, Editor(s): Alan R. Katritzky, Otto Meth-Cohn, Charles W. Rees,

Comprehensive Organic Functional Group Transformations, Elsevier Science, 1995, Pages 513-547, ISBN 9780080447056.

10. Alfred Rudin, Phillip Choi, The Elements of Polymer Science & Engineering, Third Edition, 2013, ISBN 978-0-12-382178-2, Academic Press.
11. Creamer, A., Wood, C.S., Howes, P.D. et al. Post-polymerisation functionalisation of conjugated polymer backbones and its application in multi-functional emissive nanoparticles. Nat Commun 9, 3237 (2018). https://doi.org/10.1038/s41467-018-05381-4.
12. María Rosa Aguilar and Julio San Román, Smart Polymers and their Applications, ISBN 978-0-85709-695-1, Woodhead Publishing Limited.
13. Jörg Friedrich, Metal-Polymer Systems: Interface Design and Chemical Bonding, Chapter 5, Functional Groups at Polymer Surface and Their Reactions, First published: 13 November 2017, John Wiley & Sons, Inc.
14. Jörg Friedrich, Metal-Polymer Systems: Interface Design and Chemical Bonding, Print ISBN:9783527336777 | Online ISBN:9783527679898 | DOI: 10.1002/9783527679898, 2018 Wiley-VCH Verlag GmbH & Co. KGaA.
15. Makvandi, P., Iftekhar, S., Pizzetti, F. et al. Functionalization of polymers and nanomaterials for water treatment, food packaging, textile and biomedical applications: a review. Environ Chem Lett 19, 583–611 (2021).
16. Mohamed R. Berber, James Barker, Current Advances of Polymer Composites for Water Treatment and Desalination, Journal of Chemistry, Volume 2020 | Article ID 7608423.
17. Tawfik A. Saleh, Prakash Parthasarathy, Muhammad Irfan, Advanced functional polymer nanocomposites and their use in water ultra-purification, Trends in Environmental Analytical Chemistry, Volume 24, 2019, e00067, ISSN 2214-1588.
18. Spiro D. Alexandratos, Ion-Exchange Resins: A Retrospective from Industrial and Engineering Chemistry Research, Industrial & Engineering Chemistry Research 2009 48 (1), 388-398.
19. Harry P. Gregor and J. I. Bregman, Characterization of Ion Exchange Resins. I. Acidity and Number of Constituent Cation Exchange Groups, Journal of the American Chemical Society 1948 70 (7), 2370-2373.
20. Jueun Kim, Bhupendra Kumar Singh and Wooyong Um, Effect of ion exchange resin particle size on homogeneity and leachability of Cs and Co in polymer waste form, RSC Adv., 2021,11, 2729-2732.
21. Treto-Suárez, M.A., Prieto-García, J.O., Mollineda-Trujillo, Á. et al. Kinetic study of removal heavy metal from aqueous solution using the synthetic aluminum silicate. Sci Rep 10, 10836 (2020). https://doi.org/10.1038/s41598-020-67720-0.
22. Meng Xu, Pejman Hadi, Chao Ning, John Barford, Kyoung Jin An, Gordon McKay, Aluminosilicate-based adsorbent in equimolar and non-equimolar binary-component heavy metal removal systems, Water Sci Technol (2015) 72 (12): 2166–2178.
23. Luukkonen, T., Heponiemi, A., Runtti, H. et al. Application of alkali-activated materials for water and wastewater treatment: a review. Rev Environ Sci Biotechnol 18, 271–297 (2019). https://doi.org/10.1007/s11157-019-09494-0.

24. R. W. Broach, "Zeolite types and structures," in Zeolites in Industrial Separation and Catalysis, S. Kulprathipanja, Ed., WILEY-VCH, Weinheim, 2010.

25. R. F. Lobo, "Introduction to the structural chemistry of zeolites," in Handbook of Zeolite Science and Technology, S. M. Auerbach, K. A. Carrado, and P. K. Dutta, Eds., p. 1283, Marcel Dekker, Inc., New York, 2003.

26. C. Baerlocher, L. B. McCuster, and D. H. Olson, Atlas of Zeolite Framework Types, Elsevier, Amsterdam, 6 ed. edition, 2007.

27. M. N. Rashed and P. N. Palanisamy, "Introductory Chapter: Adsorption and ion exchange properties of zeolites for treatment of polluted water," Tech. Rep., Zeolites and Their Applications, 2018, Chapter Chapter 1.

28. M.M. Hassan and C. M. Carr, "A critical review on recent advancements of the removal of reactive dyes from dyehouse effluent by ion-exchange adsorbents," Chemosphere, 209: 201–219, 2018.

29. J. Dong, Z. Xu, S. Yang, S. Murad, and K. R. Hinkle, "Zeolite membranes for ion separations from aqueous solutions," Current Opinion in Chemical Engineering, 8: 15–20, 2015.

30. Alok Kumar Meher, Sera Das, Sadhana Rayalu & Amit Bansiwal (2016) Enhanced arsenic removal from drinking water by iron-enriched aluminosilicate adsorbent prepared from fly ash, Desalination and Water Treatment, 57: 44, 20944-20956.

31. Flieger J, Kawka J, Płaziński W, Panek R, Madej J. Sorption of Heavy Metal Ions of Chromium, Manganese, Selenium, Nickel, Cobalt, Iron from Aqueous Acidic Solutions in Batch and Dynamic Conditions on Natural and Synthetic Aluminosilicate Sorbents. Materials (Basel). 2020 Nov 21; 13(22): 5271. doi: 10.3390/ma13225271. PMID: 33233399; PMCID: PMC7700620.

32. Hussain, A., Madan, S., Madan, R., 2021, 'Removal of Heavy Metals from Wastewater by Adsorption', in M. K. Nazal, H. Zhao (eds.), Heavy Metals -Their Environmental Impacts and Mitigation, IntechOpen, London. 10.5772/intechopen.95841.

33. Montserrat Iborra, Javier Tejero, Fidel Cunill, J. Felipe Izquierdo, and Carles Fité, Drying of Acidic Macroporous Styrene–Divinylbenzene Resins with 12–20 Cross-Linking Degree.

34. Yamanaka, K. (2015). Ion-Exchange Resins. In: Kobayashi, S., Müllen, K. (eds) Encyclopedia of Polymeric Nanomaterials. Springer, Berlin, Heidelberg. https://doi.org/10.1007/978-3-642-29648-2_131Industrial & Engineering Chemistry Research 2000 39 (5), 1416-1422.

35. Chaoyan Lou, Can Wu, Kai Zhang, Dandan Guo, Lei Jiang, Yang Lu, Yan Zhu, Graphene-coated polystyrene-divinylbenzene dispersive solid-phase extraction coupled with supercritical fluid chromatography for the rapid determination of 10 allergenic disperse dyes in industrial wastewater samples, Journal of Chromatography A, Volume 1550, 2018, Pages 45-56, ISSN 0021-9673.

36. M.S. Mohy Eldin, M.A. Abu-Saied, T.M. Tamer, M.E. Youssef, A.I. Hashem & M.M. Sabet (2016) Development of polystyrene based nanoparticles

ions exchange resin for water purification applications, Desalination and Water Treatment, 57: 32, 14810-14823.

37. M.S. Mohy Eldina,b, M.A. Abu-Saiedb, T.M. Tamerb, M.E. Youssefc, A.I. Hashemb, M.M. Sabetb Development of polystyrene based nanoparticles ions exchange resin for water purification applications, Desalination and Water Treatment, 57 (2016) 14810–14823.
38. K. W. Pepper, H. M. Paisley, and M. A. Young, "Properties of ion-exchange resins in relation to their structure. Part VI. Anion-exchange resins derived from styrene-divinylbenzene copolymers," Journal of Applied Chemistry, 4097–4105, 1953.
39. C.M. Wu, T. W. Xu, and W. H. Yang, "Fundamental studies of a new hybrid (inorganic-organic) positively charged membrane: membrane preparation and characterizations," Journal of Membrane Science, 216 (1-2): 269–278, 2003.
40. K.W. Pepper, H.M. Paisley, and M.A. Young, "Properties of ion-exchange resins in relation to their structure. Part VI. Anion-exchange resins derived from styrene-divinylbenzene copolymers," Journal of Applied Chemistry, 4097–4105, 1953.
41. Aref Abbasi Moud, Polymer based flocculants: Review of water purification applications, Journal of Water Process Engineering, Volume 48, 2022, 102938, ISSN 2214-7144.
42. Oluranti Agboola, Ojo Sunday Isaac Fayomi , Ayoola Ayodeji , Augustine Omoniyi Ayeni, Edith E. Alagbe, Samuel E. Sanni, Emmanuel E. Okoro, Lucey Moropeng, Rotimi Sadiku, Kehinde Williams Kupolati and Babalola Aisosa Oni, A Review on Polymer Nanocomposites and Their Effective Applications in Membranes and Adsorbents for Water Treatment and Gas Separation, Membranes 2021, 11, 139.
43. Okutucu, B., 2020, 'Wastewater Treatment Using Imprinted Polymeric Adsorbents', in A. Körlü (ed.), Waste in Textile and Leather Sectors, IntechOpen, London. 10.5772/intechopen.92386.
44. Susheel Kalia, Natural Polymers-Based Green Adsorbents for Water Treatment, 2021, ISBN 978-0-12-820541-9, 1st Edition - August 22, 2021, Elsevier, ISBN: 9780128205419, eBook ISBN: 9780128205426.
45. Pan, B., Zhang, X., Jiang, Z., Li, Z., Zhang, Q., Chen, J. (2019). Polymer and Polymer-Based Nanocomposite Adsorbents for Water Treatment. In: Das, R. (eds) Polymeric Materials for Clean Water. Springer Series on Polymer and Composite Materials. Springer, Cham. https://doi.org/10.1007/978-3-030-00743-0_5.
46. P.R. Yaashikaa, P. Senthil Kumar, S. Karishma, Review on biopolymers and composites – Evolving material as adsorbents in removal of environmental pollutants, Environmental Research, Volume 212, Part A, 2022, 113114, ISSN 0013-9351.
47. Russo, T.; Fucile, P.; Giacometti, R.; Sannino, F. Sustainable Removal of Contaminants by Biopolymers: A Novel Approach for Wastewater Treatment. Current State and Future Perspectives. Processes 2021, 9, 719.
48. Jiang, X., Li, Y., Tang, X. et al. Biopolymer-based flocculants: a review of recent technologies. Environ Sci Pollut Res 28, 46934–46963 (2021). https://doi.org/10.1007/s11356-021-15299-y

49. Subhadeep Biswas, Anjali Pal, Application of biopolymers as a new age sustainable material for surfactant adsorption: A brief review, Carbohydrate Polymer Technologies and Applications, Volume 2, 2021, 100145, ISSN 2666-8939.

50. Suprakas Sinha Ray, Rashi Gusain and Neeraj Kumar, Carbon Nanomaterial-Based Adsorbents for Water Purification, Fundamentals and Applications, A volume in Micro and Nano Technologies, 2020, ISBN 978-0-12-821959-1, Elsevier Inc.

51. Nazal, M. K., 2020, 'An Overview of Carbon-Based Materials for the Removal of Pharmaceutical Active Compounds', in M. Bartoli, M. Frediani, L. Rosi (eds.), Carbon-based Material for Environmental Protection and Remediation, IntechOpen, London. 10.5772/intechopen.91934.

52. Imran Ali, Al Arsh Basheer, X.Y. Mbianda, Alexander Burakov, Evgeny Galunin, Irina Burakova, Elina Mkrtchyan, Alexey Tkachev, Vladimir Grachev, Graphene based adsorbents for remediation of noxious pollutants from wastewater, Environment International, Volume 127, 2019, Pages 160-180, ISSN 0160-4120.

53. Ferda Mindivan, Ülküye Dudu Gül, Meryem Göktaş, Application of graphene-based adsorbents in the treatment of dye-contaminated wastewater; kinetic and isotherm studies, Journal of Vinyl and Additive Technology, 2021, 27(3): 485-496.

54. Nageswara Rao Lakkimsetty, S. Feroz, S. Karunya, L. Motilal, P. Saidireddy, G. Suman, Synthesis, characterization and application of polymer composite materials in wastewater treatment, Materials Today: Proceedings, Volume 59, Part 3, 2022, Pages 1726-1734, ISSN 2214-7853.

55. Yongshun Huang, Jiaxing Li, Xiaoping Chen and Xiangke Wang, Applications of conjugated polymer based composites in wastewater purification, RSC Adv., 2014, 4: 62160-62178.

56. Baoliang Peng, Zhaoling Yao, Xiaocong Wang, Mitchel Crombeen, Dalton G. Sweeney, Kam Chiu Tam, Cellulose-based materials in wastewater treatment of petroleum industry, Green Energy & Environment, 2020, 5(1): 37-49, ISSN 2468-0257.

57. Köse, K., Mavlan, M. & Youngblood, J.P. Applications and impact of nanocellulose based adsorbents. Cellulose 27: 2967–2990 (2020). https://doi.org/10.1007/s10570-020-03011-1.

58. Daniele C. da Silva Alves, Bronach Healy, Luiz A. de Almeida Pinto, Tito R. Sant'Anna Cadaval, Jr. and Carmel B. Breslin, Recent developments in chitosan-based adsorbents for the removal of pollutants from aqueous environments, Molecules 2021, 26, 594.

59. Ahmed M. Omer, Rana Dey, Abdelazeem S. Eltaweil, Eman M. Abd El-Monaem, Zyta M. Ziora, Insights into recent advances of chitosan-based adsorbents for sustainable removal of heavy metals and anions, Arabian Journal of Chemistry, Volume 15, Issue 2, 2022, 103543, ISSN 1878-5352.

60. John Kwame Bediako, Ji-Hye Kang, Yeoung-Sang Yun, and Soo-Hyung Choi, Facile processing of polyelectrolyte complexes for immobilization of heavy metal ions in wastewater, ACS Applied Polymer Materials 2022, 4 (4): 2346-2354.

CHAPTER

6

Advances in Ultra-porous Filters for Water Treatment

— ABSTRACT —

Conventional water purification methods including physical, chemical, and physico-chemical techniques, adsorption materials, such as grit filtration, membrane-based processes, oxidation processes, Fenton/Photo-Fenton reactions, advanced oxidation processes, ion-exchange techniques, respectively have been widely used with varying levels of success. However, all these conventional methods are either chemical-intensive, use harsh chemicals, and membranes with tight pore architecture require high applied pressure to effectively operate in water treatment applications. In order to overcome some of the limitations associated with conventional treatment techniques, highly porous, energy-less water filters are being innovated with numerous surface-active groups performing their pollutant retention characteristics. In this direction, several highly porous filters such as aerogels, electrospun nanofiber membrane, and sponge-like filters have been developed. The aerogels, electrospun nanofibrous membranes (ENMs), and sponge-like ultraporous filters prepared through cutting-edge approaches have shown substantial-high flux with a high pollutant rejection rate at less-to-no applied pressure which makes them an economical alternative for water and wastewater treatment applications.

6.1 Introduction

Conventional water purification methods including physical, chemical, and physico-chemical techniques adsorption materials, such as grit filtration, membrane-based processes, oxidation processes, Fenton/Photo-Fenton reactions, advanced oxidation processes, ion-exchange techniques, respectively have been widely used with varying levels of success. However, all these conventional methods are either chemical-intensive, use harsh chemicals, and membranes with tight pore architecture require high applied pressure to effectively operate in water treatment applications. In order to overcome some of the limitations associated with conventional treatment techniques, highly porous, energy-less water filters are being innovated with numerous surface-active groups performing their pollutant retention characteristics. In this direction, several highly porous filters such as aerogels, electrospun nanofiber membrane, and sponge-like filters have been developed. The aerogels, electrospun nanofibrous membranes (ENMs), and sponge-like Ultra-porous filters prepared through cutting-edge approaches have shown substantial-high flux with a high pollutant rejection rate at less-to-no applied pressure which makes them an economical alternative for water and wastewater treatment applications.

6.2 Critical View of Porous Filters in Water Treatment

Currently, many conventional, as well as advanced water technologies such as adsorption, ion-exchange, chemical precipitation, electrochemistry, membrane filtration, phytoremediation, and soon, have been widely employed for water purification as well as wastewater treatment applications. Most of these techniques except nanoporous membrane processes have shown several advantages for operating with simple operation procedures with relatively moderate operating cost due to their moderate energy consumption at room temperature, and have attracted attention for water purification. Even though membrane-based processes like reverse osmosis and nanofiltration which use tight nanoporous membranes have shown remarkable pollutant rejection efficiencies, however, these processes consume relatively high energy due to their high operating pressure conditions. Even though conventional microporous membranes with a pore size ranging from 0.1 to 5 μm have been widely used for separating pollutants or contaminants ranging from 1 to 10 μm, which limits their applications in water treatment applications. However, nanoporous filters or membranes with a pore size less than 0.1 μm showed high pollutant rejection performance in contaminated water treatment. Nevertheless, nanoporous membranes or filters have been effective for a large range of contaminants, even those pollutants ranging nanometer-size such as salts, metallic ions, organic molecules, and microbes from wastewater sources. [1-3]

Design and development of high-performance nanoporous membranes have been widely studied for water purification and wastewater treatment applications. Nanoporous membranes prepared using inorganic, polymeric, and hybrid polymer/inorganic materials precursors have been successfully utilized for various applications. For instance, seawater desalination is achieved through nanofiltration as well as reverse osmosis membrane having a pore size less than 5 nm. Among many methods discussed in the previous chapter, the use of nanomaterials in membrane matrix to tune both surfaces and pore properties has yielded encouraging results. Further, the use of nanoporous single-layer graphene, sub-10 nm diameter nanofibers, metal oxides, and other 2D nanomaterials have exhibited fast water permeation of water and relatively good rejection patterns for various pollutants in water and wastewater sources. [4-6]

Several studies have also evaluated the deeper relationship between the membrane structure, properties, and separation performances with the help of various mathematical models, and large data assessment. Major disadvantages associated with dense, relatively tight, or nano-porous membranes mainly used for nanofiltration, or molecular separation in the gas phase require high operating pressures for separation purposes. On the other hand, a largely dense skinned reverse osmosis membrane requires high pressure to drive the separation and electrical potential or concentration gradient to facilitate water transportation. The polymeric material in the sub-nano pore size range or relatively dense membrane skin architecture affects the permeability and selectivity. It is highly dependent on the applied pressure to induce solvent transportation in the membrane. In nanoporous membranes of filters, the separation process occurs through differences in solubility and diffusivity as well as charge-associated mechanisms. Another major disadvantage with dense or tight nanoporous membranes is a relatively lower product flux which directly affects the overall process economics. [7-9]

On the other hand, highly porous membranes or ultra-porous filters are made of a solid matrix with defined porous channels or pores which have diameters ranging from more than 0.1 to 20 μm. Even though the separation of solutes in porous membranes or filters is known to occur as a function of molecular size exclusion or through preferential rejection mechanisms. In ultra-porous filters or membranes, the degree of solute selectivity is largely based on the membrane porosity. Highly porous polymer membranes are mainly applied in the microfiltration and ultrafiltration processes. Conventionally, the microfiltration membrane with a pore regime ranging from 0.1 to 10 μm, on the other hand, pore sizes ranging from 0.001 to 0.1 μm is categorized as ultrafiltration membranes.

Recent years have seen more wastewater discharges into freshwater sources than ever before. The increasing threat of emerging pollutants, pathogenic contamination, and microplastic related accidents polluting freshwater sources and oceans daily cause severe complications to the ecosystems and environment at large. Large-scale discharges of highly contaminated wastewater are proving catastrophic to native freshwater aquatic as well as marine wildlife. For instance, modern pollutants like microplastics and oil spills that are constantly dissipating or scattering into the marine environment over a period of time have been proven deadly to the environment as well as marine life such as plants, mammals, reptiles, fish, birds, invertebrates, and algae. On the other hand, an increasing number of terrestrial activities such as the production of industrial oil contaminated wastewater discharge from oil refineries, domestic drains, automotive industrial release, shipping travel, and dumping pose a severe impact on the day-to-day life of freshwater reservoirs and groundwater sources as well. Therefore, continuing horizon oil spill related accidents and wastewater entries to the environment prompt robust, simple, and cleaner approaches. Another major problem associated with modern pollutants like microplastics and oil contaminations turns into highly stable and hazardous toxins forming stable suspension and thermodynamically stable emulsions pose serious threats and challenges to meet the desired specification for treating and producing clean water or vice versa. In such cases, many approaches have been used with limited success in the past. For example, the use of conventional deemulsifier additives for oil contaminated wastewater to attain high-quality recovery of water increases operational as well as treatment costs in addition to the added cost of post-recovery steps. [10-15]

On the other hand, in addition to conventional treatment methods including advanced membrane processes, researchers have been constantly searching for a new class of materials and techniques to address the issues associated with modern pollutants that are frightening our environment. For instance, unique nanoporous PTFE-based filters were used for the separation of two or more immiscible liquids based on the segmented flow micro-chemistry approaches. In this study, selective wettability inherently resulted in the successful separation of organic or fluorous phase from aqueous-liquid mixtures. Among many advanced approaches, membrane-based separation processes like ultrafiltration (UF) and microfiltration (MF) have been used for the separation and treatment of modern pollutants from emulsified oil/water mixtures as well as plastic effluents. In particular, MF/UF ceramic membranes have been designed and tailor-made to address separation issues associated with both complex emulsions and immiscible effluents. Among many materials, $Cu(OH)_2$ nanowire-induced filters resulted in as high as ~100000 $L.m^{-2}.h^{-1}$. On the other hand, ultrafast separation of stable emulsions, which is a difficult task as emulsions are thermodynamically stable in nature, was achieved using carbon nanotube porous (CNT)-based membranes. In this approach, porous systems showed >96% separation efficiency for oily components from the emulsion. In another approach, a simple dip-coating technique was followed to prepare super hydrophilic polyester fabric with the help of organosilicon surface modification in combination with crosslinked poly(ethylene glycol) diacrylate (x-PEGDA). As prepared hydrophilic flexible fabrics were used for the separation of contaminants both oil/water and water/oil type emulsion separation successfully.

On the other hand, electrospun polyacrylonitrile (PAN)-based fibers packed in ultra-porous filter forms have been widely used for separation and purification applications. The electrospinning technique produces fibers having diameters ranging from a few nanometers to hundreds of micrometers in diameter and several centimeters in length which can be arranged in highly ordered alignment or randomly in general. Based on the nature and type of polymer precursor electrospinning conditions can be altered to produce suitable fiber end products. Nevertheless, highly uniform nano/micro size and long-length distributed fibers are being converted into useful fibrous filter mats. The use of various electrospinning precursors in particular relatively high conductive polymer dope solutions yields highly ordered and aligned fibers that can be directly used as a packed filter for various applications. Similarly, polymer-based fibers mat can be prepared from a series of other commercially abundant synthetic polymers like polyvinylidene fluoride (PVDF), polystyrene (PS), polylactic acid (PLA), and so on and their composite versions. These fibers are then converted into filters having pore structures in the range of UF, MF, and even larger pore size range microporous (>1 μm) regime to produce reasonably high product flux.

Nonetheless, most of the electrospun fiber-based membranes or filters suffer from poor mechanical properties and being assembled material, the integrity of assembled electrospun fibers tends to be lost under applied external stimuli like pressure. Therefore, to overcome such mechanical instabilities in various ultra-porous filters, recent studies have developed ceramic coated mesh filters with relatively high mechanical strength for modern wastewater treatment. These composite ceramic filters have shown a remarkable ability to concentrate and treat contaminants like oil and microplastics from contaminated wastewater. Relatively high mechanical stability in composite ceramic filters or ceramic coated electrospun fibers has shown greater potential treatment media due to their cost-effective alternative compared to traditional treatment techniques. [16-18]

Among recent advances, polymer-based aerogels or foam membranes have been successfully tried and tested as ultra-porous filters for various water treatment applications. Ultra-porous polymer foam or aerogel systems have been mainly prepared using abundant biopolymer sources which are inherently anti-fouling, hydrophilic, and can be tailor-made to attain various forms and shapes. Many of the biopolymers have been characterized to have super-hydrophilic nature with contact angle $\theta_{water} \approx 0°$ which makes them highly selective filters for water when applied for wastewater treatment applications. For example, bio-based polycocolloid biopolymers like chitosan, sodium alginate, agarose, agar, gelatin, and so on have been successfully used for the preparation of highly porous filters in various forms. In many cases, filter precursors and other supplementary constituents have been extracted from bio-origin via easy to prepare economical extraction techniques. For instance, biopolymers require chemical crosslinking agents to stabilize their structure making them mechanically stable filters. For this purpose, several bio-origin crosslinkers have also been utilized such as genipin which is a non-toxic fruit extract maintaining all required characteristics. Another major advantage of using ultra-porous filters is that foam or aerogel-type filters require no applied pressure to obtain product water. On the other hand, the super hydrophilic nature of biopolymeric systems accelerates the selective adsorptive transportation of water from wastewater sources. In recent times, ultra-porous membranes or filters or foams tested under gravity force have shown high rejection for modern pollutants like oil from contaminated water. [19-21]

6.3. Overview of Commercial Porous Filters in Water Treatment

The increasing industrial activities have caused the uncontrolled and unregulated release of particulate matter pollutants that poses a serious concern to the global environment, ecosystem, fresh sources, and public health. As discussed in the previous section different polymeric membranes, bio-based ultra-porous filters, electrospun fibers mats, and protein-based nanofibers as renewable and sustainable filters have shown superior performance for both modern pollutants as well as particulate matters in various water samples.

Among many large-size pollutants, microbiological fluids and colloidal contaminants have proven difficult to handle and treat at low operating costs. Nevertheless, among many classes of materials tried and tested, bio-based membranes, ultra-porous membranes, or filters have shown encouraging results overcoming several drawbacks observed in conventional synthetic polymeric materials. These ultra-porous filters offer higher product permeate rates, and efficient rejection, Additionally, they offer better production rates and more efficient adsorption, and can be tailored to attain functional surfaces with the use of nanomaterial coating. Several biopolymers including chitosan, poly(vinyl alcohol) (PVA), polylactic acid (PLA), cellulose acetate (CA), and so on have been stabilized and modified following effective methods like blending, grafting, and using specially designed copolymer architectures in order to induce mechanical stability and enhance the filtration performance. [22-24]

After careful evaluation of various techniques, it is estimated that the fabrication of ultra-porous fibre membranes prepared by using electrospinning techniques is the simplest as well as highly cost-effective. Conventionally, electrospun fibrous filters with highly porous non-woven architecture have been successfully used in MF, UF, and even NF applications. However, due to their non-woven nature, fibrous filters usually exhibit fragile mechanical strength due to weak fibre/fibre packing only through physical entanglements. Nevertheless, synthetic polymer-based fibrous filters are vulnerable to bio-fouling and/or organic fouling related issues which require further surface modification. This is achieved via several approaches such as the use of nanomaterial incorporation in dope solution as well as post-spinning surface manipulation chemistry. Further, improved architecture has been adopted to improve mechanical stability in fibrous filters in which multi-layered electrospun mats have been combined to facilitate the combination of electrospun nanofibers originating from biopolymers as well as nanoparticles components. [25]

In recent advances, multi-layered fibrous filters are usually prepared with the composition of various layers in which each layer is separately fabricated to induce specific pore size and their tailor-made architecture. Initially, these loose ultra-porous fibrous filters have been used as pre-filter for clarification of highly contaminated water and wastewater. Advanced approaches also innovatively incorporate nanomaterials in the middle or top layer to induce efficient adsorptive rejection during filtration for pollutants such as viruses, bacteria, dyes, heavy metal ions, organic toxins, and so on. Nevertheless, traditionally various tightly packed highly porous filters have been used as both pre-filter and main filters for biological as well as colloidal particle retention in various contaminated sources. These include wool, polypropylene, carbon fibres, resin bound, granular carbon cartridge, string wound and so on as shown in Figure 6.1. [26-32]

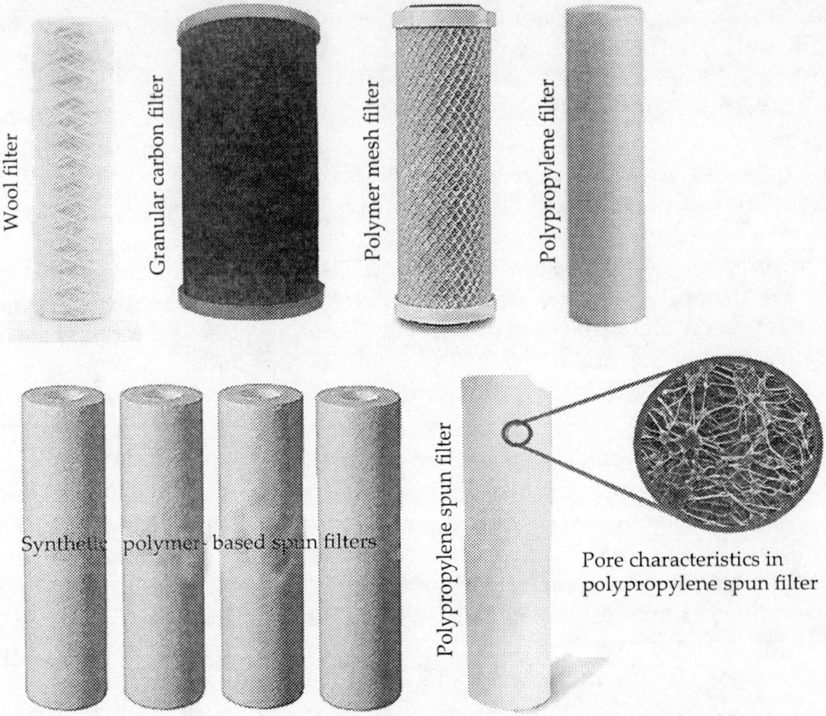

Figure 6.1. Commercially available pre-filters are prepared using different source materials and different architectures of highly porous materials used as ultra-porous pre-filters for separation and purification applications.

6.4. Advances in Aerogel for Water Treatment

Even though the first aerogel was prepared using silica as a translucent, synthetic extremely porous solid-state substance with ultra-low density and excellent thermal insulating properties. Silica-based aerogel with extremely low density and high porosity is also known as "frozen smoke." Even though silica-aerogel in its solid form has a morphology and texture like that of polymer-based foams like styrene. Recent studies have focused more on developing different precursors for various applications. For instance, synthetic polymers, biopolymers, carbonaceous precursors, iron oxide, semiconductor nanostructures, metal precursors like copper and so on have been adopted for aerogel preparation. However, aerogel within their structure carry very little solid material (~2-5%) and remaining more than 95% of the structure is nothing but air.

Among many modern pollutants, oil-in-water or water-in-oil wastewater and their emulsion form which are known to be thermodynamically stable liquid-liquid systems that pose a serious threat to the environment in the absence of appropriate treatment technique. The oil/water emulsions are typically classified into three categories depending on their characteristics and stability, namely, tight, medium loose, emulsions. While, loose and medium emulsion systems can be easily separated following phase separation technique. However, a tight or stable emulsion known to cause serious concern for the ecosystem and appropriate time with a proper de-emulsification method to break the emulsion. Recent times have seen several oil spills and petroleum transportation related accidents causing immediate and long-term catastrophy for aquatic ecosystems. On the other hand, large quantities of industrial emulsion contaminated wastewaters are discharged to freshwater bodies. These unregulated and unmonitored discharges pose greater concerns to aquatic ecosystems. High concentration of oily content in discharged water causes high chemical oxygen demand (COD) and biological oxygen demand (BOD) characteristics. Conventional chemical-based demulsifiers, so far, a popular choice, have been used to separate oil-water emulsions. However, these methods yield limited success due to several advanced attempts made to selectively separate oil from water and water from oil.

With ever increasing accidents and discharges containing wastewater from various industrial sources like petroleum, textile, tanning processing unit, steel, food, and other industries, contaminate surface and freshwater sources. On the other hand, frequent oil spill accidents of large commercial transportation vessels such as the Gulf of Mexico oil spill led to leaking of millions of tons of crude oil into the oceans which is not only a huge threat to the marine ecosystem but also a waste of valuable resources. In addition to oil contaminatios, oil spills also discharge several chemical components such as polycyclic aromatic hydrocarbons that persist in the water bodies for a long period of time and cause serious chronic negative impact on aquatic species such as molluscs and plants. This over the period leads to the thermal insulation affecting birds and mammals in the neighbouring ecosystem as well signs of which may only appear months or years later. Therefore, finding innovative and advanced materials and techniques that are quick and efficient in separate oil/water mixtures have received greater attention.

Nonetheless, direct filtration and adsorption are some of the most used techniques for oil/water separation and have been widely used in industries. Also, highly porous materials with specific affinity and wettability are commonly adopted as preliminary materials in case of smaller samples. On the other hand, larger sample volumes require continuous and sustainable materials and techniques to make solutions realistic in nature. In this direction, aerogel has shown greater potential. These aerogels can be tuned to attain specific hydrophilicity/hydrophobicity, porosity, mechanical strength, and surface characteristics to suit the nature and type of waste resource. In one such attempt, Li et al. prepared a polyurethane (PU)-based aerogel covered with graphene oxide in their bulk as well as surface structures. This composite aerogel has shown an absorption capacity for an organic component in the mixture ranging from 80 to 160 g g^{-1}, upon adsorption that can be recollected/recycled via simple mechanical compression to remove the absorbed organic constituents. [33-35]

6.5. Recent Developments in Biomaterial Aerogels

On the other hand, advanced membrane-based processes have been tested under different conditions for the removal of oil contamination from wastewater or contaminated water. Therefore, recently emerging advanced materials in different forms such as foam membranes, aerogels, surface-modified fabrics, polysaccharide agents, and inorganic meshes are successfully used for the separation of oil from water mixtures. Among them, a unique 3D network of highly porous hydrophilic aerogels preferably used for the selective separation of water over oil, similarly hydrophobic aerogels can be used to attract oil preferably over water components from contaminated mixture. The large surface area, and high porosity and easily tailor-made surface characteristics make aerogel an excellent choice for modern pollutant removal. In addition to this, bio-based aerogels having distinctive features like superhydrophilicity, biocompatibility, and sustainability aspects make them a better choice for oil/water emulsion separation. [36-38]

In recent time, highly porous foam membranes (FMs) that are like aerogels were prepared using biopolymers like agarose (Agr) and gelatin (Gel) in hybrid combination with a non-toxic crosslinking agent derived from fruit extract, genipin (G). These FMs were further stabilized with specific optimal crosslinking and gel strength and applied for oil-water separation as shown in Figure 6.2. In this study, ultra-porous aerogels prepared using agarose/chitosan colloidal type systems possess unique combinations of features that help in the robust selective separation of water from stable emulsion contaminants. Similarly, polysaccharide-based agarose and gelatine were used to prepare highly porous foam membranes. In this study also greener and sustainable sources, biopolymers like gelatine as a minor constituent, result in hydrophilic ultra-porous foam membrane. These ultra-porous filters are increasingly proving vital for sustainable and large-scale applications. Therefore, highly crosslinked major components like chitosan or sodium alginate act as support structure and minor components like gelatine or agarose induce superhydrophilicity which is significant for faster water separation. The prepared membrane was tested for selective water components from biodiesel/water emulsion containing crude vegetable oil/water emulsion, toluene/water and highly contaminated oil-spill wastewater. In all these cases, keeping sustainability in mind, all ultra-porous foam membranes were applied in crossflow filtration mode. Interestingly, all foam membranes in crossflow mode showed high purity for permeate water >99 % with remarkably high-water flux (600 $L.m^{-2}.h^{-1}$). [39]

Further, inspired by nature and the lotus leaf structure, researchers (Nguyen et al.) prepared combining engineered micro/nanostructure in the aerogel with the help of hydrophobic graphene nanomaterials. Inducing the microporous structure with the help of graphene nanosheets in aerogel resulted in rough surface characteristics in aerogel. These rough surfaces make aerogel more wettable for the organic part of the contaminants. Researchers noticed that the inclusion of graphene nanomaterials physicochemically

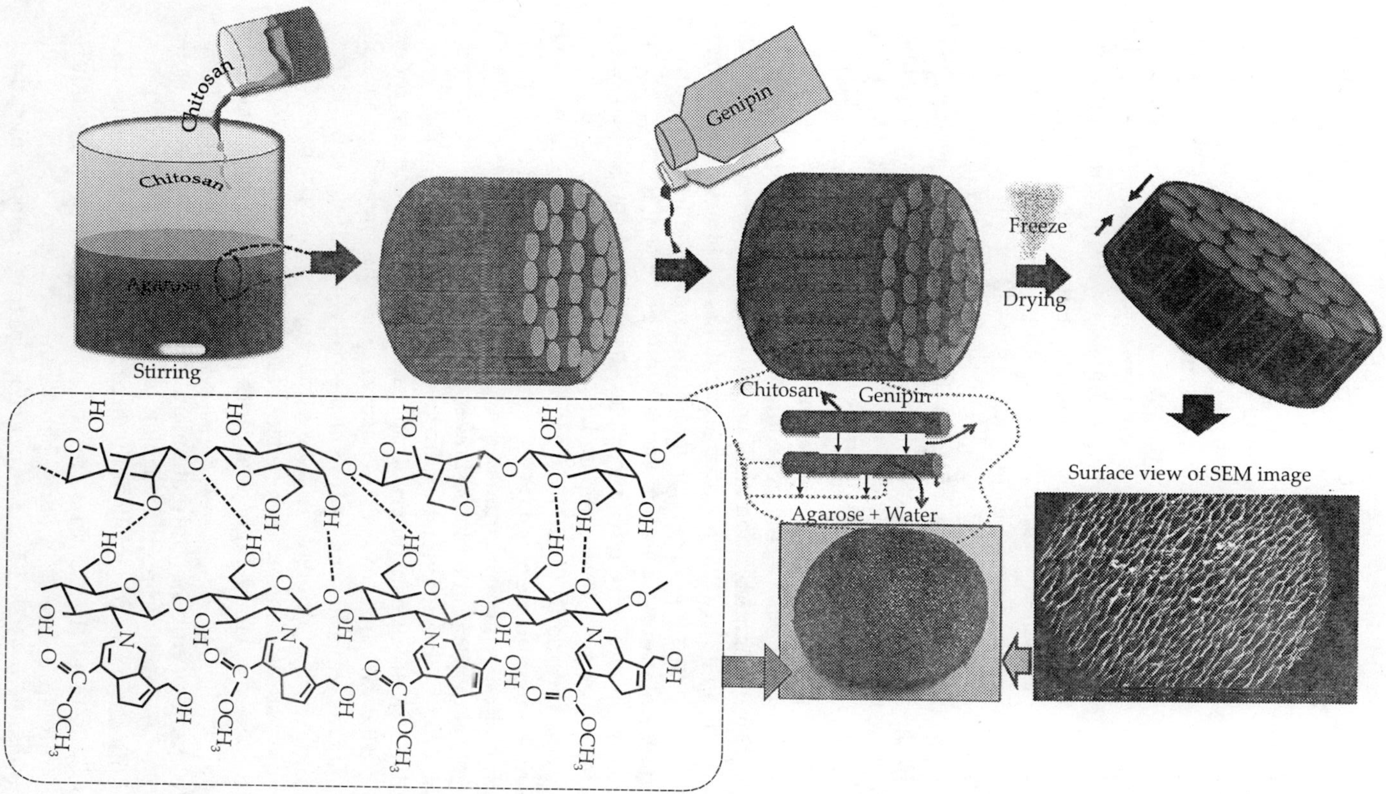

Figure 6.2. Polymer-based aerogel membranes prepared using agarose and chitosan as precursors have ultra-porous architecture useful for selective separation of modern pollutants.

turned aerogel characteristics from relatively super-hydrophilic water contact angle near 0° to super-hydrophobic contact angle 162°. However, organic or oil content separation through the aerogel pores, it is important to maintain super-lipophilicity in the aerogel. Therefore, use of graphene in aerogel successfully removes organic/oil parts maintaining super-lipophilic characteristics. This graphene/aerogel has shown excellent absorption capacity of 165 g g^{-1} absorption with high selectivity for oils and organic solvent. Further, these aerogels have shown good recyclability and reusability for next reuse. [40] Similarly, Liu et al. prepared a stable chitosan/cellulose (CS/CE) aerogel system with super-hydrophilic surface characteristics and underwater super-oleophobic greater than 150° via simple technique. As prepared CS/CE aerogels used for effective separation of oil from contaminated water mixtures and oil-water emulsions. Studies have also shown that CS/CE have great recyclability characteristics. Further, Chen et al. produced a super-hydrophobic titanium oxide (TiO_2) nanoparticles decorated cellulose aerogel. The Titania-decorated aerogel exhibits excellent super-hydrophobicity with contact angle over 170°, which selectively separates various oil-type contaminants from wastewater mixtures including toluene, chloroform, kerosene, and other contaminants. Moreover, the use of crosslinkable components in the blend system induces excellent mechanical and chemical stability to the overall porous filter. [41]

Meanwhile, in the oil from water separation process, the super-lipophilic separation media or filter can easily get contaminated or clogged because of the excessive adhesion or adsorption of oil or other organic layer particles due to their inherent lipophilicity that results in a rapid flux decline, suppression of separation efficiency thereby affecting overall membrane shelf-life. However, several cleaning solutions or agents have been proposed to subsequent surface regeneration and to remove these hard deposits of contaminants. In another study, researchers have prepared superhydrophobic polyurethane (PU) sponges as ultra-porous filters. These PU sponges further functionalized with perfluorinated graphene oxide (PGO) which was used for the separation of oil contamination from contaminated water mixtures. However, PGO functionalized PU ultra-porous filters showed gradual decrease in separation efficiency for crude oil contamination from water mixtures. Among many factors, relatively high viscosity of crude oil causes severe surface adsorption on the filter surface. In some cases, it is observed that the increase in asphalt content mainly causes enhanced crude oil viscosity, which, in turn, causes potential fouling during filtration. To assess this, another team of researchers (de Araujo et al.) studied in detail the effect of asphalt on viscosity of crude oil. This study reveals that the higher concentration of asphalt spontaneously emulsifies the crude oil-water mixtures to form many emulsion droplets. This leads to complex emulsion formation and causes difficulties in oil/water separation. Therefore, it is important to analyse the content and concentration of asphalt in oil/water mixture which influences filter media significantly. [42]

Dicyclopentadiene (DCPD) is one of the economical by-products and abundantly produced in the petroleum industry. Further, the polydicyclopentadiene (PDCPD) is produced via ring-opening metathesis polymerization (ROMP) of DCPD monomer. The PDCPD is largely used as a new class of engineering plastic having excellent mechanical properties and flexibility. On the other hand, recent advances have used PDCPD for the aerogel preparation that poses several advantages such as low density, high porosity, low cost, easy preparation, and excellent physicochemical properties, which makes them an ideal candidate for oil/water separation filter. Further, studies have used PDCPD-based aerogel filters prepared via supercritical drying process, and applied for separation of the simple oil/water mixture, oil-in-water emulsion, and asphalt containing toluene/water mixture. In addition to this, the study also extended to understand the effects of asphalt in crude oil contaminated mixtures and associated stabilizing factors on their separation were investigated. [43-45]

On the other hand, it is estimated that there are more than 10,000 different types of synthetic as well as natural dyes and pigments used for various industrial applications. These dyes amount to several hundred tons of synthetic dyes per year are produced worldwide. Major part of these dyes ends up in the environment due to their partial consumption or excessive use in the production units. This has been increasingly seen as a serious environmental concern as the large amount of dye contaminated wastewater are recently entering ecosystem without treatment. It also estimated that more than 10% of the dyes utilized for one or the other purpose worldwide are discharged directly into the environment in the form of wastewater due to intermediate losses or excessive use during the dye fixation process. In some cases, up to 20% of unutilized dyes are released unregulated as waste during textile processing. The disposal of these dyes without or partial treatment largely causes toxicity to the ecosystem and with time inhibits phot-synthetic processes causing higher COD and BOD thereby resulting in lower light penetration causing severe environmental problems worldwide.

Nevertheless, modern synthetic organic dyes used for various applications are generally characterized by their stable chemical, physical and physicochemical characteristics. Upon excessive use and release in the form of effluents rather difficult to handle and treat or remove. Because of their complex aromatic structures, which persist in aqueous environments for long times and also resist microbial degradation. Among many traditional and advanced techniques, use of adsorbents for decontamination of wastewater containing dyes is still considered an economical and technologically feasible. However, post- adsorption, separation of adsorbent is still a challenging task. In some cases, generation of several intermediate byproducts still causes considerable technological limitation. These intermittent issues have been highlighted in subsequent studies and corrective measures have been adopted to improve their efficiency in handling or removing different contaminants. This is also considered in conjunction with the existence of a variety of commercially available adsorbents that are easy to propagate with minimum generation of toxic byproducts.

On the other hand, to reduce the production of byproducts and minimize their subsequent toxic effects several researchers have shown increasing interest in developing bio-based materials or biopolymers. Biopolymers have been well researched and their production in industrial scale is also encouraging all stakeholders to adopt these materials as biodegradable, lower cost and tuneable materials stock for future. In this direction, cellulose and its derivatives have shown useful characteristics as an adsorbent due its ability to bind to different organic contaminants like dyes. However, biopolymers also intrinsically carry limited surface activities, therefore to improve their adsorption properties have been tuned with the help of similar polymeric components such as vegetable fibers that have shown improved adsorptive properties. Further, limitations associated with poor adhesion and interaction between polymer matrix and the fibers can be improved by chemical or physical modification techniques utilizing possible functionalities on fiber surface, which substantially improves the cellulose adsorption efficiency during the filtration applications.

These fibres have been surface modified using hydroxyl groups available on the backbone of fibre matrices through various chemical functionalization processes. Further, the process of maceration is carried out using the hydrogen peroxide and acetic acid, which leads to the oxidation of acetic acid via hydrogen peroxide resulting in the formation of peracetic acid. Peracetic acid is a highly selective agent in acidic medium capable of oxidizing the aromatic rings of lignin. These approaches lead to chemical modifications on the surface of fibers with isocyanates groups which can also increase their compatibility with the polymeric matrix. All these efforts resulted in a significant improvement in the performance of composites polymers and nanocomposites system filters during water treatment.

Nevertheless, among the polymeric composites the polyurethane-based systems have been distinguished set of materials due their diversity and ability to form interact with foreign polymer via chemical bonds such as hydroxyl groups present in cellulosic fibers. These efforts have greatly improved chemico-mechanical properties in advanced nanocomposite systems such as high tensile strength, elastomeric nanocomposites with strongly enhanced modulus, and high strain-to-failure. For instance, composites of cellulose and polyurethane have greatly improved tuning microcrystalline properties during the preparation of composite. In this particular study, authors attributed the improved properties in new polymeric systems to reinforcement effects due to micromechanical characteristics from stiff nanofibrils in a rubbery matrix. Also, entropic effects from polyurethane molecules interacting physically and chemically with cellulose nanofibril surfaces. Further, these composite systems have been used for adsorptive filtration of organic pollutants like malachite green. The results showed that dye composite filter showed comparatively better performance compared to polyurethane foams alone prepared using thiacalixarene instead of glycerol as the crosslinking agent.

Performance of polyurethane foam filters was further improved with the use of industrial wood residues as fillers in polyurethane composites. These efforts were also economically sustainable alternatives as the wood residues have been considered a value-added product to generate an extra income for the wood or timber industry. Use of these composite form filters has also shown improved performance for retaining pollutants from contaminated sources. [45-48]

6.6. Advances in Functional Aerogels for Water Treatment

Unregulated urbanization, rapid industrialization, and increasing mining deteriorate the environment surrounding their discharges and contamination. The major impact of contaminated water mixing is the freshwater resources due to excessive contamination carrying various modern pollutants such as pharmaceutical wastes, surfactants, dyes, and heavy metal ions, which are the main cause for concern. In such a situation, availability of freshwater for daily household consumption severely affects and forces artificial water scarcity. According to recent estimates, >650 million people globally lack access to clean and safe water. Higher levels of modern pollutants in fresh and surface water also induce water-borne diseases. Among many industries, tannery processing units, textile dyeing operations, paper and paint industries release over 50,000 tons of dyes to the aquatic ecosystem annually and cause considerable water pollution and waste of millions of tonnes of valuable resources. On the other hand, recent studies have recorded enormous amounts of persistent or emerging pollutants (EPs) in various water sources. Until recently, many commonly used substances were considered harmless, however advanced detection and analytical techniques have characterized several daily used substances such as pharmaceuticals, surfactants, cosmetics ingredients, and natural/synthetic hormones have increasingly posed threat to human health as well as environment. This continuous adulteration of freshwater resources thus threatens the integrity of the aquatic environment at large. In this situation, several provincial, national as well as international monitoring and regulatory authorities have commissioned a series of programmes to study and regulate the release and treatment of persistent contaminants in various water sources.

Removal of modern and persistent EPs using conventional detection as well as treatment methods have shown limited success. This is necessitating the development of simple and robust advanced techniques to handle the worsening situation. This challenging task is addressed via the development of smart functional materials that are tailor-made to suit the type of contaminant. In addition to the surface water contamination due to industrial activities, several countries are facing increasing concentrations of high concentration of pollutants such as fluoride, arsenic and other heavy metals which severely affect the quality of groundwater sources used for drinking purposes. In addition to human activities, natural processes are also enforcing the exacerbated level of contamination thereby leading to deadly diseases when used or consumed above their permissible level. According to the World Health Organization (WHO) maximum permissible concentration limit for arsenic and fluoride is 50 $\mu g.L^{-1}$ and 1.5 $mg.L^{-1}$, respectively. This situation is expected to worsen the situation in near future as the geographical depositions of arsenic and fluoride minerals are increasingly leaching into the groundwater resources.

More so, these geographical mineral deposits are largely distributed in many countries such as Indian earth crust, China, Argentina, Pakistan, Mexico, and Bangladesh. Also, the coexistence of arsenic and fluoride were also found to be several folds higher than the permissible concentration limits in groundwater. [49-52]

Thus, several efforts are being made to develop advanced and efficient treatment materials and technologies for the removal of emerging or persistent pollutants like pharmaceuticals, organic dyes, fluoride, and arsenic from contaminated sources. Nevertheless, some of the conventional techniques are also being modernized to address the issue associated with the global water crisis. In this direction, adsorption, advanced oxidation processes, ion exchange, precipitation coagulation, electrodialysis, membrane technologies and photo-thermal-assisted technologies like photo-Fenton reactions have been notified as potential techniques to further consider. Among these, simple physical methods like membrane-based technologies in their advanced outlook have shown as "state-of-the-art" methods for such applications. However, even advanced membranes have shown some limitations while treating modern persistent pollutants. Membrane processes also suffer from several other issues like high capital cost, requires large plant footprint, expensive setup, and inadequate rejection EPS of small and neutral nature. Traditionally, adsorption techniques have been the widely used and most reliable technology used for wastewater of low to moderate contamination. Adsorption technique is relatively simple and low-cost operation which produces high quality reusable water.

Traditionally, many carbonaceous materials such as activated carbon and sand beds have been used for this purpose. However, recent advances in nanomaterials and nanotechnology-assisted processes are yielding diverse material stock which can directly be used as the future of water disinfection media. These advanced and functional materials and their processes can be tailor-made to suit the unique end wastewater or contaminated water characteristics. Meanwhile nanotechnology promises to provide smaller, however more effective materials, devices, or processes. In this, immobilizing active nanomaterials or functionalities having greater affinity towards the pollutants can be enhanced, besides tailor-made distribution in various substrates or carriers and achieved simple regeneration. Also, tuning physicochemical properties, active nanomaterials can be easily recycled, which broadens their applicability toward real wastewater or contaminated water treatment. Interestingly, active functional nanomaterials can be prepared not only using inorganic precursors, but also by engineering the biomaterials to attract modern pollutants like dyes, fluoride, arsenic, pathogens and microplastics from water. Biopolymers are a green and ubiquitous source widely used as advanced water purification materials in futuristic technologies owing to their exceptional properties. In this direction, several studies have used a series of biopolymers including chitosan, sodium alginate, agarose, gelatin, cellulose and their derivatives as individual and hybrid systems for retaining modern pollutants from contaminated sources.

In one such attempt, chitosan-based aluminum (CS/Al), chitosan-iron (CS/Fe) and combination of CS/Al/Fe composites were explored as efficient and sustainable adsorbents for fluoride, arsenic, and other emerging pollutant removal from an aqueous medium. However, standalone composite nanomaterials have shown limitations to discrete water systems. Therefore, advanced approaches use adsorptive membranes or hybrid membranes as means for addressing water contamination related issues. Interestingly, this approach with a combination of nanomaterials decorated on an ultra-porous substrate acting as a high- flux membrane retains several pollutants in series and simultaneously like dyes, fluoride, arsenic, active pharmaceutical ingredients, artificial hormones and so on. [53]

However, high-cost of nanomaterial preparation processes, inflexibility for diversified pollutants, relatively low water permeability, nanomaterial leaching scare and nanotoxicity restrict their large-scale utility. Therefore, it is important to develop cost-effective procedures and protocol for robust and versatile water purification. With these expectations, several researchers have attempted to work on aerogel and their functionalized composites as an alternative system. Specific and targeted functionalization on various aerogels were initially hypothesized to yield improved affinity toward diverse pollutants. On the other hand, high porosity was utilized to get higher product flux. In this direction, several studies revealed that functionalized aerogels as advanced filters have exhibited superior adsorption capacity, high fluxes, and regeneration characteristics against a series of contaminants.

As it is done with pristine polymer aerogels, composite functionalized aerogels are usually prepared through gelation of precursors comprising three-dimensional (3D) hybrid network and subsequent freeze-drying of hydrogel. During the process of freeze-drying stable 3D network structure with highly porous macrostructures are retained via sublimation of solvent components in the form of ice. In advanced approaches, several extensive research efforts have been carried out to design novel composite like MOF induced aerogel recipes and substrate precursors utilizing combinations of polymer and nanomaterial filters. These advanced functional aerogels have been further utilized for various modern pollutant treatments such as emerging pollutants, heavy metals, artificial hormones, pharmaceutical active ingredients, oil-water separation and so on. More recently, these advanced separation media are being converted to technologically feasible filter media. [54, 55]

In this direction, synthetic polymer-based fibrous aerogel, polyvinylidene difluoride-based porous filters, and polyvinyl alcohol/graphene hybrid aerogel were employed for moderate wastewater like oil/water separation with remarkably high product water flux in a cross- continuous flow method. Similarly, biopolymer-based aerogel membranes, viz., agarose/gelatin, chitosan/agarose, alginate/graphene oxide, sodium alginate/cellulose nanofibril (CNF), and agarose-/bovine serum albumin-based hybrid aerogel membranes, have been reported as sustainable ultrafilter barrier for robust selective separation of water from thermodynamically stable oil/water emulsion. Whereas aerogels are characterized by exceptional porosity and

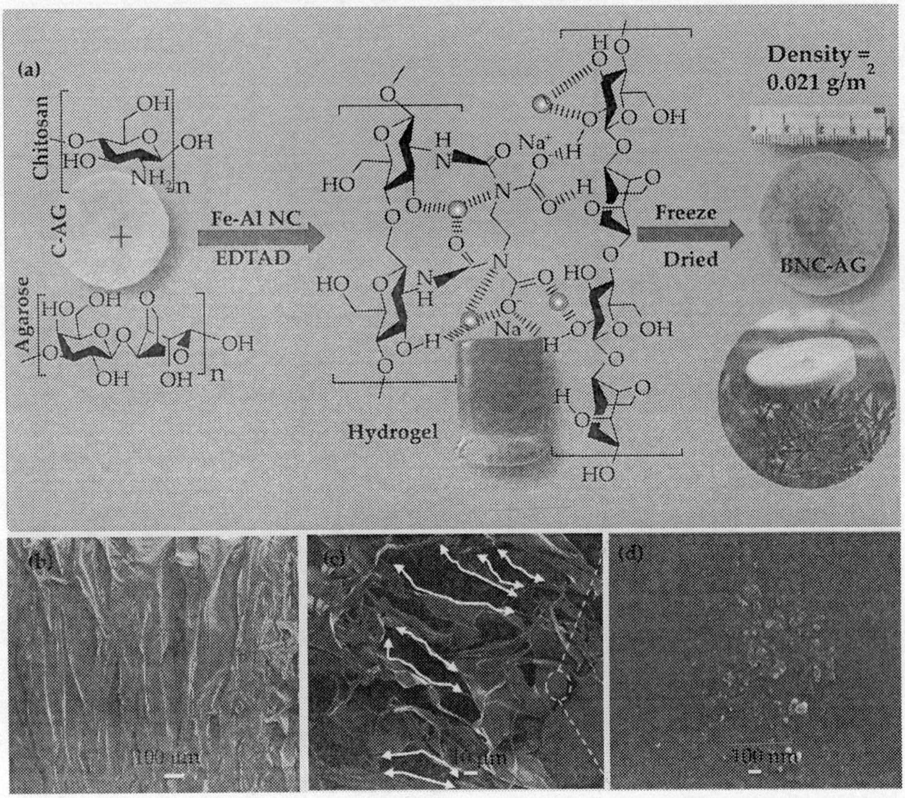

Figure 6.3. Advanced Al/Fe induced composite aerogels used for the treatment of emerging pollutants from various contaminated water sources and real water samples.

ultralow density, pristine polymer networks are relatively inert for any interactions. Therefore, to create active functional sites on the relatively inert polymer surfaces thereby improving interactive properties use of smart nanomaterials have been adopted for specific targeted applications. These efforts have also been mainly considered to induce a task-specific functionality on naturally abundant, non-toxic, and biodegradable ultra-porous 3D aerogel substrates. The design of functionality on the polymer substrate can be followed based on targeted application and nature of polymer and available active sites. In this direction, several nanomaterials with metal oxides, hydroxides, amine, carboxylic and oxyhydroxide end chemical nature have been induced or decorated on polymer backbone. For instance, a novel chitosan/agarose-based aerogel ultra-porous filters were decorated with γ-AlOOH and α-FeOOH nanomaterials as sustainable active sites which facilitate selective retention of modern pollutants with ultrafast water permeability as shown in Figure 6.3.

Similarly, several polymeric systems have also been adopted to surface modification such as chitosan/agarose, chitosan/sodium alginate, chitosan/agar, chitosan/cellulose and so on. Likewise, the main polysaccharide polymer agarose is largely used as a gelation agent which undergoes extensive intermolecular hydrogen bonding, resulting in double-helical type architecure that aggregate into thick bundles. For instance, chitosan in combination with agarose and assisted by ethyl-enediaminetetraacetate dianhydride (EDTAD) results in a stable superhydrophilic highly stable and ultra-porous aerogel. These hydrogels followed by aerogels have been stabilized to suit the various extreme stimuli such as pH and temperature. Even here chitosan is a major component and agarose is a minor but plays an important role in stabilizing hydrogel and deriving the final aerogel system. Further, the aerogel system was surface coated with silver nanoparticles to induce antimicrobial properties making it nanomaterials composite (BNC-Ag) having enhanced antibacterial properties to filter. On the other hand, both AlOOH and FeOOH-induced bionanocomposites aerogels have been applied for selective separation of fluoride and arsenic from the aqueous contaminated water. These state-of-the-art materials have also shown excellent adsorption capacity for other emerging pollutants like organic dyes, surfactant, pharmaceuticals, hormones along with arsenic (V) and fluoride with high fluxes and excellent adsorptive capacity. More importantly, the functional aerogels have shown large-scale applicable potential and excellent recyclability for real contaminated water. In most of the cases, functional aerogel filters have shown enormous potential as next-generation multifunctional water treatment systems to mitigate the global water scarcity.

6.7. Nanocomposite Aerogels

In view of the extensive use of modern-day chemicals and prevalence in various industries across the sector worldwide, there is a high proficiency of these chemicals leach into the water resources or infiltrate through wastewater discharges. On the other hand, oil contamination is increasingly creating a major concern to the environment. Therefore, high-efficiency modern chemical pollutants and oil filtration is essential to avoid these substances or oil wastage to reach the environment. Oil/water separation is a vital step in various industrial processes such as tertiary oil extraction, oil spill clean-ups, oil drilling in exhausted oil fields, and oil-bearing industrial wastewater separation and purification.

In this context, several researchers are engaged in developing effective and sustainable nano composite aerogel as an alternative separation and purification system for industrially important substances. Even though it is highly expected to develop ecofriendly and sustainable separation or absorption systems such as foam-type or mesh-type biopolymers with nanostructured active sites. Separation of complex modern pollutants like oil in water can largely be treated using nanostructures assemblies of sodium silicate and TiO_2 nanoparticles in two-dimensional (2D) architecture coated on stainless steel meshes. Likewise, cone structured arrays have also been prepared using polypyrrole functionalized assemblies of graphene oxide to selectively collect micron-sized oil micro-droplets from contaminated mixtures. In another study, 2-dimensional stainless steel meshes were modified with polyacrylamide hydrogels having specific affinity towards the component from the oil/water mixture.

However, limited volume capacity of conventional 2D meshes is due to the weight of larger mesh substrates. Furthermore, two-dimensional meshes are susceptible to mechanical damage which limits their robust filter efficiency for vapor oil than three-dimensional (3D) meshes. Indeed, large volumes are readily filtered through thicker 3D meshes without increasing the size of the substrate. More significantly, 3D meshes are reliable and do not compromise with the filtering ability as they possess well organized interconnected network structure which prevent collapse of compression due to repeated use. In addition to this, the larger surface which provides sufficient interface to reaction to take place on 3D meshes traps significantly larger droplets of oil vapor. Nevertheless, keeping the sustainability aspect in mind, a combination of many 2D and 3D filters is being adopted overcoming inherent drawbacks in modern pollutant separation.

On the other hand, graphene having superior thermal, electrical and mechanical characteristics to traditional inorganic materials can be of great interest. Graphenaceous materials have also been known to induce significantly better antibiotic properties which can inhibit the formation of bacteria during oil/water separation applications. These properties synergistically act to make particular ultra-porous filters for a widespread use for the selective adsorption of modern pollutants like volatile organic pollutants, and heavy metal ions. In this direction, 3D structured graphene-based nanocomposite ultra-filters have been prepared. In particular, graphene/Co_3O_4 and graphene/NiO with 3D graphene networks in the foam architecture have been utilized as macroporous systems in selective gas, water and oil separation applications. Further, several forms of graphene foams have also been prepared with significantly improved hydrophobicity through surface treatment methods such as oxygen plasma treatment or decoration of nanomaterials following self-assembled monolayer coating.

Interestingly, detailed surface characteristic analysis using field-emission scanning electron microscopy (FE-SEM) reveals a 3D structure of ultra-porous foam with pores size ranging from 150–200 μm diameter well suited for larger pollutant separation. Interestingly, a higher magnification SEM micrograph also revealed the graphene-coated area shows that the graphene nanosheets had been well organized with both laminated and rough surface pattern. Further, detailed surface characteristics also revealed that nanosheet assembled ultra-porous filters were omniphilic in nature which is due to the surface O_2-plasma treatment. Also, pristine hydrophobic graphene foam filters have been turned to omniphobic foam following self-assembling modification procedure using (heptadecafluoro-1,1,2,2-tetrahydrodecyl) trichlorosilane (HDF-S). With this, as prepared 3D structural graphene is hydrophobic; however' upon O_2-plasma treatment process generates defects on the surface of graphene backbone rendering omniphobic characteristic to the end product. Further, O_2-plasma treatment resulted in creating hydroxyl groups (–OH) on the graphene surface inducing omniphobic characteristic and these functioanal active groups along with trichlorosilane–$SiCl_3$ of HDF-S form covalent interaction from which a strong siloxane network could be introduced making it a suitable surface barrier for selective filtering of liquid/liquid or gas separation applications. [56]

On the other hand, emerging pollutants such as active pharmaceutical ingredients (APIs) like Dichlophenac, Ciprofloxacin and surfactant constituents present in various personal care products (PCPs) in high concentration generally leach into municipal or domestic wastewater sources. These categories of substance are dumped into freshwater resources without treatment daily. This is due to lack of efficient advanced treatment techniques and awareness at various levels. Also, most of the conventional techniques applied for treatment of such pollutants showed limited success. However, these substances upon consumption or entering the environmental life cycle are known to promote antibiotic resistance in various species which is also a major health hazard. For instance, excessive release of Ciprofloxacin into the environment may cause serious health hazards. Ciprofloxacin, being a third-generation fluorinated quinolone-based pharmaceutical ingredient, is a highly stable chemical and exhibits broad-spectrum antibacterial activity in humans when consumed. Recent studies show the high concentration of ciprofloxacin in hospital discharge water, and even effluent treatment plants that are connected to domestic wastewater discharge concentrations ranging from 31 mg/L in drug formulation facility wastewater discharge to 150 g/L in direct hospital wastewater discharges. On the other hand, a non-steroidal anti-inflammatory medicine (NSAID), diclofenac, generally used as a treatment drug for inflammation has been identified as one of the most commonly found pharmaceutical residue in freshwater bodies due to its widespread use and disposal. More worryingly, diclofenac has been determined in various wastewater treatment plants and effluent concentrations up to μg/L. Also, these APIs do not undergo degradation and most of the traditional treatment techniques have proven limited success, which necessitates the development of advanced hybrid treatment techniques such as advanced ozonation, advanced oxidation processes in combination with nanomaterials-based advanced adsorption methods. Although some combinations of techniques have been effective in treating modern pollutant issues temporarily, however, degradation or adsorption techniques of these processes generate secondary pollutants such as highly toxic degraded smaller molecules or oxidation by-products which pose further complexity to further water treatment steps.

To address these complexities, several advanced materials and methods like physical treatment process involving membrane technique, chemical method coagulation, flocculation, advanced oxidation and electrochemical oxidation, catalytic degradation, precipitation, and other technologies, photo-Fenton reactions, have been developed. However, increasing awareness in adopting greener and sustainable processes, end users are now demanding appropriate materials combined with simple processes for emerging or persistent pollutants treatment. Therefore, in this direction advanced ultra-porous adsorptive filtration techniques have shown promising results to be an efficient, cost effective, accurate, low-cost method for the removal of emerging pollutants in a crossflow or continuous flow configuration. This approach is advantageous over the batch-wise technique which avoids potential generation of secondary pollutants due to unexpected interaction between adsorbent and pollutant and intermediate compound production due to elevated wastewater or contaminated source conditions that are stored for a longer period.

Among many other adsorbent materials evaluated, bio-origin polymeric substrates offer sustainable alternative solutions due to their large hydrophilic pores, high surface area, abundance, facilitate easy surface functionalization like FeOOH and AlOOH coating at low-cost modification methods, long durability. They also offer chemical as well as mechanical robustness upon suitable crosslinking. Natural and sustainable materials also offer a major advantage in repeated recyclability and reusability. These highly porous systems selectively separate high quality water from contaminated mixture retaining all pollutants in their pore architecture. Also, these functionalized advanced filters do not produce secondary pollutants making them one of the cleanest separation and purification media. With minimum to no recovery loss, functional nanocomposite aerogel ultra-filters offer high quality remediated water product water. This makes use of natural or biopolymers usage as future water treatment techniques a significant innovation for the future. Nevertheless, even though biopolymer-based substrate meshes or filters have been widely used in water treatment applications, the use of hazardous biopolymer crosslinkers, extensive reusability concerns, or scale-up limitation which requires entirely a new infrastructure which is estimated to be relatively high-cost are the main disadvantages of the biopolymer-based systems. As a result, some researchers have demonstrated the use of chitosan and sodium alginate in combination with metal complex system (CS:SA-M) generally referred to as functional bio-meshes as effectively and sustainably scavengers for various emerging pollutants such as fluoride (F^-), chromium (Cr^{6+}), artificial hormones, dye molecules, drug molecules along with other emerging pollutants wastes such as surfactants from contaminated water. In one such effort chitosan/agarose-based aerogel surfaces were functionalized with AlOOH and FeOOH active sites for series of pollutant retention and recovery from the contaminated sources as shown in Figure 6.3. Further, recent studies have also extensively used different natural biopolymers originated from abundant polysaccharides sources that are low-cost and provide highly reactive sites for functionalization have been converted into modern pollutant capturing meshes. This is due to biogenic polymers having an ability to interact with a wide range of multivalent salt precursor present on polymeric backbone such as Zn^{2+} , Al^{3+}, Fe^{3+} La^{3+}, and Zr^{4+} to form stable potential pollutant capturing sites through their electrostatic interactions. [57, 58]

Similarly, efforts are also being made to utilize naturally abundant fibers as potential substrate in creating ultra-porous barriers. In this direction, natural cellulosic fibers, cotton fiber, macromolecular proteinaceous nanofibril materials like amyloid, ferritin, and silk nanofibrils are abundant in nature. These low-cost and sustainable fibers can also be obtained for the purpose via simple extraction processes and can be directly used as filtration mesh in separation and purification applications. Various proteinaceous nanofibrils, mechanically robust silk proteins have shown exceptional characteristics for applications like water filtration where chemical inertness as well as mechanical robustness is significantly desired. Nevertheless, silk-based proteneous nanomaterials have attracted significant attention and have experienced rapid growth in the materials science research sector due to their exceptional properties like high tensile strength, mechanical robustness and toughness even compared to man-made fibers derived from steel, Kevlar and so on.

Further, one-dimensional silk nanofibrils (SNFs)-based ultra-porous membranes having high nanomaterials loading efficiency, substantial mechanical robustness as well as flexibilities that can be prepared at low production costs have been functionalized as an effective barrier for a series of modern pollutants. Thus, SNFs-based functional materials have been fabricated with high performance active sites on their surface to mitigate the hazardous pollutants of impaired waters around the world. Initially, a group of researchers reported use of SNFs in the form of ultrathin membrane to separate organic dyes from colloidal nanoparticles with ultrahigh flux permeance. These ultra-porous membranes tested in a long run showed promising results in both high retention capacity as well as high fluxes with added advantages of biodegradability and greener alternative for modern problems. The affordability of these ultra-porous membranes have been tested at various levels and these point of use sustainable membranes are expected to provide an economical alternative for the production of safe and clean drinking water in developing countries that lack larger water treatment infrastructure.

Despite the extensive recent progress, protein-based ultra-thin filters generally suffer from relatively low adsorption, gradually declined flux due to large molecule fouling propensity such as pathogenic pollutants. These limitations restrict protein-based filters for universal applications mainly due to low active surface area and robust affinity for a variety of pollutants. Even though studies have adopted a multilayer filter preparation strategy, which relatively enhanced their mechanical robustness maintaining high ultrahigh flux but limited molecular loading, the selectivity toward particular pollutant separation and the lack of regeneration associated with organic/bio-pollutant fouled membranes or fiber filters remains challenged. Further studies also proved that the use of nanomaterials on their surface both in situ or post-nanomaterials coating techniques can be effectively used to control the pore size and their distribution in fiber membranes. Thus, incorporating nanomaterials following various methods and techniques have been adopted in order to develop nano functionalized silk filters for several potential applications. Therefore, it is necessary to combine synthetic or natural inorganic nanomaterials with natural fibers resulting in nanocomposite fibril-based membranes. Nonetheless, several previous attempts have shown the benefits of active nanomaterials in combination with fibrous protein such as metal chalcogenides, hydroxyapatites, graphene oxide, cellulose nanocrystal, activated carbon, and Kevlar, etc., resulting in exceptional surface properties. This prompted new researches on utilizing SNFs for potential surface functionalization to expand their scope and make a potential adsorptive remover of every kind of pollutant from contaminated water sources. However, these functionalization processes involve multi-step preparation approaches and have generally shown limited selectivity for molecular separation application.

On the other hand, advancement in nanotechnology protocols have led to the preparation of non-precious metal-based nanocomposites for functionalization of SNFs. Metal oxyhydroxides have been studied as surface-active adsorbents for a wide range of modern contaminants owing to their high surface area and active binding sites via various interactions. As discussed earlier, metal oxyhydroxide functionalized biomaterial-based aerogels were successfully tested as 'point-of-use' ultra-porous filters for the treatment of contaminated water. Nonetheless, these materials and approaches inherently exhibited several drawbacks such as low ion selective, low-permeability, weaker stability under extreme pH conditions and more importantly nanomaterials have shown susceptibility for leaching from the polymer substrate which is a major concern of nanomaterials and are unable to withstand acidic conditions. On the contrary, use of systematic nanocomposite fiber preparation strategy was expected to yield a highly efficient ultra-high flux filter through the integration of chemically and mechanically robust protein nanofibrils with highly active multi-functional sites exposed to contaminants.

Further, advanced studies have also revealed the extraordinary coordination ability of SNFs to metal ions through their nitrogen functionality can be significantly utilized to achieve high pollutant treatment efficiencies. In this regard, various metallo-proteins such as metallothioneins, ferritin, and so on have demonstrated that the metal ion can effectively bind or interact with various mechanisms to amino acids in protein fibrils. In one such attempt, silk-based protein fibers were used as support to zero-valent palladium metal nanocatalysts for asymmetric hydrogenation. These hybrid nano/catalysts showed exceptional inorganic/macromolecular interface to be effectively used for catalytic hydrogenation of oximes and oxazolones. Inspired by the unique features, metallo-protiens potentials have been further explored whether in SNFs have been surface modified using task-specific active nanomaterials. Keeping technological feasibility in mind, surfaces modified with AlOOH and FeOOH on SNFs surface were configured in heterogeneous multilayered architecture following a simple layer-by-layer assembly approach as shown in Figure 6.4. This unique multi-layered filter not only provided better porosity arrangements with significantly high permeate flux but also task-specific active sites captured series of emerging pollutants such as pharmaceutical ingredients, fluoride, heavy metal ions, waste, macromolecules, dyes, and oil-water emulsion with high adsorption capacity, which, in turn, remarkably enhanced the purification efficiency. Further, catalytic features of nanomaterials were utilized to regenerate the surface of modified SNFs. This is important to address the potential fouling issues that occur during the treatment of contaminants either originating from surface water, groundwater or wastewater sources. In this study, multi-layered filters with photo-Fenton active nanomaterial were further utilized to clean the surfaces exposed under controlled UV or visible light. Overall, the presence of emerging pollutant scavengers with task-specific adsorption affinity towards target molecules

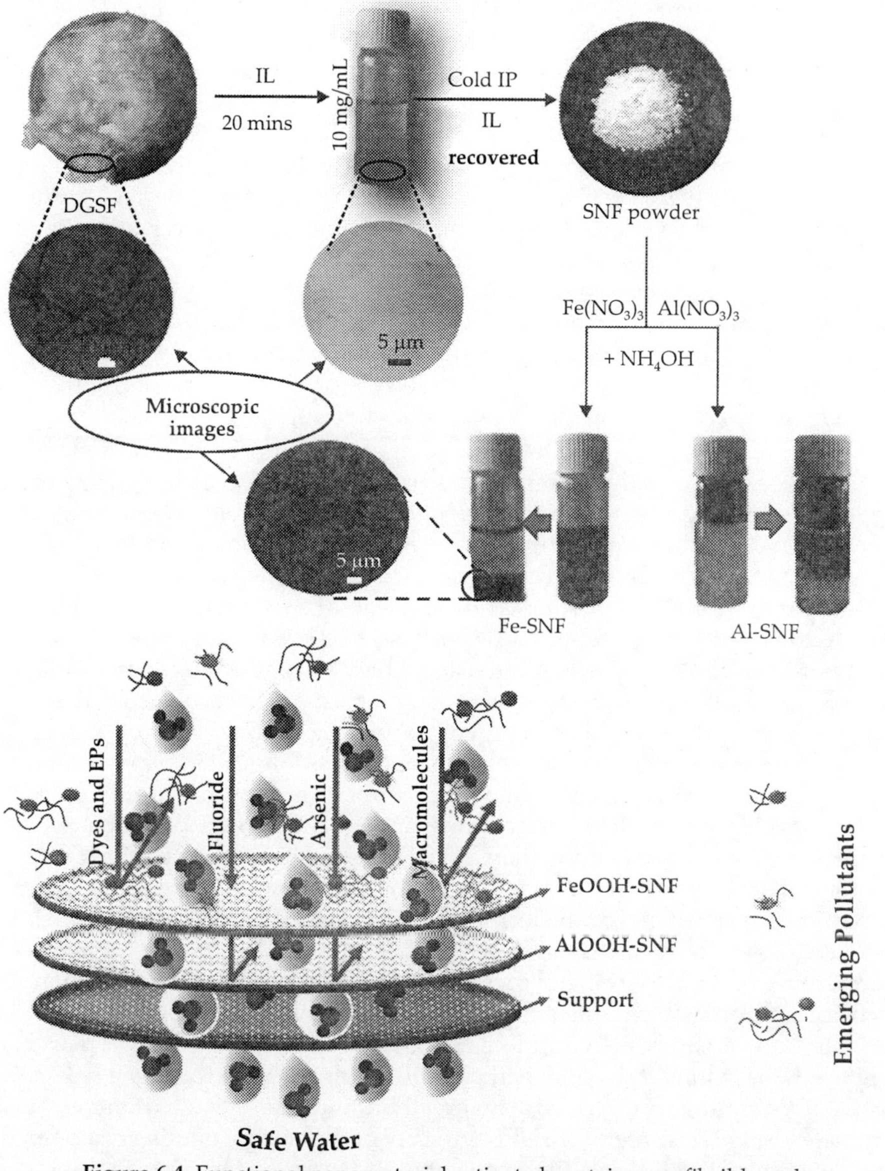

Figure 6.4. Functional nanomaterial activated protein nanofibril-based advanced ultra-porous filters used as universal water treatment applications.

and additional photo-catalytic activity makes these advanced ultra-porous filters a sustainable and green alternative. In addition to this, recent advances in nanotechnology protocols ideally recommend the water treatment process equipped with self-cleaning properties via photocatalytic oxidation degradation of pollutant post retention is the most effective and reliable method to treat fouled filters or membranes.

6.8. Advances in Electrospun Fiber Filters

Electrospinning technique offers unique polymer fibers arranged both in random and well-ordered architecture. Electrospun nanofibers have attracted substantial interest in several applications such as wound dressing, self-healing, encapsulation for drug delivery operation, tissue engineering and water filtration due to their simple facile preparation technique at cost-efficient procedures. Electrospun fiber mats also can be fabricated with tailor made spinning routes as well as can be easily incorporated the special feature depending on their end use. Electrospinning techniques offer versatile fiber preparation routes with fiber diameters ranging from 10 nanometers to several micrometers. Electrospinning can also be successfully achieved using different types of raw materials from a variety of polymers, inorganic precursors, and composites of both polymer/inorganic dope solutions. Moreover, electrospun fibers can be easily turned into water filtration membranes via simple random assembly or ordered assembly methods.

Several recent studies have revealed that electrospun fibers have several advantages to be an effective water filtration membrane due to their high degree of porosity which can provide up to 80% random or ordered voids, which include both interconnected as well as open pore structures. In addition to this, high specific surface area originated by nano-sized fibers and their loose packing can be used to retain a series of pollutants on the surface. The typical electrospinning setup involves a syringe pump which pushes precisely feed or dope solution either polymer or composite solution using a high voltage power source on to the collector, an electrically conductive spinneret, a grounded collector, all the dope polymer solution are electrospun on to rotating or stationary collector as shown in Figure 6.5. While the electrospinning process follows simple operation in generating uniform sized nano/microfibers through an electro-hydrodynamic mechanism. However, rheological properties, interaction and conductivity parameters of the dope solution play a vital role in determining the fiber size and their distribution. On the other hand, the combination of dope solution properties and process parameters make the overall process complicated. The electrospinning processing parameters such as polymer dope solution concentration, conductivity of dope solution, molecular weight of dope solution, viscosity, and process parameters such as pumping rate, the applied voltage, feed rate, and air gap which controls solvent evaporation characteristics play a significant role in tuning size and their distribution on the collector. Nonetheless, the electrospinning process is also greatly influenced by environmental parameters such as temperature and humidity of the electrospinning chamber which deterministically drives the spinnability and their end fiber characteristics.

Interestingly, most of the polymers can be in one or the other form and can be electrospun into fibers either as individual polymers or composites of polymers. These electrospun nanofibers can be assembled into either MF, or UF, or NF, with controlled pore size and their distribution for various potential applications. In one such earliest attempt, researchers have prepared randomly arranged electrospun nanofiber mats with membrane filter porosity at ca. 80 vol%. In this attempt, the pore size due to the random arrangement of the nanofibrous mats

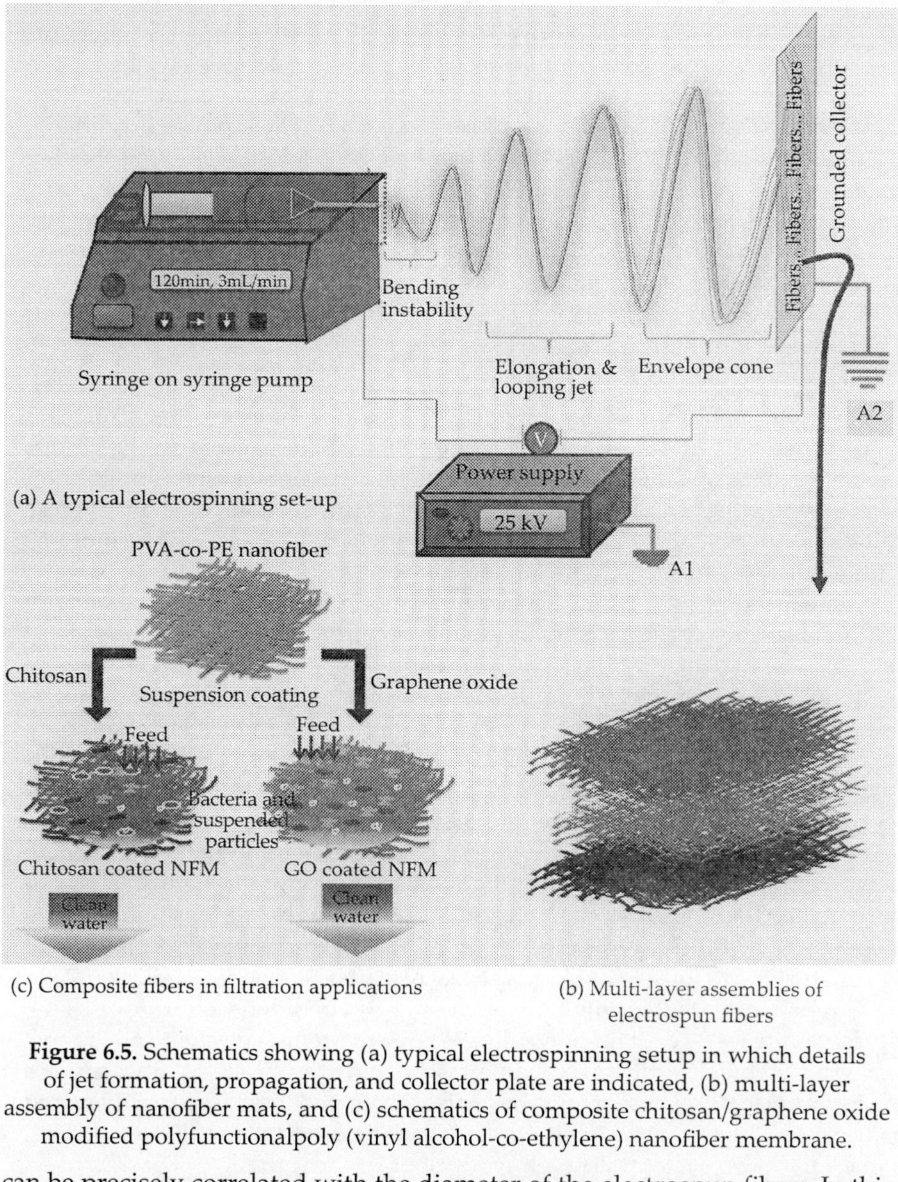

Figure 6.5. Schematics showing (a) typical electrospinning setup in which details of jet formation, propagation, and collector plate are indicated, (b) multi-layer assembly of nanofiber mats, and (c) schematics of composite chitosan/graphene oxide modified polyfunctionalpoly (vinyl alcohol-co-ethylene) nanofiber membrane.

can be precisely correlated with the diameter of the electrospun fibers. In this study, researchers have obtained the average pore size of the nanofibrous filters to be ca. 3±1-fold the average nanofiber diameter. Furthermore, the maximum pore size of the randomly arranged nanofibrous filter was noted to be ca. 10±2 times the nanofiber average diameter. Therefore, simple modification methods can also yield advanced classes of electrospun nanofibers with improved surface characteristics and separation properties following the process of polymer blending, composite formation, or functionalization processes.

In addition to this, electrospun nanofiber mats can be post-surface modified to obtain specific ligand or molecule immobilization onto the electrospun fiber assemblies. These post-surface modification approaches have induced fiber ultra-porous filters a specific affinity to selectively capture targeted pollutants from various contaminated water or wastewater feeds. Interestingly, these post-immobilization processes make fiber filter surfaces attain specific separation characteristics to purify the pollutant molecules based on their physical, chemical, and physicomechanical properties. Nonetheless, modern pollutants having specific biological functionalities rather than molecular size can also be easily retained or filtered out using nanofiber assemblies. For such applications, several proteins, ligands, and enzymes have been utilized as immobilizing agents on the range of biodegradable natural polymers such as polyglycolic acid (PGA), polyvinyl alcohol (PVA), and polylactic acid (PLA) depending on their end applications and requirement.

Recent advances have led to the innovations in preparations of layered electrospun nanofibrous filters in which tailored ultra-porous filter structures can be tuned to suit the end applications. Nevertheless, electrospun fibrous filters have shown some limitations such as low mechanical stability, inferior chemical stability, and thermal instabilities. However, to overcome these limitations, researchers have adopted nanomaterials-based dope solutions to tune both the surface characteristics as well as pore properties. Largely, polymer nanocomposite-based electrospun nanofibers have been prepared with additional surface characteristics as the use of different nanomaterials such as metal oxides, hydroxides, oxyhydroxides and biological active ingredients in the spinning solution yield tailor-made fiber mats. However, the presence of a high volume of reinforced fibers creates complexities in the electrospinning process of dope solution which significantly decreases the properties of the end-prepared nanofibers. Therefore, multi-layered electrospun filters (Figure 6.5(b)) were further prepared as a potential alternative for the simple combination of electrospun nanofibers with bio-origin nanoparticles and nano-whiskers to overcome the above-mentioned limitations such as low mechanical strength and enhanced anti-fouling characteristics. Interestingly, all these fibers were electrospun in multi-layer format to enhance the mechanical properties of fiber mats. On the other hand, advanced research has widely adopted non-biodegradable synthetic polymers such as polyvinylidene fluoride (PVDF), polyacrylonitrile (PAN), and polysulfone (PSU) to prepare mechanically robust and highly stable fiber filters. Nonetheless, keeping the sustainability aspect as a major consideration, researchers have also developed bio-based multi-layered electrospun filters as sustainable alternatives to be used as ultrafiltration as well as microfiltration systems. Therefore, most of the advanced approaches use a combination of synthetic and bio-polymers for effective spinning of multi-layered electrospun filter mats such as poly(ethylene terephthalate) (PET) and PAN, and bio-based polymers like chitosan, in particular chitosan/graphene oxide modified functional poly(vinyl alcohol-*co*-ethylene) as shown in Figure 6.5(c).

Therefore, the electrospinning process offers a simple, flexible, and rapid route to fabricate nanofibrous scaffolds or mats or filters packed in various pore size ranges. These ultra-porous packings of fibers upon appropriate assembly and packing can result in multilayered fibrous filters usually composed of various active nanomaterials. The design of specific pore size and surface characteristics can yield fibrous filters with specific mechanical flexibility during the water filtration applications. Furthermore, induction of nanomaterials and deposition of the functionalized active sites on nanofiber both on the mid or top layer provides efficient adsorptive characteristics which help in eliminating contaminants such as heavy metal ions, bacteria, dyes molecules, viruses, emerging pollutants, and other modern toxins from various contaminated sources. Even though several researchers have attempted to scale up or industrialize the electrospinning process as the most versatile technique to produce nanofibrous filters. However, some of the technological limitations in designing multiple needle assemblies and poor fiber packing efficiencies in final packed filters remain pose serious challenges. Nevertheless, assessing several technological potentials, researchers, innovators, and technologists are engaged in overcoming technological limitations of the electrospinning process. [59-61]

6.9. Challenges and Prospects of Ultra-porous Filters

The development of multi-functional ultra-porous filters originating from aerogels, composite aerogels and electrospun membrane filters offers an opportunity for advanced filtration applications, as an effective alternative to conventional NF, UF and MF processes to be used in various water treatment applications. Interestingly, both aerogel as well as electrospun nanofiber filters offer highly porous structure with nearly 80% interconnected and bicontinuous pore structure offer higher flux, which considerably reduces the energy consumption during filtration processes compared to conventional polymer membranes. On the other hand, incorporation of nano-sized inorganic as well as biomaterials such as metal oxides, hydroxides, layered materials, cellulose nanofibrils and biomacromolecules with an average diameter of 5–100 nm yields a wide pore size distribution and surface characteristics in both aerogels as well as electrospun fiber mats. These task-specific active sites help in capturing highly toxic and conventionally complex pollutants from contaminated sources. In addition, the bio-origin and hydrophilic nature of aerogel and composite nanofiber mat precursors considerably decreases the organic as well as biofouling of the membrane filter making them more biocompatible and biodegradable in real applications. However, extensive research is required to develop highly reliable, industrially feasible, and completely sustainable ultra-porous filters having biodegradable characteristics post-treatment of real contaminated water and wastewaters. From the economic and sustainability point of view, it is important to explore the precursors for ultra-porous membrane or filter preparation that are cost efficient and easily adoptable with the currently available infrastructures. In preparing tailored ultra-porous and functional aerogels and nanofibers, following lyophilization and solution blowing or co-axial electrospinning, or centrifugal spinning, respectively, minimizing process cost tremendously contributes to the overall economics and sustainability of the technology.

Accumulate impact of all process parameters consequently help in addressing the concerns of the various end users regarding the development of non-toxic, low-cost, and environmentally benign ultra-porous filters. During this process, it is also important to comprehensively explore innovative engineering modules to make filtration performance of such ultra-porous filters affordable and effective in addressing the global concerns on widespread drinking water as well as industrial wastewater treatment.

6.10. Conclusion

Advanced water treatment techniques are being developed considering sustainable and affordability factors. Conventional water treatment techniques have always been characterized by their high capital and operational costs. In some cases, advanced water treatment technologies use environmentally harsh chemicals and substances to operate effectively. In such situation, ultra-porous water filters have shown great potential in reducing energy consumption and affordability in effectively treating various types of water contamination. This chapter has given detailed account on the design and development of ultra-porous water filters and their potential applications in various types of contaminated water treatment.

References

1. Tan X, Rodrigue D. A Review on Porous Polymeric Membrane Preparation. Part I: Production Techniques with Polysulfone and Poly (Vinylidene Fluoride). Polymers (Basel). 2019 Jul 8; 11(7): 1160. doi: 10.3390/polym11071160. PMID: 31288433; PMCID: PMC6680680.
2. Fujikawa, S., Selyanchyn, R. & Kunitake, T. A new strategy for membrane-based direct air capture. Polym J 53, 111–119 (2021). https://doi.org/10.1038/s41428-020-00429-z
3. Golovkov V., Sokhoreva V., Sigfusson T. Formation of chemically resistant track membranes based on polyvinylidene fluoride. Pet. Chem. 2012; 52: 462–464.
4. Güell C., Ferrando M., López F. Monitoring and Visualizing Membrane-Based Processes. John Wiley & Sons; New York, NY, USA: 2009.
5. Kanellopoulos N. Nanoporous Materials: Advanced Techniques for Characterization, Modeling, and Processing. CRC Press; Boca Raton, FL, USA: 2016.
6. 16. Abdelrasoul A., Doan H., Lohi A., Cheng C.H. Morphology control of polysulfone membranes in filtration processes: A critical review. Chembioeng Rev. 2015; 2: 22–43. doi: 10.1002/cben.201400030.
7. C. M. Kao, B. M. Yang, R. Y. Surampalli, and Tian C. Zhang Limitation of Membrane Technology and Prevention of Membrane Fouling, Membrane Technology and Environmental Applications. 2012.
8. Shanxue Jiang, Yuening Li, Bradley P. Ladewig, A review of reverse osmosis membrane fouling and control strategies, Science of The Total Environment, Volume 595, 2017, Pages 567-583, ISSN 0048-9697.
9. Haiqing Lin,Yifu Ding Polymeric membranes: chemistry, physics, and applications Journal of Polymer Science, 15 September 2020, Volume 58, Issue 18 p. 2433-2434.
10. Hossein Miraji, Asha Ripanda, Edward Moto, A review on the occurrences of persistent organic pollutants in corals, sediments, fish and waters of

the Western Indian Ocean, The Egyptian Journal of Aquatic Research, Volume 47, Issue 4, 2021, Pages 373-379, ISSN 1687-4285.

11. Gong, P., Xu, H., Wang, C. et al. Persistent organic pollutant cycling in forests. Nat Rev Earth Environ 2, 182–197 (2021).
12. Varjani SJ, Gnansounou E, Pandey A. Comprehensive review on toxicity of persistent organic pollutants from petroleum refinery waste and their degradation by microorganisms. Chemosphere. 2017 Dec;188:280-291. doi: 10.1016/j.chemosphere.2017.09.005. Epub 2017 Sep 4. PMID: 28888116.
13. Tiphanie Deblonde, Carole Cossu-Leguille, Philippe Hartemann, Emerging pollutants in wastewater: A review of the literature, International Journal of Hygiene and Environmental Health, Volume 214, Issue 6, 2011, Pages 442-448, ISSN 1438-4639.
14. Deblonde T, Cossu-Leguille C, Hartemann P. Emerging pollutants in wastewater: a review of the literature. Int J Hyg Environ Health. 2011 Nov; 214(6): 442-8. doi: 10.1016/j.ijheh.2011.08.002. Epub 2011 Aug 31. PMID: 21885335.
15. Morin-Crini, N., Lichtfouse, E., Fourmentin, M. et al. Removal of emerging contaminants from wastewater using advanced treatments. A review. Environ Chem Lett 20, 1333–1375 (2022).
16. Subrahmanya T, Ahmad Bin Arshad, Po Ting Lin, Januar Widakdo, Makari H. K., Hannah Faye M. Austriaa, Chien-Chieh Hua, Juin-Yih Laia and Wei-Song Hung, A review of recent progress in polymeric electrospun nanofiber membranes in addressing safe water global issues, RSC Adv., 2021, 11, 9638-9663.
17. Kerianne M. Dobosz, Christopher A. Kuo-Leblanc, Tyler J. Martin, and Jessica D. Schiffman, Ultrafiltration Membranes Enhanced with Electrospun Nanofibers Exhibit Improved Flux and Fouling Resistance, Industrial & Engineering Chemistry Research 2017 56 (19), 5724-5733.
18. Haisheng Chen, Manhong Huang, Yanbiao Liu, Lijun Meng, Mengdie Ma, Functionalized electrospun nanofiber membranes for water treatment: A review, Science of The Total Environment, Volume 739, 2020, 139944, ISSN 0048-9697.
19. Svetlana I. Semenova, Haruhiko Ohya, Khantong Soontarapa, Hydrophilic membranes for pervaporation: An analytical review, Desalination, Volume 110, Issue 3, 1997, Pages 251-286, ISSN 0011-9164.
20. N. A. Ahmad, C. P. Leo, A. L. Ahmad & W. K. W. Ramli (2015) Membranes with Great Hydrophobicity: A Review on Preparation and Characterization, Separation & Purification Reviews, 44: 2, 109-134.
21. Zhaohe Huang,Shumeng Yin,Jianzhong Zhang,Na Zhang Recent advances in membrane hydrophilic modification with plant polyphenol-inspired coatings for enhanced oily emulsion separation, Journal of Applied Polymer Science, 24 February 2021, Volume 138, Issue 25 50587.
22. Saheed Mustapha, Tijani Jimoh Oladejo, Ndamitso Muhammed Muhammed, Abdulkareem Ambali Saka, Ajala Adewale Oluwabunmi, Mohammed Abdulkabir, Oke Omotunde Joel, Fabrication of porous

ceramic pot filters for adsorptive removal of pollutants in tannery wastewater, Scientific African, Volume 11, 2021, e00705, ISSN 2468-2276.

23. Adrian Kaiser, Wendelin J. Stark, and Robert N. Grass, Rapid Production of a Porous Cellulose Acetate Membrane for Water Filtration using Readily Available Chemicals Journal of Chemical Education 2017 94 (4), 483-487.
24. Yanqiu Lu, Wei Liu, Jiangtao Liu, Xue Li, Sui Zhang, A review on 2D porous organic polymers for membrane-based separations: Processing and engineering of transport channels, Advanced Membranes, Volume 1, 2021, 100014, ISSN 2772-8234.
25. Qi Sun, Briana Aguila, Yanpei Song, and Shengqian Ma, Tailored Porous Organic Polymers for Task-Specific Water Purification, Accounts of Chemical Research 2020 53 (4), 812-821.
26. Aluigi, Annalisa; Vineis, Claudia; Tonin, Claudio; Tonetti, Cinzia; Varesano, Alessio; Mazzuchetti, Giorgio, Wool Keratin-Based Nanofibres for Active Filtration of Air and Water, Journal of Biobased Materials and Bioenergy, Volume 3, Number 3, September 2009, pp. 311-319(9).
27. Andreas MautnerNanocellulose water treatment membranes and filters: a review, Polymer International, Volume 69, Issue 9, September 2020, Pages 741-751.
28. Pasichnyk, M., Gaálová, J., Minarik, P. et al. Development of polyester filters with polymer nanocomposite active layer for effective dye filtration. Sci Rep 12, 973 (2022).
29. Marek Gryta, Application of polypropylene membranes hydrophilized by plasma for water desalination by membrane distillation, Desalination, Volume 515, 2021, 115187, ISSN 0011-9164.
30. Wu, J., Cao, M., Tong, D. et al. A critical review of point-of-use drinking water treatment in the United States. npj Clean Water 4, 40 (2021).
31. Millen HT, Gonnering JC, Berg RK, Spencer SK, Jokela WE, Pearce JM, Borchardt JS, Borchardt MA. Glass wool filters for concentrating waterborne viruses and agricultural zoonotic pathogens. J Vis Exp. 2012 Mar 3; (61): e3930. doi: 10.3791/3930. PMID: 22415031; PMCID: PMC3466626.
32. Anna Cescon and Jia-Qian Jiang, Filtration Process and Alternative Filter Media Material in Water Treatment Water 2020, 12, 3377.
33. Beatriz Merillas, Judith Martín-de León, Fernando Villafañe, and Miguel A. Rodríguez-Pérez, Transparent Polyisocyanurate-Polyurethane-Based Aerogels: Key Aspects on the Synthesis and Their Porous Structures, ACS Applied Polymer Materials 2021 3(9), 4607-4615.
34. Parwani M. Rewatkar, Adnan Malik Saeed, Hojat Majedi Far, Suraj Donthula, Chariklia Sotiriou-Leventis, and Nicholas Leventis, Polyurethane Aerogels Based on Cyclodextrins: High-Capacity Desiccants Regenerated at Room Temperature by Reducing the Relative Humidity of the Environment, ACS Applied Materials & Interfaces 2019 11(37), 34292-3430.
35. Suja P. Sundaran, Reshmi C. R. and Sujith A, Tailored design of polyurethane based fouling-tolerant nanofibrous membrane for water treatment, New J. Chem., 2018, 42, 1958-1972.

36. Zheng L, Zhang S, Ying Z, Liu J, Zhou Y, Chen F. Engineering of Aerogel-Based Biomaterials for Biomedical Applications. Int J Nanomedicine. 2020 Apr 3; 15: 2363-2378. doi: 10.2147/IJN.S238005. PMID: 32308388; PMCID: PMC7138623.

37. Anina James, Deepika Yadav, Bioaerogels, the emerging technology for wastewater treatment: A comprehensive review on synthesis, properties and applications, Environmental Research, Volume 212, Part A, 2022, 113222, ISSN 0013-9351.

38. Loredana Elena Nita, Alina Ghilan, Alina Gabriela Rusu, Iordana Neamtu and Aurica P. Chiriac, New Trends in Bio-Based Aerogels, Pharmaceutics 2020, 12, 449.

39. Jai Prakash Chaudhary, Nilesh Vadodariya, Sanna Kotrappanavar Nataraj, and Ramavatar Meena, Chitosan-Based Aerogel Membrane for Robust Oil-in-Water Emulsion Separation, ACS Applied Materials & Interfaces 2015 7 (44), 24957-24962.

40. D. D. Nguyen, N. H. Tai, S. B. Lee and W. S. Kuo, Energy Environ. Sci., 2012, 5, 7908–7912.

41. H. Peng, J. Wu, Y. Wang, H. Wang, Z. Liu, Y. Shi and X. Guo, Appl. Phys. A, 2016, 122 .

42. S. B. de Araujo, M. Merola, D. Vlassopoulos and G. G. Fuller, Langmuir, 2017, 33, 10501–10510

43. Z. Yao, L. W. Zhou, B. B. Dai and K. Cao, J. Appl. Polym. Sci., 2012, 125, 2489–2493.

44. K. Cao, Q. Fu, L. Zhou and Z. Yao, Prog. Chem., 2012, 24, 1368–1377 CAS.

45. E. Oluwatuyi, Oluwapelumi O. Ojuri, Arman Khoshghalb, Cement-lime stabilization of crude oil contaminated kaolin clay, Journal of Rock Mechanics and Geotechnical Engineering, 12, 1, 2020, Pages 160-167.

46. Ramasamy Ganesamoorthy, Vinod Kumar Vadivel, Rajnish Kumar, Omkar S. Kushwaha, Hadas Mamane, Aerogels for water treatment: A review, Journal of Cleaner Production, Volume 329, 2021, 129713, ISSN 0959-6526.

47. Uastaferro M, Reverchon E, Baldino L. Agarose, Alginate and Chitosan Nanostructured Aerogels for Pharmaceutical Applications: A Short Review. Front Bioeng Biotechnol. 2021 May 12;9:688477. doi: 10.3389/fbioe.2021.688477. PMID: 34055766; PMCID: PMC8149959.

48. Jai Prakash Chaudhary, Sanna Kotrappanavar Nataraj, Azaz Gogda and Ramavatar Meena, Bio-based superhydrophilic foam membranes for sustainable oil–water separation, Green Chem., 2014,16, 4552-4558.

49. Xiangze Jia,Mohammad Peydayesh,Qiang Huang,Raffaele Mezzenga, Amyloid Fibril Templated MOF Aerogels for Water Purification, Nano Micro Small, Volume18, Issue4, January 27, 2022, 2105502.

50. Jie Wei, Shi-Han Gui, Jun-Hua Wu, Dan-Dan Xu, Yun Sun, Xiao-Ying Dong, Yang-Yong Dai, Yong-Feng Li, Nanocellulose-Graphene Oxide Hybrid Aerogel to Water Purification, Applied Environmental Biotechnology (2019) - Volume 4, Issue 1, p. 11.

51. Hao Kong, Yun Chen, Guozheng Yang, Bin Liu, Lei Guo, Yan Wang, Xin Zhou and Gang Wei , Two-dimensional material-based functional aerogels for treating hazards in the environment: synthesis, functional tailoring, applications, and sustainability analysis, Nanoscale Horiz., 2022,7, 112-140.
52. Jatoi, A.S., Hashmi, Z., Mazari, S.A. et al. Recent developments and progress of aerogel assisted environmental remediation: a review. J Porous Mater 28, 1919–1933 (2021).
53. Manohara Halanur Mruthunjayappa, Vibha T. Sharma, Kalpana Dharmalingam, Nataraj Sanna Kotrappanavar, and Dibyendu Mondal, Engineering a Biopolymer-Based Ultrafast Permeable Aerogel Membrane Decorated with Task-Specific Fe–Al Nanocomposites for Robust Water Purification, ACS Applied Bio Materials 2020 3 (8), 5233-5243.
54. Pijush Kanti Chattopadhyay, Nayan Ranjan Singha, MOF and derived materials as aerogels: Structure, property, and performance relations, Coordination Chemistry Reviews, Volume 446, 2021, 214125, ISSN 0010-8545.
55. Martin R. Lohe,a Marcus Rosea and Stefan Kaskel, Metal–organic framework (MOF) aerogels with high micro- and macroporosity, Chem. Commun., 2009, 6056-6058.
56. Bong J, Lim T, Seo K, Kwon CA, Park JH, Kwak SK, Ju S. Dynamic graphene filters for selective gas-water-oil separation. Sci Rep. 2015 Sep 23;5:14321. doi: 10.1038/srep14321. PMID: 26394930; PMCID: PMC4585788.
57. Glenita Bridget D'Souza, Anshu Kumar, Smitha V. Kamath, Ashok Shrishail Maraddi, Sanna Kotrappanavar Nataraj, Designing engineered biopolymer mesh filter for robust sequestration of chromium (VI), fluoride and other emerging pollutants: A sustainable approach, Chemical Engineering Journal, Volume 443, 2022, 136462, ISSN 1385-8947.
58. Manohara Halanur Mruthunjayappa, Nataraj Sanna Kotrappanavar, Dibyendu Mondal, Bioinspired engineering protein nanofibrils-based multilayered self-cleaning membranes for universal water purification, Journal of Hazardous Materials, Volume 424, Part C, 2022, 127561, ISSN 0304-3894.
59. Yujie Chen, Muhammad Shafiq, Mingyue Liu, Yosry Morsi, Xiumei Mo, Advanced fabrication for electrospun three-dimensional nanofiber aerogels and scaffolds, Bioactive Materials, Volume 5, Issue 4, 2020, Pages 963-979.
60. Nayl AA, Abd-Elhamid AI, Awwad NS, Abdelgawad MA, Wu J, Mo X, Gomha SM, Aly AA, Bräse S. Review of the Recent Advances in Electrospun Nanofibers Applications in Water Purification. Polymers (Basel). 2022 Apr 14; 14(8): 1594. doi: 10.3390/polym14081594. PMID: 35458343; PMCID: PMC9025395.
61. Wee-Eong Teo, Ryuji Inai & Seeram Ramakrishna (2011) Technological advances in electrospinning of nanofibers, Science and Technology of Advanced Materials, 12: 1, DOI: 10.1088/1468-6996/12/1/013002.

CHAPTER

7

Advances in Block Copolymer-based Membranes

— ABSTRACT —

Pressure-driven membrane separation processes like reverse osmosis (RO), nanofiltration (NF), ultrafiltration (UF), and microfiltration (MF) have been widely used for potable water production and domestic as well as industrial wastewater treatment applications. These membrane-based separation processes have been known to yield quick and high-quality product water with a high percent of pollutant rejection at high flux rates. These processes generally use synthetic polymer-based filtration barriers for the effective removal of various types of pollutants and contaminants from water sources. Even though membrane processes have shown several advantages, these processes suffer from minor but important limitations like fouling, durability, gradual decline in their permeability, etc., which causes a loss in overall separation efficiency and life of the membrane. Nevertheless, advanced approaches to membrane materials development are focusing on the synthesis and development of membrane precursor materials that possess highly film forming properties and can result in ordered pore structures. In this direction designer polymers such as block copolymeric systems have been known to possess exceptional properties. Block copolymers have unique pore-forming characteristics via their self-assembly properties which result in forming ordered nanopores in selective layers. Upon processing, these ordered nanopores provide much needed superior high-water permeability, antifouling characteristics, thermal resistance, chemical resistance, and durability during their applications. Therefore, this chapter focuses on elucidating advances in block copolymer-based membranes useful for pressure-driven MF, UF, NF, and RO processes as low-energy and low-cost alternatives.

7.1 Introduction

Pressure-driven membrane separation processes like reverse osmosis (RO), nanofiltration (NF), ultrafiltration (UF), and microfiltration (MF) have been widely used for potable water production and domestic as well as industrial wastewater treatment applications. These membrane-based separation processes have been known to yield quick and high-quality product water with a high percentage of pollutant rejection at high flux rates. These processes generally use synthetic polymer-based filtration barriers for the effective removal of various types of pollutants and contaminants from water sources. Even though membrane processes have shown several advantages, these processes suffer from minor but important limitations like fouling, durability, gradual decline in their permeability, etc., which causes a loss in overall separation efficiency and life of the membrane. Nevertheless, advanced approaches to membrane materials development are focusing on the synthesis and development of membrane precursor materials that possess highly film in nature and can result in ordered pore structures. In this direction designer polymers such as block copolymeric systems have been known to possess exceptional properties. Block copolymers have unique pore-forming characteristics via their self-assembly properties which result in forming ordered nanopores in selective layers. Upon processing, these ordered nanopores provide much-needed superior high-water permeability, anti-fouling characteristics, thermal resistance, chemical resistance, and durability during their applications. Therefore, this chapter focuses on elucidating advances in block copolymer-based membranes useful for pressure-driven MF, UF, NF, and RO processes as low-energy and low-cost alternatives.

7.2 Limitations with Conventional Porous Membranes

Currently used conventional and commercially successful filtration membranes are made from numerous synthetic origin polymers as well as biopolymers. Commercial membranes are used in the processes like microfiltration, ultrafiltration, nanofiltration, reverse osmosis, electrodialysis, gas separation, and pervaporation. The critical evaluation of the current status of various types of membranes used in different processes suggests the inherent limitations associated with the conventional preparation techniques. Conventionally, several simple techniques have been practiced for the preparation of porous as well as non-porous or dense membranes. Among many, solution casting, non-solvent induced phase separation, thermal induced phase inversion, vapor-induced phase separation, track etching, electrospinning, and sintering have been prominent choices for turning synthetic polymers into useful membranes. Interestingly, most of the membranes prepared to originate from thermoplastics in nature that are also known as rubbery or glassy polymers. These possess unique physicochemical characteristics like solubility and glass transition temperature responses. These properties are useful in obtaining good and stable film-forming properties. Upon heat treatment to a temperature above, an amorphous polymer known as rubbery transition temperature (T_g), the polymer turns from the glassy state to the rubbery state. These temperature ranges in various polymers can be used for inducing optimal porosity and physicomechanical properties in membranes. [1-3]

Nonetheless, preparing a dense membrane is the simplest among all methods which involves two steps. The first step involves the preparation of a suitable porous substrate or support which is generally prepared via the phase inversion technique, and the second step is the preparation of the selective or dense barrier layer on the surface of the support. The phase inversion technique on the other hand is widely used for the preparation of highly porous polymeric membranes. In this technique, initially, the polymer solution is prepared which is a polymer-rich phase, and when the casted polymer solution is placed in a polymer-poor phase and thermodynamically unstable coagulation bath, the presence of a non-solvent, polymer phase separation occurs. In this process, polymer solubility, solvent vapor pressure, solution evaporation, temperature, humidity, solvent, and non-solvent miscibility play a vital role in obtaining optimal pore size and their distribution in final membranes.

Therefore, phase inversion is a simple solvent/non-solvent demixing process in which an initially homogeneous polymer solution which is also known as the dope solution is transformed from a liquid state to a solid state via controlled evaporation and coagulation step. This method is the most commonly used and it helps to fabricate polymeric porous membranes with a large size pore structure. A coagulation bath or medium also known as the non-solvent bath is generally nonsolvent to polymer but is miscible with solvent. The solvent and non-solvent exchange in the coagulation bath facilitates phase-inversion in which polymer solidification results in a porous membrane. This method is used largely to prepare most of the porous membranes that find direct applications as MF, and UF membranes. These membrane pores vary from 5 nm to 100 nm and further with advanced approaches, pore size can be tuned to below 5 nm and up to micrometer ranges.

Due to the similar permeation and rejection characteristics, microfiltration and ultrafiltration are considered similar processes. These processes are also related except for pore size changes slightly making one of the major distinctions. For instance, in the MF process isotropic micro-porous membranes are widely used, while in the UF process, membranes are generally made by the Loeb and Sourirajan process and have anisotropic pore-structures comprising of a finely porous selective or surface skin layer and a relatively open micro-porous substrate. On the other hand, most conventional UF and MF membranes have a single-layer barrier and are prepared via a single-phase inversion process. For decades now, MF and UF membranes pore size and their morphologies have experienced little or no change even after utilizing several advanced materials and innovative preparation protocols. [4-6]

This limitation originates largely from the limited scopes associated with the phase inversion techniques, for instance, the non-solvent-induced phase inversion technique involves the immersion of a solution cast film comprising individual polymer or composite into a coagulation or non-solvent bath usually water in most cases which result in solid polymer phase separation. In this solvent exchange process, ultimately the polymer-poor phase solidifies as a relatively large void structure resulting in the formation of pores, while the polymer-rich phase results in the solid membrane matrix. In this process, between the solution casting, and coagulation process, a variety of factors plays a vital role in inducing selective layer pore size and their distribution such as intrinsic features of polymers, polymer concentration, solvents, solution casting, evaporation timing, and coagulation step which can lead to obtaining varieties of final membrane morphologies.

Even though recent advances in nanotechnology-based protocols have immensely changed the selection of the material and refurbished all the membrane preparation parameters depending on the end user demand for different applications. In this direction, electrospun nanofibrous mats or membranes (ENMs) are one such end product that has been adopted as alternative material and morphologies for conventional polymer-film-based UF and MF membranes. These electrospun nanofibrous membranes have shown relatively improved permeability and rejection performance, relatively flexible pore tuning characteristics at low environmental impact materials stock, and preparation routes at low energy consumption. Major advantages of ENMs give flexible pore packing features in which different pore sizes can be achieved using appropriate diameter nanofiber packing. Therefore, these ENMs are considered ideal alternative candidates for film-based membranes for future water purification applications. Also, electrospinning technology also has evolved over the years since its first rediscovery in 1993 (Reneker et al.). Extensive research has led to the innovation in simple single nozzle spinning to large-scale multi-step preparation techniques to produce nanofiber scaffolds or mats or membranes. Further, electrospinning technique involving coaxial electrospinning has also been adopted as a modified version of the conventional electrospinning process which comprises the arrangement of multiple dope solutions as feed systems to simultaneously electrospun two or more polymer dope solutions from coaxial capillaries. Also, 3D printing technology also provided an innovative electrospinning opportunity with specific fiber morphologies and their packing features resulting in a simple mat or membrane format for structured materials useful for a variety of separation and purification applications.

Advanced versions of ENMs have also shown improved porosities compared to commercially successful polymer-film-based membranes. While polymer-film-based membranes show 60–80 vol% porosity whereas electrospun nanofibrous membranes can give a higher porosity as 80 to 95 vol%. This is because pore size or voids due to the packing of fibers can be controlled directly by altering the fiber diameter and their combination. Moreover, another major advantage of using nanofiber scaffold is that electrospun fibers provide a large surface area and a combination of composite dope solutions can induce abundant functional active sites on their surfaces. All these synergic effects thus provide ENMs with high retention characteristics for contaminants via both exclusive rejections as well as functional adsorption. Nevertheless, ENM's exceptional pore volume percentage which is close-void in their packing morphology provides nanofiber scaffolds with an interconnected network structure thereby permitting exceptionally high permeation flux and low hydraulic resistance. This also results in lower fouling propensities in ENMs. [7]

However, ENMs also suffer from some major disadvantages when used in the processes like UF and MF under externally applied pressure conditions. The major drawback associated with ENMs is that nanofibrous assemblies undergo mechanical deformation under high pressure in UF and MF processes. Nonetheless, advanced approaches have shown improved mechanical stability by adjusting important electrospinning process parameters, forming composites of nanofibers, and post-spinning arrangements. All these improved properties allow the nanofibrous membrane to attain well-organized packing to result in stable MF/UF absorptive membranes. Further, coaxial electrospinning approaches have resulted in improved pore packing which can be extended to RO, FO, and NF applications. Also, by adjusting the electrospinnability of synthetic polymers, hydrophobic ENMs can be prepared for membrane distillation (MD) applications. Disadvantages associated with mechanical deformation can also be overcome by arranging ENMs fibrous layer into one-, two-, multiple- layered arrangements depending on the intended applied pressure and application.

Multiple studies have been conducted to address the issues discussed above. In addition to poor fibrer packing abilities which make them inferior to mechanical strength which is detrimental to load-bearing applications, recent detailed studies have also revealed that the toxicity associated with the polymers and their composites and residual solvent in the spun fibres restrict their use in direct drinking water purification applications.

Therefore, studies are still majorly focusing on overcoming inherent drawbacks of the conventional electrospun nanofibrous membranes by focusing on improved dope solution preparation involving biopolymer precursors, improving their spinnability, and improving their barrier layer properties with enhanced packing densities with emphasis on reinforcement and post-treatment protocols. On the other hand, the non-woven nature of electrospun also restricts their immediate use, therefore, researchers are adopting several functionalizations to induce selective adsorptive sites to make them better pollutant retention barriers.

7.3 Advanced Membranes in Water Treatment: Block Copolymer Membrane

Overexploitation of natural resources and their unregulated discharge into the environment is causing environmental degradation. Currently, to purify natural and freshwater sources like ponds, lakes, rivers, and increasingly now oceans that are extensively polluted due to the contamination into drinking water quality, membrane-based treatment technologies are being favored. Membrane processes follow simple and robust physical treatment mechanisms by exclusively removing the impurities or contaminants from contaminated domestic, industrial, and/or agricultural wastewater. Other major advantages associated with membrane processes come from relatively small footprint, simple physical separation, relatively low energy consumption, ease of membrane manufacturing, scalability, and their engineering into a module to suit the end user requirements. However, membrane process performances are characteristically evaluated considering their unique inherent trade-off norms between permeability or product flux rate and solute rejection selectivity. Another main practical challenge which is concerning the shelf-life of a membrane in use is its organic and/or biofouling propensity and subsequent degradation of parent polymer.

Nevertheless, commercially practiced trade-off norms set between membrane permeability and their solute selectivity have raised intense discussion among researchers and innovators in recent times. These interests are mainly directed not only toward addressing trade-off restrictions but also efforts intensified to enhance permeability and water quality. Interestingly, in porous membranes, the trade-off norms greatly influence and in some cases impair filtration performance evaluation. Ideally, all surface or morphological modification efforts have been to increase the water permeability without compromising solute or pollutant rejection or their selectivity via various strategies involving surface porosity tuning and/or considerably reducing the selective or active layer thickness. However, in non-porous or dense membranes the impact of the trade-off criteria on membrane performances can be eliminated possibly due to the uniform polymer network structures which experience limited external stress. Nevertheless, in all polymer-based membrane systems, both organic and biofouling, the major challenge persists during real operations, which leads to a drastic decline in performance. [8, 9]

To address issues associated with flux, rejection, and fouling in conventional membranes, several advanced and innovative approaches have been adopted considering two main considerations, namely: (1) to create a materials stock or prepare new and advanced materials having long durability and good filtration properties in order to prepare cost-effective and sustainable alternatives for membrane preparation, and (2) to design and optimize structure-specific morphologies which the help in selective separation of pollutants.

As shown in Figure 7.1, two important aspects of novel membrane materials stocks have been in detail evaluated to sort out optimum performance. However, the focus has always been to institute detailed studies to evaluate inherent trade-off limitations between selectivity and permeability. In this direction, several combinations of materials involving inorganic,

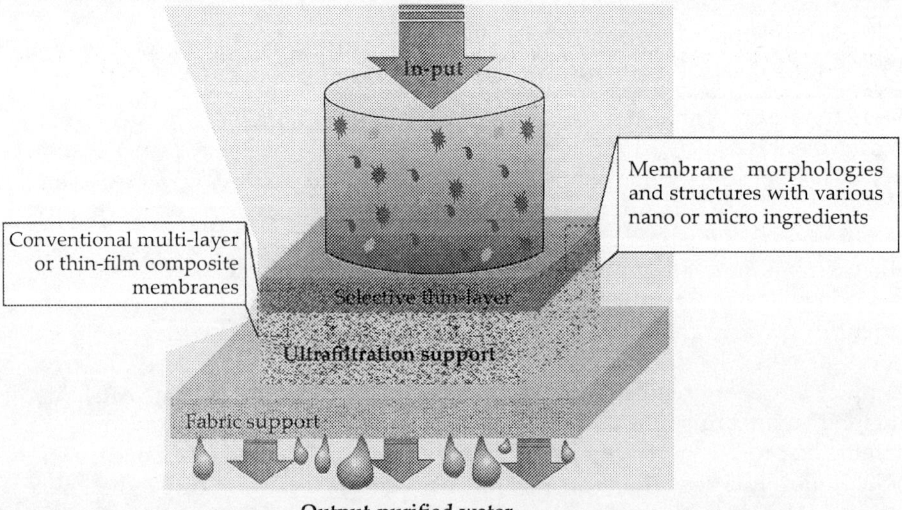

Figure 7.1. Schematic representation of parameters considered for selecting materials stocks and structural tuning considered during the design of conventional membranes.

organic, or polymeric, and polymer/inorganic hybrid materials have been tried and tested in many combinations and configurations. Among them, polymeric membranes have been considered superior to inorganic systems due to their mechanical properties, chemical resistance, and flexible processability under cost-effective processing conditions. In addition to this, typical polymers are extensively used for membrane applications at the commercial level including polysulfone (PSF), polyethersulfone (PES polyacrylonitrile(PAN), polyvinylidenefluoride(PVDF), and polyetheretherketone (PEEK). [10-12]

On the other hand, a new class of designer polymer block copolymers has shown immense potential as a new membrane material having great potential in controlling pore size and their unique distribution characteristics. Block copolymers have a rich and well-defined morphology in the melt, guided by the interaction between different blocks. These advanced classes of polymeric materials when applied as membrane barriers offer the potential of precisely tuning selective blocks of copolymer at the nanoscale. This unique feature of creating and optimizing nanopores comes from the physical removal of specific blocks in novel systems.

Nevertheless, it is important to understand conventional membranes in the context of trade-off performances made up of polymers, and hybrid polymer systems having inorganic fillers like ceramics, zeolites, glass, and metals. Before moving forward, it is also important to evaluate the characteristics of commercially successful combinations of materials adopted widely for various applications. It is well understood that composite and nanocomposites of polymers have exhibited excellent materials properties as well as separation and purification characteristics in water filtration and wastewater treatment. This is whilst their existence is relatively a high cost, complex fabrication process, and involves solvent systems that are classified as environmental concerns. These are some of the limiting factors now bothering the future membrane market and researchers at large. Nonetheless, inducing nano-sized fillers or active sites in the form of composite additives into the polymer matrices has exponentially improved their overall performances and has positively impacted the overall commercial market. These composite membranes have been designed and fabricated following different strategies such as composite dope formation, in situ growth of nanomaterials in a polymer matrix, post-surface modification, intermediate functionalization process, and so on. Also, nano-fillers dispersed in polymers in diverse positions can form nanocomposite polymers of uniform physicochemical characteristics leading to remarkable enhancement in hydrophilicity, high product water permeability, improved rejection rates, and anti-fouling properties, and flexible mechanical characteristics. Besides simple flat-sheet and hollow fibre configurations, polymer-based membranes have also been prepared in electrospun nanofibrous, aerogels, ultra porous mesh filters, and functionalized filter cartridges or beds that offered cost-effective, more efficient, affordable solutions for water purification.

Nonetheless, conventional polymer-based thin-film composite (TFC) and thin-film nanocomposite (TNFC) have been widely used and commercially successful configurations that popularize the technology across the sectors.

In these configurations, membrane structure, morphology, pore size, and pore size distribution played a vital role in achieving maximum output in terms of purified product water. Among many commodity and synthetic polymers, PS, PES, PEEK, PAN, and composites of these have been used at the commercial level. However, among many TFC membranes, polyamide (PA)-based TFC membranes have been considered the gold standard for RO, NF, and FO operations due to their robust filtration efficiency and affordability. Even in electrospun mats in nanofibrous membrane format, fibrous scaffolds have shown great potential as ultra-porous filters. Further, there have been several attempts to replace polyamide with relatively more sustainable biopolymers. Nevertheless, served as an alternative materials stock have been explored for the conventional support as well as selective layers in TFC membranes and to fabricate advanced thin-film composite (ATFC) membranes with unique and target-specific properties, such as surface affinity, hydrophilicity, high porosity, and interconnected pores. In this direction biopolymers like chitosan as an alternative to polyamide have shown improved separation and purification applications. [13-16]

Even though biopolymer-based TFC membranes have shown higher permeability and lower fouling propensity and moderate energy consumption which effectively contribute to the overall performance of traditional TFC membranes, the mechanical robustness of synthetic polymers has proven to be unmatchable yet. In addition to this, the introduction of multifunctional nanomaterials into the polymer matrices has also greatly improved the properties of membranes. Nanomaterials of various origins and types such as ceramics, zeolites, metal-organic framework (MOF), covalent organic framework (COF), and polymeric nanoparticles have provided much-needed permeability enhancements and improved life span to conventional polymer membranes. These nanomaterials have been either used in between an interlayer or interfaces of support and selective layer or mixed in the polymer doped solution resulting in the mixed matrix membranes decorating the selective layer surface making them highly exposed to the contaminated sources. In most cases, conventional membranes have shown more adoptable engineering benefits such as flexibility, and processability, providing a better module fabrication. More so, composites of polyamide layer and the improved engineering designs of support layer in advanced TFC membrane systems have substantially improved permeability and anti-fouling properties. Nevertheless, modern pollutants including emerging pollutants, persisting chemicals, complex wastewater sources, and ever-transforming intermediates pose serious challenges to all the existing membrane processes. Therefore, nature-inspired sustainable solutions are being researched to overcome limitations with conventional membranes. [17-19]

Therefore, this chapter critically overviews the progress in conventional polymer membranes, commercially successful composite and nanocomposite membranes, and limitations of the commercially successful membrane as well as their modified versions. This chapter also overviews the new class of membrane materials used for designer membranes with the specific surface as well as pore characteristics. Moreover, it is important to assess the properties of the precursors or raw materials used for the preparation of an advanced class of membranes from a perspective of membrane processability, module preparation, and their compatibility with engineering components. These synergetic aspects induce effective membrane barrier properties both in bulk and selective layers.

In this direction, several nature-inspired concepts have also been adopted such as extreme wettability characteristics like super hydrophilicity from biopolymers, super hydrophobicity like the lotus leaf effect, and natural architectures found in biomimetic membranes. From all these efforts, several unique features have been achieved in membrane surface as well as functional properties. In addition to this, controlling pore sizes and their distribution in the membrane surface has always been the greatest challenge for researchers both in the academic and industrial sectors. Even though conventional solution casting (SC), solvent-induced phase inversion (NIPS), thermal induced phase inversion (TIPS), and track etching (TE) methods follow simple and easy to prepare protocols, in almost all cases pore size control and their distribution are very broad. [20, 21] This aspect of the conventional membranes is a relatively daunting task. To overcome these inherent issues, researchers have deliberately attempted to design novel membrane precursors or raw materials to result in extensive control over pore tuning. In this direction, a new class of copolymers with well-defined molecular weights and their specific molar fractions have been converted to membranes. A block copolymer is classified as a polymer comprising a molecular chain in which there is a linear arrangement of dissimilar polymer blocks, a block being defined as a portion of a polymer molecular chain in which the monomeric units have at least one constitutional or configurational feature absent from the adjoining portions as shown in Figure 7.2. Among many advantages, block copolymers undergo self-assembly transformation which greatly depends on their types of polymers, molar fraction, and nature of molecular fractions. Upon self-assembly, block copolymers possess a characteristic and well-defined morphology above their melting point subsequent cooling (in the melt) which is guided by the interaction between constituent blocks linked through covalent bonding. These special characteristics turn out to be very useful when used for membrane preparation. Each block in a polymer can be used or eliminated with precise physicochemical processes to create a well-defined pore structure. Therefore, block copolymers offer enormous possibilities in tuning nanoscale pores precisely and can be further combined with hybrid materials to induce a variety of synergic properties in one single membrane. Generally, there are two main types of block copolymers, namely: (a) linear block copolymers, and (b) branched or graft copolymers. [20]

The main difference between a linear block copolymer and a graft copolymer is that a linear block copolymer has blocks of repeating units arranged alternatively in order or random, on the other hand, a graft copolymer contains branches of repeating units, a largely dissimilar nature. A polymer

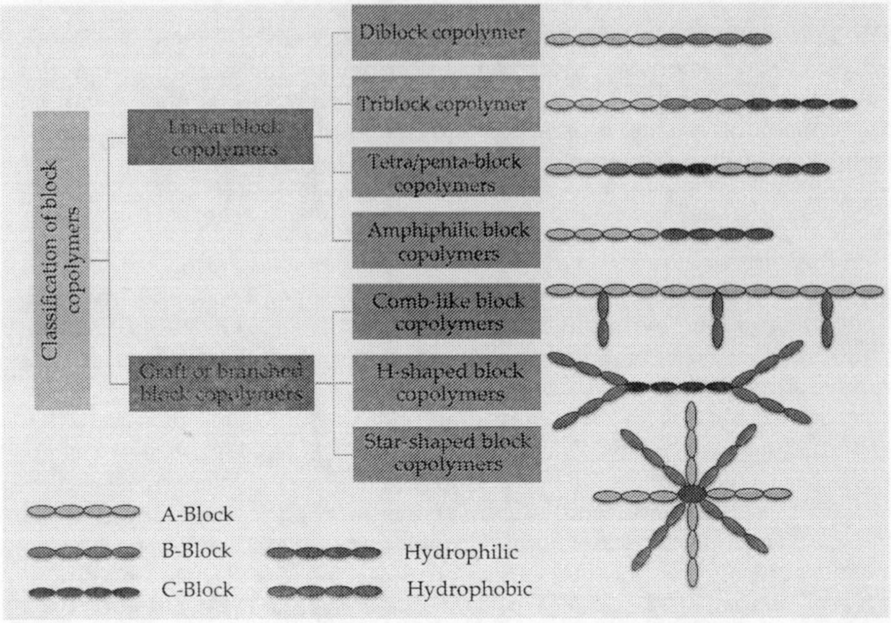

Figure 7.2. Detailed classification of the block copolymer family.

is a macromolecule that can exist in many different chemical properties in the form of various repeating units linked to each other via covalent chemical bonds. On the other hand, block copolymers are prepared following controlled polymerization of one monomer (for instance, A) followed by chain extension with a different monomer (for instance, B) to form AB diblock copolymer. Similarly, if the third monomer is covalently linked using third block (C) to result in a tri-block copolymer as ABC block copolymers. These block copolymers can be further extended to tetra- and penta-block copolymers. Generally, block copolymers are made up of blocks of dissimilar polymerized monomers and covalent linkages to each block are denoted as -*b*- where *b* = block. For instance, polystyrene -*b*-poly(methyl methacrylate) or PS-*b*-PMMA, is made up of first polymerizing styrene, and then subsequently polymerizing methyl methacrylate (MMA) from the reactive end of the polystyrene chains. Similarly, Polyethylene–*b*-vinyl acetate (PEVA), Polyacrylonitrile-*b*-butadiene–*b*-styrene (ABS), PS-*b*-poly(ethylene-*alt*-propylene) (PE-*b*-PP), Polystyrene-*b*-polyisoprene (PS-*b*-PI), Poly(*n*-butylacrylate)-*b*-poly(acrylonitrile)(PBA-*b*-PAN), polystyrene-*b*-poly(ethyleneoxide) -*b*-polystyrene(PS-*b*-PEO-*b*-PS), Polystyrene-*b*-poly(ethylene-*b*-propylene) (PS -*b*-PEP), nitrile rubber and so on have been prepared with specific targeted end properties. [22-26]

7.4 Importance of Self-assembly in Block Copolymers

Block copolymers are comprised of polymer segments of different repeating units linked covalently together in a polymer chain as a typical block copolymer of A as the first monomer and B as the second monomers would have the structure in the resulting copolymer as AAAAA-BBBBB-AAAAA–

BBBBB. This type of repeating alternative block copolymer is typically known as a 'diblock' copolymer. However, copolymers can also be prepared using many topologies that are referred to as block copolymers. On the other hand, a typical triblock copolymer has three blocks, one central block flanked by two blocks A and B containing a different repeating unit C, for example, AAAA-BBBB-CCCC-AAAA-BBBB-CCCC. Similarly, alternative block copolymers having segmented blocks consist of various alternating small blocks of two or more different types of repeating units. Graft or brush-like block copolymers on the other hand have one long main backbone chain of one type of repeating unit and hanging or grafted to it are several smaller blocks as second repeating units branching off the main polymer chain backbone. Figure 7.2. gives a detailed illustration of several types of major block copolymer architectures. [27, 28]

Further, following step-growth polymerization proceeds having difunctional monomers, the block copolymers are prepared using one B–B monomer and similarly two monomers having 'A' functionality: A–A and A'–A'. As selected monomers carrying difunctional nature, A–A and B–B follow step-growth polymerization, it is difficult to synthesize a stable diblock copolymer. Therefore, in order to prepare diblock copolymers following step-growth chemistry, limited synthetic techniques can be explored to prepare diblock polymeric systems. For instance, the utilization of A–B-type monomers such as 6-amino hexanoic acid can result in a well-defined diblock copolymer. In this technique, after reaching a specific percentage conversion of targeted molecular weight which is estimated using Carothers's equation to know the converted monomer into specific molecular weight, a selective end-capping or terminating agent is introduced to remove one of the functionalities i.e., either A or B, based on the desired end block polymer. At this moment, each polymeric chain or oligomeric containing one functionality either B or A, respectively can be terminated. Further to this, a second A–B-type monomer can be subsequently introduced to couple with the initially prepared oligomer or low molecular weight polymer to result in the formation of a diblock copolymer. A similar technique can be further adopted to prepare triblock copolymers. This is carried out using the reaction between one difunctional monomer like A–A with a different difunctional monomer system like B–B resulting in polymeric chains with A end or terminal groups. The subsequent addition of A–B-type monomers can result in the two outer blocks. Interestingly, both techniques are complex and difficult to regulate to obtain a well-defined end product. [29]

On the other hand, linear multi-block copolymers also referred to as 'segmented' block copolymers can be well controlled to obtain designer copolymers. Interestingly, block copolymers possess several advantages due to their ability to induce synergistic properties such as thermal and mechanical characteristics of the individual blocks from their parent monomers like polymer blends. As block copolymers are the assemblies of dissimilar polymer blocks having individual physicochemical characteristics and different repeating units that are covalently linked together, however, the phase separation behavior and morphology dramatically vary from polymer blends. If the constitutive units or blocks are immiscible, the melt upon gradual cooling turns into phase-separated islands.

On the other hand, block copolymers are prepared following controlled polymerization of one monomeric unit followed by polymer chain extension using a different monomeric unit. Block copolymers are normally prepared in alternative or linear and graft or branched configuration via controlled polymerization of one monomer (A), followed by chain growth using a different monomer (B) to form AB or similarly extension with a third monomer (C) to yield ABC block copolymers. In most cases, atom transfer radical polymerization (ATRP) is the preferred polymerization sequence followed in order to synthesize block copolymers with well-defined block length and size. ATRP is carried out mechanistically using transition metal-mediated atom transfer radical addition (ATRA) reactions. Generally, ATRP follows a decreasing order of activity: acrylonitrile (AN)>methacrylates>styrene~acrylates monomers. However, during a chain growth reaction it continues from a less active monomer to a more reactive monomer unit. For instance, chain growth from polyacrylate-based macro-initiator effectively further extensively initiates the ATRP of MMA. To be an effective ATRP, the end-group on the macro-initiator preferably should be –Br and the catalyst like CuCl, which halogen exchange is expected to take place forming a –Cl-terminated extending polymer chain and a CuBr catalyst complex. This type of halogen interchange or exchange requires a 1:1 ratio of macro-initiator to catalyst/transferable halogen. Whereas, when a low reactivity monomer is to be attached as the second segment in such a situation lower concentrations of catalyst can be used for the completion of copolymer formation. Similarly, multifunctional initiators can be used in the process of copolymer preparation of A-B-A type alternative copolymer or AB-star or multi-armed block copolymers. [30]

Interestingly, advanced approaches in polymer synthesis protocols have made it possible to extend the series of monomers of different chemical and physicochemical nature to be incorporated as a block in the copolymer. This can be achieved via sequential RAFT and ATRP polymerization processes following a dual functional catalyst bromoxanthate which acts as both initiator–transfer and agent–terminator (iniferter). For instance, poly(vinyl acetate)-*b*-PMMA, poly(vinyl acetate)-*b*-PS, and poly(vinyl acetate)-*b*-poly(methyl acrylate) block copolymers with a relatively low polydispersity of $Mw/Mn<1.25$ were prepared following successive RAFT and ATRP polymerization technique with the help of bromoxanthate iniferter as a single catalyst complex for both steps.

Similarly, several macro-initiators can be used for the preparation of copolymers via different polymerization processes including conventional radical polymerization (RP), cationic or anionic polymerizations like cationic ring-opening polymerization (CROP), anionic ring-opening polymerization (AROP), ring-opening metathesis polymerization (ROMP), condensation polymerization, and post-metallocene catalysis. The only condition is that copolymer chain should have at least one terminal functionality or be converted to an ATRP-initiating moiety. In such a situation, most of the ATRP catalyst/initiation system can be used to extend the copolymer block as an extension step. [31]

Recent advances have also led to the discovery of "click-chemistry" and its use in the preparation of multi-segmented block copolymers. "Click chemistry" was first introduced by K. B. Sharpless in 2001 to elucidate the reactions of high yielding nature that are used to create only byproducts that can be removed without any sophisticated techniques like chromatography. Generally, these byproducts are stereospecific and chemistry is simple to perform. In copolymer preparation, the copper-catalyzed cycloaddition reaction can be used to produce macromolecular architectures like diblock and triblock copolymers. For instance, α,ω-diazido-terminated PS-b-PEO-b-PS triblock copolymer can be prepared using propargyl ether in DMF in the presence of a catalyst, copper bromide/N,N,N',N",N"-pentamethyldiethylenetriamine (CuBr/PMDETA) in click chemistry with the help of microwave irradiation. Interestingly, recycled catalysts from these reactions can be recycled and reused for both the formation of the first precursor block in a copolymer backbone as well as in the chain-extended multi-block copolymer. [32]

In another block copolymer preparation method, difunctional oligomers are covalently linked to prepare high molecular weight block copolymers. For instance, in order to synthesize a copolymer of a poly(siloxane imide), poly(dimethyl siloxane) (PDMS), prepolymer or oligomers are prepared with end-capped aminopropyl functionality. This relatively large macromonomer can be further reacted with another diamine and a dianhydride to result in a larger macromolecule. As a result of this, the portions of the macromolecule that resulted from the reaction of second diamine and the dianhydride would be referred to as 'polyimide block', whereas the siloxane block would be called the 'siloxane block'. In another approach, the multiblock copolymers synthesis technique is commonly used in the preparation of polyether polyols such as poly(ethylene glycol) (PEG) or poly(propylene glycol) (PPG) as relatively low T_g and soft segment blocks. Following this technique, PPG or PEG diols are reacted with an excess of diisocyanate to form a 'pre-polymer' that are then subsequently reacted with a 'chain extender' such as butane diol (BD) and hexane diol orto form high molecular weight copolymers.

Interestingly, the ability of different dissimilar polymer oligomers, pre-polymers, or the monomeric blocks to combine to result in tailor-made end copolymer can be utilized to create fascinating block copolymer patterns. In this approach, the properties of the constitutive blocks offer tuneable thermal, physicochemical and mechanical properties. The primary advantage in designing block copolymers arises from the immiscibility nature of monomeric blocks due to their unique thermodynamics of phase mixing. Similarly, even though the blends of PDMS and polyimide homopolymers can also be phase separated, copolymers composed of PDMS block covalently linked with polyimide block would have significantly reliable phase separation. In addition to this, the phase separation process in block copolymers yields unique morphological sequences as a function of block composition which also dictates many of the physicochemical, thermal, and mechanical properties in the resulting block copolymer. Further, at temperature treatment between the low- and high-T_g blocks, which is also known as the service temperature window, the high glass transition temperature blocks function as physical crosslinks and the low-glass transition temperature locks act as more flexible segments resulting in flexible elastomeric properties. Due to this, even thermosets can be melt-processed at temperatures above the highest thermal transition.

7.5 Self-Assembly and Pore Formation in Block Copolymers

In block copolymers, the direct covalent linkage of two dissimilar blocks or units in macromolecular chains facilitates physical properties that considerably diverge from the homopolymer blend counterparts. Therefore, in diblock copolymers, the polymer chains no longer consist of isolated block molecules of type A and B respectively, and the miscibility of the two block components is still highly unfavorable. In this situation, the covalent connectivity confines facilitate the phase separation of dissimilar blocks at the molecular level. This phenomenon in block copolymer systems is known as micro-phase separation as described in Figure 7.3 offering several potential patterns or micro-sectional arrangements useful for several potential applications.

Interestingly, the phase behaviour in self-assembled linear-shaped diblock copolymer largely governed by three main experimentally controllable factors, namel:, (a) the molecular weight and the degree of polymerization (*N*), (2) the controllable volume fraction of one block (*f*), and (3) the Flory-Huggins interaction parameter (χ). The Flory–Huggins parameter, χ, describes the

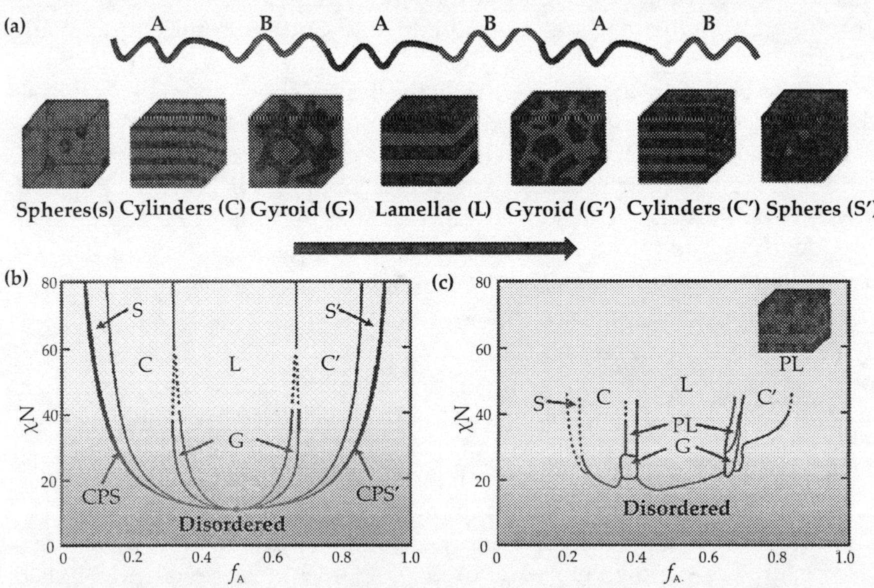

Figure 7.3. Schematic representation of (a) a symmetric A-B diblock copolymer having different morphologies inits melt states as a function of volume fraction of block A (f_A) resulting in sphere (S), cylinders (C), gyroid (G) lamellae (L) and and their inverse volume fractions, gyroid (G′), cylinders (C′), and sphere (S′). (b) Mean-field phase diagram, and (c) showing additional characteristic stable morphologies.

excess free energy of mixing which in both polymer blends as well as block copolymer systems govern phase separation behaviors. Interestingly, in chemically-distinct non-polar polymers, the value of interaction parameter, χ, is dominated by the mismatch in cohesive energy densities of the individual monomeric units. These interaction parameters provide direct measure of the interaction extent between molecules, for instance, polymers or surfactants , or between the polymeric molecule and solvent in which solution is prepared. [33]

When block copolymers are heat treated above their glass transition temperature (T_g), the polymer melt morphology turns into solution form well-defined block copolymers based on their narrow polydispersity in connection with the volume fraction of block copolymers. In film or membrane preparation, block copolymer can be prepared following solution casting. As in case of other polymers, the solution-casting techniques require specific solvent selectivity and optimal solution concentrations which greatly influences the polymer bulk morphology. The morphologies of copolymers in film or membrane form plays a vital role in applications such as water purification, fuel cells, high-performance thermoplastic elastomers, and so on. In high-performance thermoplastic applications, thermoplastic elastomers possess low-modulus polymer block characteristically having low-T_g which is covalently linked to high-modulus polymer block. Further, using inherent incompatibility of dissimilar blocks of copolymer selective micro-phase separation is achieved to impart the extraordinary mechanical properties such as tensile characteristics such as high reversible elongation, high tensile strain, and low strain hysteresis.

Pore sizes in block copolymers can be uniquely tuned and altered using the characteristic physicochemical properties of individual blocks in the systems. Interestingly, unique nano-sized structures can be spontaneously prepared when the interfacial area between the two immiscible phases is minimized in the block copolymer systems. For example, a lamellar morphology in final self-assembled block copolymer can be achieved with symmetric diblock copolymer having length of the A-block is equal to the B-block, while the size of the lamellae is determined by the total molecular weight as it is clear from Figure 7.3, which displays detailed phase behavior of diblock copolymers which is also summarized in the phase diagram. In this manner changing molar fractions of the blocks microstructures in block copolymer can be aligned as sphere (S), cylinder (C), gyroid (G), lamellar (L) and in special cases a bicontinuous micro-domains in the bulk assemblies or to relatively thick films. [34]

Microphase separation is a physicochemical phenomenon largely found in the polymer systems prominently studied in block copolymers composed of immiscible or incompatible blocks or chemical components, where they tend to instantaneously form macro- or micro-phase-separated structures or patterns with microscopic length scales because of intramolecular phase separation in bulk or in concentrated solutions. During the phase separation, the spatial scales are governed and driven by their chain lengths, whereas their morphologies can be altered through their relative chain length compositions. This phenomenon is called micro-phase separation, and the subsequent structures are named micro-phase separation structures or microphase-separated structures. In most of the designer copolymers the prefix "micro-" is used simply to refer to "micro-size or smaller" patterns and their realistic periodicity ranges from nearly few-nanometers to several hundred nanometers. This micro-phase separation directly depends on the length of a block in the polymer system or molecular weight of a unit or block or component of copolymers.

In this direction, a typical dimension of the chain length (*d*) can be in the range of 10 to 100 nm. This is the unique physical characteristic which is generally only found in block copolymers which can be easily designed to achieve nano-structured patterns. These characteristic patterns can be well controlled both by theoretical predictions and designing experiments. In addition to their ability to form self-assembled structures, conformational properties in copolymers can also be explored to control critical phase separation behavior compared to homopolymers. It is estimated that the critical point of symmetric homopolymer blends is generally denoted in volume fraction ($f_1 = f_2 = \phi c = 0.5$), however in block copolymer is theoretically found at $(\chi N)c = 2$. On the other hand, the critical point of a symmetric diblock copolymer seems to be substantially higher, which is $(\chi N)c = 5.2$. Interestingly, due to the inevitable A-B linkage in block copolymer with reduced gain in entropy of mixing in block copolymer melts accounts for this factor. In block copolymers, often a value of micro-phase separation factor is found to be 10.5 in the literature, but it should be noted that in this case *N* represents the block or segment length of the total copolymer, instead the individual blocks. Finally, for strongly and equally phase separated lamellar structures the chains are highly extended in their phase separation as the long period *d* scales to *N*2/3, whereas homopolymer melts simply follow random-walk statistics, that is, end-to-end distance or radius of gyration which is proportional to *N*1/2. On the other hand, minimization of the A-B blocks interfacial tension in dissimilar diblock copolymers is f_1, $6 = f_2$, driving the system to form unique but multiple other phases. In addition to lamellae architecture as discussed previously, the strong microphase regime of $\chi N>100$ can lead to self-assembly of spherical as well as cylindrical structures. In this case, both phases contain essentially pristine components and the A-B interfaces are very sharp. Although the interface in the phase separation broadens at intermediate separation strengths of $10<\chi N<100$, which results in additional bicontinuous gyroid network patterns or morphologies. These network patterns have also been known to result in stability in this volume fraction regime.

7.6 Controlling Pore Size in Block Copolymer Membranes

Block copolymers are a fascinating class of macromolecules which can provide a rich and well-defined morphologies in the melt and upon self-assemblies driven by the physicochemical interaction between various dissimilar blocks. When these unique macromolecules are used in membrane applications, block copolymers offer enormous possibilities in precisely controlling and tuning their degradable or easily removal block at the nanoscale level. These block copolymers can also be combined with different polymers as well as foreign materials with varying properties to make them composite in a single membrane. In addition to designer-type macromolecules can also be found in large scale commodity forms used in automobile tires made up of large scale rubby polymer such as poly(acrylonitrile-*b*-butadiene-*b*-styrene) block copolymer commercially known as SBS rubber. Similarly, nitrile and ethylene-vinyl acetate have also been combined to form commodity copolymers. Interestingly, block copolymer micelles are largely formed via the self-assembly of either amphiphilic or with oppositely charged blocks in an aqueous medium. In this configuration, a hydrophilic block forms a corona whereas hydrophobic blocks form the core of the micelles. In this way, block copolymers in bulk can be effectively assembled using restricted mechanical shear. [35]

Further, several researchers have used combinations of block copolymer systems to achieve microphase separation in the film or membrane form. For examples Angelescu et al. used shear flow to align block copolymers of polystyrene-*block*-polyisoprene (PS-*b*-PI), PS-*b*-poly(ethylene-*alt*-propylene) (PS-*b*-PEP) and poly(*n*-butyl acrylate)-*block*-poly(acrylonitrile) (PBA-*b*-PAN) resulting in PEP in the cylindrical and spherical microdomains. Interestingly, viscous PDMS oil or a cross-linked PDMS elastomer pad is used to impart a shear stress on the thin films with shear being employed at temperatures ranging between the T_g and the order/disorder temperature of block copolymer. To achieve PEP cylinder conformation in a single-layer film, the PEP cylinders were assembled or aligned parallel to the shear direction and a high degree of lateral ordering was achieved over cm^2 areas after imparting shear. However, study revealed that a monolayer of PEP spheres in (PS-*b*-PEP) cannot be shear aligned because of insufficient mechanical anisotropy of hexagonal lattices. Hence, a minimum of two or more layers of spheres in the self-assemblies required to obtain relatively good lateral ordering of shear-aligned spherical micro-domains. Study also demonstrated that the presence of defects in the shear-allied assemblies like dislocations poses an intrinsic limitation of this overall technique. As a consequence, shear aligning techniques play a vital role for the fabrication of block copolymer-based devices which does not require perfect lateral ordering. [36]

In another study, Thomas et al. developed a rapid 'directional solidification' process for alignment of selected segments in block copolymer domains, which depends on the controlled directional crystallization of solvents such as anthracene (AN) or benzoic acid (BA). For instance, semi-crystalline block copolymers containing polyethylene (PE) blocks were homogeneously mixed with *n*-butyl acrylate (BA) or AN above their melting temperature. Further, upon cooling, the solvents crystallize during which it acts as substrates on which epitaxial crystallization of PE blocks. This synergic combination of both epitaxy of PE blocks and directional solidification of solvents drives the crystalline substrates leading to highly ordered micro-domains that are assembled parallel in the direction of the rapidly grown solvent crystals. Moreover, directional solidification process can be utilized to align microstructures in diblock copolymers that are amorphous in nature such as PS-*b*-PMMA, poly(*n-butyl acrylate*)-*b*-poly(acrylonitrile) (PBA-*b*-PAN), and polystyrene-*b*-polyisoprene (PS-*b*-PI). In this case, only the directional solidification of organic solvents leads to alignment of AN or BA involving segregation of micro-domains along with the growth direction of solvent crystals. Interestingly, alignment, orientation or order and their behavior in block in copolymer system to form specific microstructures can be well controlled based on the molecular weights, polydispersity index, concentration of the solution and film thickness.

Therefore, several advanced studies reveal the potentials of block copolymer in controlling pore structural control in the range of 10 to 100 nm length scale is crucial for the functioning of many separation and purification applications. Besides ordered porosity, the mean pore size and their dispersion are few significantly important parameters that block copolymers provide for a range of applications. These broad but controllable pore characteristics are best suited in mass transport-limited situations even governed by the Knudsen diffusion along with size exclusive separation mechanisms. [37]

Technologically, block copolymer based self-assembly can provide a promising bottom-up approach to produce large scale ordered micro/mesoporous structures with controlled, tuneable pore size and morphologies. This process proceeds with a persistent micelle templating approach where core and corona sizes can be well ordered and regulated following the selection of appropriate block or fraction size. On the other hand, inorganic precursors typically derived from sol–gel methods can also be embedded into the corona fraction of block copolymer micelles through preferential supramolecular interactions. Following this, the hybrid block copolymer composite can be transformed into a well ordered inverse opal-type mesoporous structure via either heat treatment above their Tg or direct thermal calcination, or chemical treatments or photocatalytic reactions or various physical including UV-ozone degradation, oxygen etching. The use of block copolymer as a structure-directing template or agent which offers reliable and definitive control over pore size and porosity. Even though pore size and overall porosity can be easily tuned via simple the mixing ratio between different block sizes and in case of composites, organic and inorganic precursor, however, the pore size is mainly determined by the molecular weight of the sacrificial block which forms the micelle core, thereby resulting in a final pore dimensions and morphologies in the range of 5 to 50 nm as shown in Figure 7.4. [38]

Interestingly, another set of recent studies have revealed the close relationship between the dispersity of block copolymers and their size distribution during their self-assembly and co-assembly feature during phase separation. This helps in highlighting the significance of precise control over the macromolecular characteristics of the structure-directing template or agent. In general, the preparation of block copolymers with specific molecular weight and narrow dispersion (Đ) is a significant and elaborate process that involves a multi-step preparation route, some of which are difficult to control and terminate. Therefore, it is difficult to synthesise monodisperse copolymer blocks in general conditions. Therefore, block copolymer synthesis to achieve microdomain sizes resulting in MF or UF or NF is either conducted

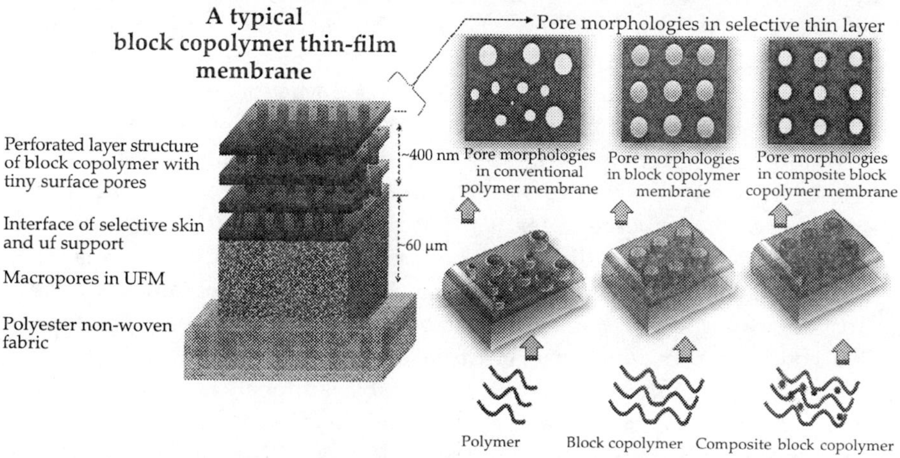

Figure 7.4. Schematic of a typical thin-film composite membrane and pore structures, morphologies and their orientation in thin-film selective layer originated from their source polymer precursors, namely, homopolymer, block copolymer and composites block copolymer.

following sequential addition of monomers via living or controlled chain-growth polymerization procedures or polymerization that relies on coupling reactions between different end-functionalized chains blocks or units or segments. In the first case, close control must be achieved over reaction kinetics and conversion efficiency in order to synthesise well-ordered and pre-defined molecular weights and narrow polydispersity block copolymer. For this purpose, cationic or anionic polymerization or controlled radical polymerization have been applied for the synthesis of narrow polydisperse block copolymers. In another strategy, effective coupling via end-group functionalization is followed to obtain well defined copolymer for example, Thiol–ene, Diels–Alders, or cycloaddition reactions. However, these coupling reactions have been designed as rapid and selective to ensure effective linking between the polymer chains. Nevertheless, in each case synthesis procedure and conditions are highly sensitive and vulnerable to form impurities and therefore, precise reaction conditions are highly desired to achieve narrow dispersive copolymer.

Further, purification of end block polymers from their impurities are achieved either by size exclusion chromatographic (SEC) fractionation or high-performance liquid chromatography (HPLC) along with the separation of main block copolymer by their molecular weights. For instance, fractions containing polystyrene-*block*-polyisoprene (PS-*b*-PI) copolymers with narrower distribution in molecular weight and chemical composition can be fractionated and collected using a HPLC via two step separation process.

7.7 Block Copolymer-based Microfiltration Membrane

Conventional microfiltration membranes are generally prepared using polymer solution cast or phase inversion techniques and in the form of electrospun nanofiber mats or scaffolds arranging them in the pore range typically between 0.1 μm and 1 μm. These polymers or membrane precursors are generally selected based on their high porosity, low density/weight, and highly flexible film forming property. However, MF membranes are also broadly classified film or fiber mat with pore packing or pore size ranging from 0.05 to 10 μm which make suitable as barrier or separation application as both filter and adsorptive membrane. With these pore characteristics, drinking water sources such as surface water, wastewater and other wastewater sources having emerging pollutants like natural microbial pathogens, organic matter, and complex surfactants can be easily treated to produce to obtain clean and safe drinking water. These membranes can be of film or flat caste or electrospun scaffold configuration which can be used effectively removing micron- or submicron-size contaminants. Also, by tuning surface characteristics of polymers, adsorbing metal ions and organic compounds can also be achieved. However, conventional membranes prepared from both plain film form or electrospun nanofibers can have either a single-self-supported layered structure or a multi-layered composite structure. [39, 40]

Even though self-supported MF membranes prepared using PVDF, PAN, PES, PSF, and so on have commercially been successfully used in various applications, border pore size distribution in conventional polymer membranes make overall performance inferior in the long run. MF membranes with microporosity (0.1–1 um) work on the principle of size exclusion, selectively allowing the smaller molecules to pass through and filtering out larger molecules. On the other hand, broader pore causes mechanical deformation in self-supported as well as electrospun nanofibrous membrane resulting in drastic decline in permeability, enables foulant intrusion in to the bulk of interconnected pore voids or nanofibrous polymer matrix in multi-layer nanofibrous layers.

In all cases, further to improve the solute rejection, permeability and overall performance of the MF membranes, in situ or post-functionalization is followed by grafting large-chain polymers on the backbone. In addition to compositing PAN, PVDF, PES and PSF, other commercially available membrane substrates like cellulosic, polyacrylates, polycarbonates, and polyolefins, have also been functionalized via surface pore activation process including UV-radiation-induced grafting, chemical grafting and plasma-induced grafting. However, the chemical, electron-induced excitation, and ionization processes cause dissociation due to which electrons and ions react with the monomer and the subsequent polymeric substrate through undesirable side reactions. This may cause difficulties in overall membrane characterization due to generation of a wide variety of surface-active sites with complex chemistries and functionalities.

Nonetheless, advanced approaches using block copolymers to attain very well-ordered pore structures with narrow pore size distribution. Block copolymers also provide high percentage of porosity in the membrane resulting in large accessible pores for solvent to permeate, useful for both

MF and UF separation applications. Overall, nanostructured pores can easily be achieved by tuning dissimilar block in the copolymers linked together covalently into the periodic arrays of spherical, cylindrical, gyroid and lamellar morphologies with typical feature sizes of 10-100 nm in which one component can be selectively removed via phase separation physicochemical parameters to generate the nanostructures. On the other hand, the block copolymers used for UF membrane applications, the self-assembly of relatively low molecular weight blocks are selected to achieve narrow pore size and their distribution along with higher pore densities. In this regard, researchers have made numerous advanced attempts to achieve well-defined cylindrical or 3-dimensional nanostructures from block copolymer films or membranes to produce narrowly distributed UF range pores.

7.8 Block Copolymer-based Ultrafiltration Membranes

Like MF membranes, several advanced but innovative protocols have been adopted for the preparation of block copolymer-based on UF membranes. Some of these strategies involve selective solvent evaporation, selective-etched monoliths, and non-solvent induced phase separation (NIPS). For instance, Cussler et al. used polylactide (PLA)- containing block copolymer was transformed to generate porous templates following rapid solvent evaporation or shear-aligned monoliths procedure. This is followed by the selective hydrolysis of the PLA microdomains to result in highly ordered ultra-porous structures in the resulting membrane. On the other hand, Peinemann and group prepared a new type of nanoporous ranging in UF membrane range using non-solvent induced phase separation with polystyrene-*b*-poly(4-vinylpyridine) (PS-*b*-P4VP). Through this approach the cylindrical channels were introduced on the top of a random porous-sponge-like layer that contains large porous structures. With this, the size selectivity during separation of biomolecules type solutes can be well controlled by altering the pH-responsiveness in which characteristics of P4VP can easily be controlled to perform separation and transportation. In another study Kim et al. developed nanoporous UF membrane selective etching of the cylindrical microdomains of PMMA were generated and selectively etched to create an ultra-sized continuous porous network. This method is particularly well controlled in generating microdomains which can be isolated in the range of 30 nm. These UF membranes were used for the removal of human rhinovirus type 14 (HRV-14). This work also inspired further reconstruction of polystyrene-*b*-poly(methyl methacrylate) (PS-*b*-PMMA)-based series of ultrafiltration membranes for various applications.

In most of the studies PS-*b*-PMMA with different mole fractions have been fabricated to result in UF range membranes. These results demonstrate the ability to fabricate nanoporous UF membranes from self-assembled BCP morphologies via tuneable pore size appropriate for the selective separations of a large range of chemical species, in particular biomolecules. However, separating smaller divalent ionic species, dye molecules and biological contaminants is still a challenge. However, in all cases, it is vital to avoid or nullify the component-degradative routes which can reduce the mechanical stability of the membrane, even when it retains overall membrane performance. Therefore, to induce mechanical stability to UF membranes, a relatively high concentration of PSb-PMMA block copolymer solutions were casted to prepare thick films followed by thermal annealing on substrates with well-balanced interfacial interactions to prepare the membrane. This also allows the microdomains orientation normally aligned to the substrate and air interfaces, however the bulk of the copolymer membrane would still experience random orientation in the bulk. On the other hand, use of appropriate solvent which is selective to the degradable PMMA in both for sphere and cylindrical microdomains can also be well designed in which excessive solvent could cause simple osmotic swelling of the PMMA microdomains. Subsequent rinsing of swollen microdomains in selective solvent can be washed out to result in nanopores that penetrate to the bulk and through the membrane resulting in UF membrane with pores ranging from 6 to 17 nm. These exceptionally well-ordered pore in the UF membranes can be used for selective separation of colloidal pollutants such as nanoparticles, for instance Au NPs.

Therefore, several studies presented that UF range pore generation can be achieved and be controlled in the block copolymer with sphere and cylindrical microdomains orientations at both interfaces. On the other hand, relatively thick nanoporous UF membranes can easily be produced via a simple swelling and rinsing of the minor component in a matrix that remains glassy or degraded in nature. Also, using designer or appropriate solvent which is selective towards PMMA trapped in spherical or cylindrical microdomains in which the restricted or confined swelling of PMMA microdomains can be rapidly removed to generate highly uniform nanopores architecture that penetrated through the membrane. [41-43]

7.9 Recent Advances in Block Copolymer Membranes Preparation Strategies

Recent advances have mainly adopted heat treated or thermally driven physical pore forming techniques in various block copolymer-based membranes. Like the directional solvents crystallization process, temperature and concentration gradients processes have been used to achieve ordered microphase separation in long-ranged patterns in the block copolymers. In this direction, Hashimoto et al. followed a thermal gradient technique to control the macroscopic ordering and orientation of microphase separation of block copolymer microdomains in bulk of membranes. This is achieved in block copolymer following order-disorder transition (ODT) temperature of the overall system. An ODT occurs when a macromolecular system containing block copolymers changes between one of the ordered phases to the disordered phase, vice versa. Therefore, through heat treatment, the entire block copolymers membranes above the ODT and subsequently cools with a slow interval in a thermal gradient, first the coolest part of the membrane gets microphase separated. Further, with gradual time, the thermal gradient reaches across the membrane resulting in the ordered microphase separation in line with the template existing microdomains. Therefore, microphase separation and their unique order orientation is induced in the block copolymers with controlled morphological arrangements following thermal gradient which is moved across the membranes slowly and steadily.

Further, Angelescu et al. followed an advanced method of microdomains creation in block copolymer membrane. In this method, block copolymer film was first casted on silicon wafer which is facilitated by microfabricated heater and sensor to control and monitor the heating of block copolymer film on the wafer and further creation of microphase separation. This method is known to induce well controlled and accurately placed microdomains in a block copolymer due to the accuracy in temperature gradient and cooling which helps in well-ordered self-assemblies. When the temperature gradient is applied on block film coated on silicon wafer, layer-by-layer (monolayer) cylindrical microdomains generate further heating that results in a high degree of lateral order is induced along with parallel cylinders in the bulk of the membranes or film. Study also revealed that an oscillatory temperature gradient could decrease the potential defect density in the lateral ordering of the block copolymer films. Nevertheless, the orientation of cylindrical domains was not well correlated to the direction and the propagation of temperature gradient that induced in the film or membrane. Further to understand on this pattern Berry et al. extended the thin film-zone refinement process to polymer system where the block copolymer was heat treated to temperatures well below the ODT however, it was found that there was sufficient mobility in the blocks to have a local reorientation or reorganization in the block copolymer which minimizes the possibilities of removing packing or orientation defect in the resulting membranes or films. Therefore, the study noted the significance of slow and steady heat treatment of block copolymer and gradual but slow cooling across the membrane or film results in well aligned and ordered cylindrical microdomains with microscopic distances. This would be vital in creating well-ordered and well denied pore structures in the block copolymer membranes useful for water treatment applications.

In another study, a zone-casting method in which polymer solutions are casted in to form thin films or membrane on a mobile substrate. This technique has been adopted to create large-scale arrangement of nanoscale domains in microphase separated domains in block copolymers. Further, Kowalewski et al. demonstrated use of poly(*n*-butyl acrylate)-*block*-poly(acrylonitrile) (PBA-*b*-PAN) to create a large-scale and long-range ordered lamellae in block copolymer thin films or membrane using a simple zone casting technique. In this technique, copolymer solution was prepared in *N,N*-dimethylformamide (DMF) and following zone-casting method membrane onto a moving substrate with the help of a flat nozzle syringe equipment. In this technique, solvent evaporation rate plays a vital role in obtaining ordered microphase separated domains. Therefore, to obtain well-ordered porous systems in block copolymers, solvent evaporation rate is well controlled via the temperature gradient and alterations in the copolymer solution as well as substrate temperature. It also should be noted that, by the orientation and order of the microdomains of block in the film or membrane can also be controlled by altering solution casting parameters as well. [44]

In another study, Kimura et al. followed a simple gradient method to prepare long-ranged and ordered cylinder-type phase separation in PS-*b*-PB block copolymer through the flow of solution on a tilted stationary as well as flexible substrates followed by subsequent solvent evaporation and annealing. Following this, the parallel cylindrical phase separated domains were observed aligned in the direction of solution flow on substrate with a high degree of lateral order over large surface areas. This is significantly noticed near the top edge where the polymer solution flow field is strongest. Further, Tang et al. followed a similar to zone-casting technique type method to prepare aligned lamellar-forming PBA-*b*-PAN block copolymer membrane. In this design, zone-casting strategy type protocol was followed to inject the block copolymer solution through the nozzle on the substrate in slow moving condition which controlled the solvent evaporation facility. Through this process, the microphase separation resulted following solvent evaporation leading to perpendicular lamellae assembly in the macroscopic size. [45]

In another method, ultraviolet light irradiation assisted simultaneous block copolymer cross-linking and selective block degradation techniques have been developed to create highly ordered porous structures in the membranes. For example, UV radiation on copolymer film/membrane was induced to induce *cis–trans* isomerization, which can transform conformational change either in the range of volume change of specific block or even an alteration in the segmental interaction parameter. In either case, the block copolymer film or membrane can be navigated through an ODT of the polymer system. In this technique, wavelength of UV radiation can be altered to reverse the isomerization in case of a particular wavelength radiation is not sufficient to drive a *trans–cis* isomerization tins the block copolymer to produce ordered state self-assembly. Through this via sweeping of light between two bands across the surface, the ordered block microdomains separation in block copolymers can be arranged in order with fixing several defects into the disordered state and back into the ordered state. However, as the order in block of a copolymer system appears in the form of microphase separation is achieved through the template mechanism via the existing microphase-separated morphology. This technique is exactly like that of thermal gradient process, however, only change in the process is absence of physical contact driving force with the block copolymer sample. This is an added advantage compared to the previous techniques. However, realizing this photo-assisted process in real condition has proven to be difficult, since technique firstly requires the synthesis of block copolymers that are photoactive and procedures to prepare photo-active block copolymers are very less in number as of today.

7.10 Conclusion

Future and challenges in developing block copolymer membranes

In summary, block copolymers have been considered as versatile and speciality polymers that can provide several fascinating properties due to their unique and specific repeating units, which gives them an outstanding candidate characteristic to be a polymer-based membrane. Since, modern-day polymeric membranes demand a well ordered pore structure on an symmetric substrate, the block copolymers can be one of the best candidates to be an advanced class of membranes useful for separation and purification technologies, in particular water treatment applications. Ever since their first use in 2007 as asymmetric porous structure produced through phase separation techniques, block copolymers have attracted tremendous interest to be used as MF or UF membranes. Since then, several researchers from polymer and membrane science background have adopted this new field of materials research to develop thin-film form porous structures that are stable on a variety of substrates.

Further, several polymer modification methods have also been followed to fabricate isoporous flat-sheet membranes like that of conventional technique such as non-solvent induced phase separation (SNIPS) in combination with evaporation-induced self-assembly of block copolymers. Following this new concept, a controlled solvent evaporation is achieved with the help of controlled gas flow. This novel approach is known as gas-flow-induced polymers self-assembly (gSNIPS). Further systematic studies following the gSNIPS concept, several designer pore morphologies achieved through the process parameter altering such as the concentration of the block copolymer and gas flow. These parameters have shown to be useful in drastically reducing the overall cost of the membrane fabrication process while maintaining the uniqueness of porosity and their morphologies.

In another recent advancement, large scale block copolymer-based membranes have been prepared using polymer spraying technique. Here, controlling concentration of block copolymers and subsequent spray parameters, economical technological advantages are being achieved compared to conventional solution-casting doctor-blade technique followed by phase inversion protocols. Here, the scale-up of the spray coating process can be easily achieved by designing simple approaches, namely, one-fluid nozzles, two-fluid nozzles, and air-brush that are commonly used in the coating industries. Nonetheless, recent advancements in the spray coating technologies can easily be adopted to achieve optimum membrane thickness on various compatible substrates. One such attempt has been the use of spray technique to coat polystyrene-block-poly(4-vinylpyridine) diblock copolymer solution and their subsequent conversion into ordered nanoporous membranes. In this study, among many parameters, two-fluid nozzles approach has been identified as the most promising and potential route to achieve efficient end product. This approach allows the coating of membrane or polymer film ~150 nm homogeneously on various flexible substrates which can facilitate further iso-porosity. One of the major limitations with spray coating technique, when low concentration polymers are used on relatively porous substrates, coating solutions tend to penetrate and lead to pore blocking. In such a situation, membranes have shown low permeances 50 L/(m2.h.bar), and in some cases complete dense film formation due to pore blocking has been observed. To avoid such a situation, researchers have further developed an advanced method of tuning porosity in block copolymer self-assembled structures via template-assisted nano-catalysis. In this approach, all types of block copolymers including diblock, triblock, and so on with tailor-made and tuneable block periodicity, domain order, orientation, and degree of lateral orders are achieved.

Therefore, exploration of block copolymers through their versatile self-assembly potentials using various types and combinations of block copolymers led to the discovery of several MF and UF membranes that have shown great potential in various applications. More recently, anti-fouling propensity characteristics of block copolymers have attracted growing attention with many promising applications other than UF/MF such as nanofiltration, controlled drug delivery, cell separation, template-directed membrane catalyst, optics, and gas separation. Further, recent advances in the field of block copolymer-based *membranes* have broadened their reach not only restricted to applications but also to design, develop and synthesise block copolymers of narrow polydispersity index. This is believed to define porosity, structure, and morphologies in the target membrane in both flat sheet as well as hollow fiber conformation to be applied as membranes under various real-life conditions. Overall, the advanced research in the field of block copolymer membranes and their ordered nanostructure has especially abled separation and purification technologist and innovators to adopt new class of materials that are well defined and organized for specific real-time applications with great potentials forecasted for further growth.

References

1. Shanthana Lakshmi Duraikkannu, Roberto Castro-Muñoz, Alberto Figoli, A review on phase-inversion technique-based polymer microsphere fabrication, Colloid and Interface Science Communications, Volume 40, 2021, 100329, ISSN 2215-0382.
2. Ayesha Kausar (2017) Phase Inversion Technique-Based Polyamide Films and Their Applications: A Comprehensive Review, Polymer-Plastics Technology and Engineering, 56: 13, 1421-1437.
3. Agnieszka K. Hołda, Ivo F.J. Vankelecom, Understanding and guiding the phase inversion process for synthesis of solvent resistant nanofiltration membranes. Appl. Polym. Sci. 2015, doi: 10.1002/APP.4213042130.
4. Kiso, Y. Factors affecting adsorption of organic solutes on cellulose acetate in an aqueous solution system. Chromatographia 22, 55–58 (1986).
5. Editors: Hui-Hsin Tseng, Woei Lau, Mohammad Al-Ghouti, Liang An, 60 Years of the Loeb-Sourirajan Membrane, Principles, New Materials, Modelling, Characterization, and Applications, 1st Edition - April 2, 2022, Paperback ISBN: 9780323899772, eBook ISBN: 9780323886246.
6. T. Matsuura, P. Blais, S. Sourirajan, J. of Applied Polymer Science, 20, 1515 (1976).
7. AbdElAziz A. Nayl 1,* , Ahmed I. Abd-Elhamid 2 , Nasser S. Awwad 3 , Mohamed A. Abdelgawad 4 , Jinglei Wu 5 , Xiumei Mo 5 , Sobhi M. Gomha 6,7, Ashraf A. Aly 8 and Stefan Bräse Review of the Recent Advances

in Electrospun Nanofibers Applications in Water Purification, Polymers 2022, 14, 1594.

8. Rini John, Kaushik Pal, Jitha S Jayan, Saritha Appukuttan, Kuruvilla Joseph, New emerging review on advances in block copolymer based water purification membranes, Journal of Molecular Structure, Volume 1231, 2021, 129926, ISSN 0022-2860.
9. Leiming Guo, Yong Wang and Martin Steinhart, Porous block copolymer separation membranes for 21st century sanitation and hygiene, Chem. Soc. Rev., 2021,50, 6333-6348.
10. D. M. Warsinger, S. Chakraborty, E. W. Tow, M. H. Plumlee, C. Bellona, S. Loutatidou, L. Karimi, A. M. Mikelonis, A. Achilli, A. Ghassemi, L. P. Padhye, S. A. Snyder, S. Curcio, C. Vecitis, H. A. Arafat, J H. Lienhard, A review of polymeric membranes and processes for potable water reuse, Progress in Polymer Science, 81: 209-237, June 2018.
11. Hieu Trung Nguyen, Minh-Thuan Pham, Truc-Mai Thi Nguyen, Ha Manh Bui, Ya-Fen Wang & Sheng-Jie You (2022) Modifications of conventional organic membranes with photocatalysts for antifouling and self-cleaning properties applied in wastewater filtration and separation processes: A review, Separation Science and Technology, 57: 9, 1471-1500.
12. Najmul Haque Barbhuiya, Utkarsh Misra and Swatantra P. Singh, Synthesis, fabrication, and mechanism of action of electrically conductive membranes: a review, Environ. Sci.: Water Res. Technol., 2021,7, 671-705.
13. W.J. Lau, A.F. Ismail, N. Misdan, M.A. Kassim, A recent progress in thin film composite membrane: A review, Desalination, Volume 287, 2012, Pages 190-199, ISSN 0011-9164.
14. Zhe Yang, Peng-Fei Sun, Xianhui Li, Bowen Gan, Li Wang, Xiaoxiao Song, Hee-Deung Park, and Chuyang Y. Tang, A Critical Review on Thin-Film Nanocomposite Membranes with Interlayered Structure: Mechanisms, Recent Developments, and Environmental Applications, Environmental Science & Technology 2020 54 (24), 15563-15583.
15. haskar, V.V., Kaleekkal, N.J. Next-generation thin-film composite nanofiltration membranes for water remediation: a review. emergent mater. (2021).
16. Ismail MF, Islam MA, Khorshidi B, Tehrani-Bagha A, Sadrzadeh M. Surface characterization of thin-film composite membranes using contact angle technique: Review of quantification strategies and applications. Adv Colloid Interface Sci. 2022 Jan; 299: 102524. doi: 10.1016/j.cis.2021.102524. Epub 2021 Sep 24. PMID: 34620491.
17. Qihui Qian, Patrick A. Asinger, Moon Joo Lee, Gang Han, Katherine Mizrahi Rodriguez, Sharon Lin, Francesco M. Benedetti, Albert X. Wu, Won Seok Chi, and Zachary P. Smith, MOF-Based Membranes for Gas Separations, Chemical Reviews 2020 120 (16), 8161-8266.
18. Xin Li, Yuxin Liu, Jing Wang, Jorge Gascon, Jiansheng Li and Bart Van der Bruggen, Metal–organic frameworks based membranes for liquid separation, Chem. Soc. Rev., 2017,46, 7124-7144.

19. Die Ling Zhao, Susilo Japip, Yu Zhang, Martin Weber, Christian Maletzko, Tai-Shung Chung, Emerging thin-film nanocomposite (TFN) membranes for reverse osmosis: A review, Water Research, Volume 173, 2020, 115557, ISSN 0043-1354.
20. Jun Tae Jung, Jeong F. Kim, Ho Hyun Wang, Emanuele di Nicolo, Enrico Drioli, Young Moo Lee, Understanding the non-solvent induced phase separation (NIPS) effect during the fabrication of microporous PVDF membranes via thermally induced phase separation (TIPS), Journal of Membrane Science, Volume 514, 2016, Pages 250-263, ISSN 0376-7388.
21. Tasselli, F. (2014). Non-solvent Induced Phase Separation Process (NIPS) for Membrane Preparation. In: Drioli, E., Giorno, L. (eds) Encyclopedia of Membranes. Springer, Berlin, Heidelberg.
22. Thomas H. Epps, III *a and Rachel K. O'Reilly , Block copolymers: controlling nanostructure to generate functional materials – synthesis, characterization, and engineering, Chem. Sci., 2016, 7, 1674-1689.
23. Hongbo Feng , Xinyi Lu , Weiyu Wang , Nam-Goo Kang and Jimmy W. Mays Block Copolymers: Synthesis, Self-Assembly, and Applications, Polymers 2017, 9, 494.
24. Sajjad Husain Mir2, Gaulthier Rydzek, Larry Akio Nagahara, Ajit Khosla and Parvaneh Mokarian-Tabari, Review—Recent Advances in Block-Copolymer Nanostructured Subwavelength Antireflective Surfaces, Journal of The Electrochemical Society, Journal of The Electrochemical Society, 2020 167 037502.
25. Lei Chen, Shuo Wang, Qianqian Yu, Paul D. Topham, Changzhong Chen and Linge Wang, A comprehensive review of electrospinning block copolymers, Soft Matter, 2019,15, 2490-2510.
26. Ho-Cheol Kim, Sang-Min Park, and William D. Hinsberg, Block Copolymer Based Nanostructures: Materials, Processes, and Applications to Electronics, Chemical Reviews 2010 110 (1), 146-177.
27. Yiyong Mai and Adi Eisenberg, Self-assembly of block copolymers, Chem. Soc. Rev., 2012,41, 5969-5985.
28. Zhen Wang,Chun Lam Clement Chan,Tianheng H. Zhao, Richard M. Parker, Silvia Vignolini, Recent Advances in Block Copolymer Self-Assembly for the Fabrication of Photonic Films and Pigments, Advanced Optical Materials, Volume 9, Issue 21 2100519.
29. Liwen Xing, Ji-Ren Liu, Xin Hong, Kendall N. Houk, and Christine K. Luscombe, An Exception to the Carothers Equation Caused by the Accelerated Chain Extension in a Pd/Ag Cocatalyzed Cross Dehydrogenative Coupling Polymerization, Journal of the American Chemical Society 2022 144 (5), 2311-2322.
30. Krzysztof Matyjaszewski and Jianhui Xia, Atom Transfer Radical Polymerization, Chemical Reviews 2001 101 (9), 2921-2990.
31. Angelika E. Neitzel, Thomas J. Haversang, and Marc A. Hillmyer, Organocatalytic Cationic Ring-Opening Polymerization of a Cyclic

Hemiacetal Ester, Industrial & Engineering Chemistry Research 2016 55 (45), 11747-11755.

32. Neal K. Devaraj and M. G. Finn, Introduction: Click Chemistry, Chemical Reviews 2021 121 (12), 6697-6698.
33. Alberto Alvarez-Fernandez, Barry Reid, Jugal Suthar, Swan Yia Choy, Maximiliano Jara Fornerod, Niamh Mac Fhionnlaoich, Lixu Yang, Benjamin Schmidt-Hansberg and Stefan Guldin, Fractionation of block copolymers for pore size control and reduced dispersity in mesoporous inorganic thin films, Nanoscale, 2020,12, 18455-18462.
34. Nunes, S. P. (2016). Block Copolymer Membranes for Aqueous Solution Applications. Macromolecules, 49(8), 2905–2916.
35. Chen Li Qian Li Yusuf Valentino Kaneti, Dan Hou, Yusuke Yamauchi and Yiyong Mai, Self-assembly of block copolymers towards mesoporous materials for energy storage and conversion systems, Chem. Soc. Rev., 2020, 49, 4681-4736.
36. Alberto Alvarez-Fernandez, Barry Reid, Jugal Suthar, Swan Yia Choy, Maximiliano Jara Fornerod, Niamh Mac Fhionnlaoich, Lixu Yang, Benjamin Schmidt-Hansberg, Stefan Guldin, Nanoscale , Volume 12, Issue 35, Online publication date: 2020.
37. Maryam Radjabian, Volker Abetz, Advanced porous polymer membranes from self-assembling block copolymers, Progress in Polymer Science, Volume 102, 2020, 101219, ISSN 0079-6700.
38. Zhang, S., Zhou, J., Wang, Z. et al. Preparation of polysulfone-based block copolymer ultrafiltration membranes by selective swelling and sacrificing nanofillers. Front. Chem. Sci. Eng. 16, 745–754 (2022).
39. Yan Luo, Xiaoteng Wang, Ren Zhang, Maninderjeet Singh, Ali Ammar, Diana Cousins, Mohammad K. Hassan, Deepalekshmi Ponnamma Samer Adham, Mariam Al Ali Al-Maadeed and Alamgir Karim, Vertically oriented nanoporous block copolymer membranes for oil/water separation and filtration, Soft Matter, 2020,16, 9648-9654.
40. Jayraj K. Shethji,Stephen M.C. Ritchie, Microfiltration membranes functionalized with multiple styrenic homopolymer and block copolymer grafts, Journal of Applied Polymer Science, Volume132, Issue36, September 20, 2015.
41. Burts, K.S., Plisko, T.V., Bildyukevich, A.V. et al. Modification of polysulfone ultrafiltration membranes using block copolymer Pluronic F127. Polym. Bull. 78, 6549–6576 (2021).
42. Jiemei Zhou and Yong Wang, Selective Swelling of Block Copolymers: An Upscalable Greener Process to Ultrafiltration Membranes? Macromolecules 2020 53 (1), 5-17.
43. Dongwei Ma, Jiemei Zhou, Zhaogen Wang, Yong Wang, Block copolymer ultrafiltration membranes by spray coating coupled with selective swelling, Journal of Membrane Science, Volume 598, 2020, 117656, ISSN 0376-7388.
44. Maciej Kopeć, Rui Yuan, Eric Gottlieb, Carlos M. R. Abreu, Yang Song, Zongyu Wang, Jorge F. J. Coelho, Krzysztof Matyjaszewski, and Tomasz

Kowalewski, Polyacrylonitrile-b-poly(butyl acrylate) Block Copolymers as Precursors to Mesoporous Nitrogen-Doped Carbons: Synthesis and Nanostructure, Macromolecules 2017 50 (7), 2759-2767.

45. Fujiwara, T., Kimura, Y., & Teraoka, I. (2000). Tailoring of block copolymers based on the stoichiometric control of the end-functionality of telechelic oligomers and the utilization of large-scale fractionation by phase fluctuation chromatography: a synthetic strategy for the preparation of end-functionalized poly(L-lactide)-block- poly(oxyethylene). Journal of Polymer Science, Part A: Polymer Chemistry, 38 (13), 2405-2414.

CHAPTER

8

Recent Advances in Nanotechnology-based Water Filtration

— ABSTRACT —

With recent innovations in the field of nanoscience and nanotechnologies, newer properties of materials have been discovered and evaluated for various applications including water treatment applications. Consequently, there have been several efforts to develop nanomaterials that are capable of treating, sequestering, scavenging, or absorbing harmful substances, chemicals, and pollutants, from contaminated watebodies. Among many, ceramic, inorganic nanomaterials, natural clay-based nanomaterials, and nano-sized polymer systems include both synthetic as well as biopolymers. Several initial studies in this direction have revealed that nanomaterial-based water filters can be a revolutionary alternative to all the conventional water treatment techniques. In the recent past, the use of various nano-sized materials in the form of nanoparticles, carbon composites, filter beds, ion-exchange cartridges, and composites in the polymer matrices that are converted into aerogels, fibrous filters as well as natural fibres have shown proficient potential in achieving long-term sustainable goals in water treatment sector. In particular, polymers and their blend systems such as chitosan/sodium alginate-based have been used to induce surface functionality using various inorganic nanomaterials for the rapid removal and retention of hazardous and emerging pollutants from contaminated water sources. This chapter provides in detail information on the recent advances in nanomaterials-based technologies aiming to fulfil ever-increasing futuristic demand in the water purification and wastewater treatment sectors.

8.1 Introduction

Globally, the water crisis is forcing researchers, innovators, technologists, and policy-makers to formulate stricter guidelines for water resource management, water usage, wastewater generation, and discharges. As it is a historically well-known fact that water is crucial for life and due to overexploitation of resources, the number of freshwater sources available for human consumption is diminishing or shrinking by the day. On the other hand, increasing the variety of pollutants and their concentration levels makes conventional water treatment technologies either ineffective or irrelevant. This is adding more to the already severe problem. Meanwhile, attempts are being made to bring down the cost of treated drinking water pumped from various sources. In this direction, membrane-based technologies have been now used as cost-effective and quick treatment techniques which have drastically improved the lives of larger populations in recent times in underdeveloped countries. [1]

Nevertheless, there is estimated over 80% of the world's wastewater generated through various sources like municipal, industrial, and agricultural activities are dumped or discharged into the environment untreated or partially treated. These contaminated wastewaters are generally characterized by high concentrations of hazardous organic and toxic inorganic compounds, which in the due course of time transform into a complex system and enter the food chain thereby endangering the wildlife, humans, and aquatic organisms. Upon consumption of traces or higher concentrations of modern pollutants can cause immunological disorders, cancer, developmental retardation, kidney damage, and so on. According to recent studies, less than 1% of the world's freshwater resources are accessible, and it is estimated that the challenges will remain worse by 2050. According to some studies, global demand for freshwater resources remains predicted to be more than one-third of what it is now. The ever-growing population and growing demand for high-quality potables can be satisfied in near future, only if water resources and wastewater generation and discharges are managed efficiently, effectively, and sustainably through the appropriate treatment and management techniques. Therefore, water resources and waste management become a prerequisite and top priority for a better future. [2]

On the other hand, to address the issues associated with accessible freshwater and water scarcity, for decades, several traditional methods, materials, and treatment techniques have been tried and tested. The most used water treatment and management methods have been ion-exchange, adsorption, precipitation, coagulation, membrane-based filtration, oxidation, electro-oxidation, floatation, and several recent innovations which are yet to be tried at the community as well as commercial levels shown in the form of infographic in Figure 8.1. These techniques over the years have shown several advantages, but many of these methods have also shown technological limitations to achieve total environmental remediation. Among all, membrane technologies have been relatively quick and clean and used in widespread domestic water treatment at low-cost operations. Some of the other major advantages associated with commercially successful membrane technologies are: (a) the membrane process operates and produces quality water without heating, (b) pollutant separation is purely physical in nature, and (c) membrane separation processes do not require additional chemical or any other addictive substances to facilitate the treatment process. These factors make overall membrane processes widely acceptable. Some of these techniques are quick in producing clean water such as adsorption and advanced oxidation processes used in the treatment of a variety of contaminated water as well as wastewaters. [3-6]

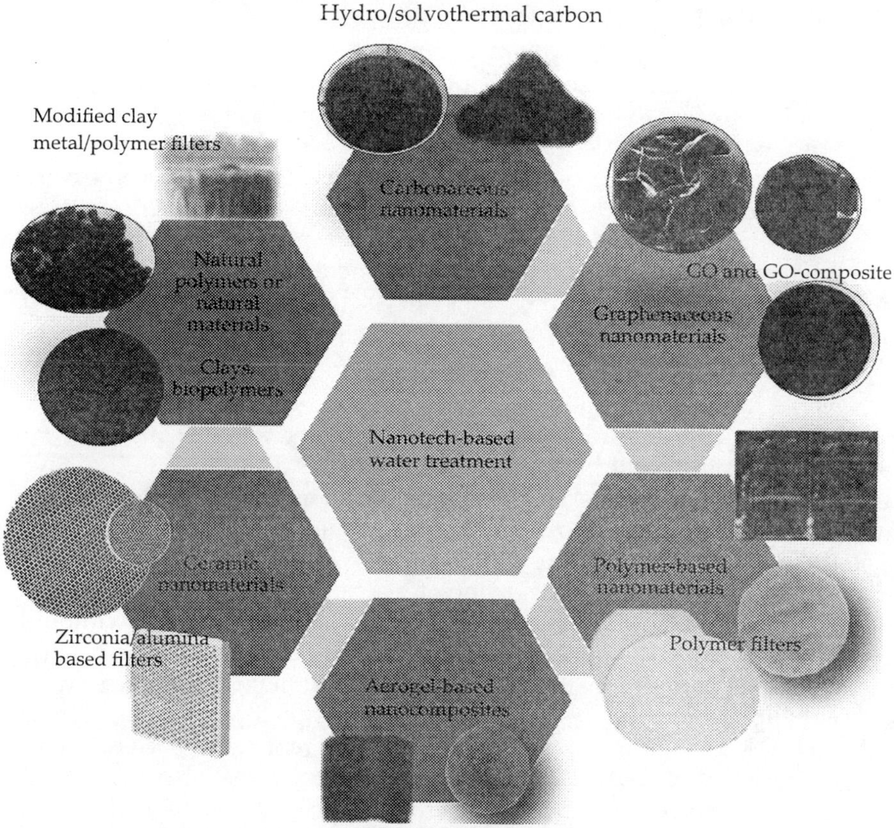

Figure 8.1. Infographics of several conventional as well as advanced water filters prepared using various precursors and their modified versions, including polymers, carbons, graphene, and ceramics.

Nonetheless, conventional treatment materials and methods have lately shown some negative impression on overall treatment processes such as the production of byproducts, extensive use of synthetic materials shadowing the issues of sustainability, and cost of input energy in both preparations as well as during their operation. For instance, conventional adsorbents and reagents used in oxidation processes are known to produce several by-products of much higher toxicity and complexities than the targeted pollutants themselves. Likewise, organic, and bio-molecular fouling on membrane surfaces limits their life span to such an extent that the discarded membranes themselves have been considered an environmental hazard. Thus, many conventional water treatment materials and methods need a complete overhaul in addition to developing sustainable maintenance protocols. However, additional maintenance materials and their usage protocols, in turn, increase the overall operational cost of the treatment process. In general, an ideal pollutant retention media or adsorbent or membrane, or reagents should have robust and high productivity arrangements. Widely used membranes should be turned into high permeability barriers with enhanced contaminant rejection characteristics maintaining a long shelf-life.

Even though recent studies have adopted several advanced protocols, these conventional materials still suffer from severe issues like poor flux permeability, ineffective separation, retention and/or selectivity, restricted porosity and regeneration, poor mechanical robustness, and declining activity with time. In addition to this, membrane technology is ineffective to convert large amounts of resources into purified water which makes the overall process inferior. Both at the domestic level as well as industrial scale, recovery is restricted for close to 70% of the feed water during the treatment process. Further, the presence of newer and persistent or emerging pollutants such as drug molecules, microplastics, artificial hormones, perfluoro compounds, and heavy metal ions restrict their efficiency. Therefore, limited separation performance in addition to high capital investments involved with membrane technologies restricts their overall use in domestic applications. [7-9]

Nevertheless, recent advances in membrane development using nanotechnology-assisted protocols have shown greater potential in improving membrane life and water recovery efficiencies. Advanced studies have revealed that the nanomaterial-based protocols majorly governed by the physicochemical interactions are gaining prominence globally because of their efficiency, reliability, and ease of processing at a low cost and with minimal inexpensive infrastructure. These nanomaterials possess a large surface area-to-volume ratio and tuneable surface functions which help them to withhold pollutants selectively. Nanomaterials can also be prepared and used in batches and large-scale quantities making them potential candidates for future water treatment applications. Interestingly, many precursors for the nanomaterials can be selected from natural sources which are inherently anti-bacterial, anti-viral, and resistant to organic as well as bio-fouling. On the other hand, with the suitable tuning of size, shape, and morphologies of the nanomaterials, materials' properties can be enhanced to their theoretical upper limits.

With these extraordinary optimized properties, nanomaterials can be packed in a device and can be prepared with specific selectivity to pollutants with high permeability, controlled wettability, catalytic activity, antifouling characteristics, and remarkable anti-bacterial activity through their regulated structures. Thus, nanomaterials have been evaluated to be an effective filter media in removing a series of pollutants of different nature such as organic and inorganic origin from an aqueous medium thereby potentially exhibiting an economically feasible and realistic alternative for water treatment in near future. Even though conventional membranes that are widely available with instant product generation capabilities for a wide range of applications will play a vital role in addressing issues associated with contaminated water, however, search for sustainable materials stock that goes for the preparation of membrane, membrane module, and engineering device will remain the priority.

Thus, various nanomaterials with a large volume to surface area and many surface-active functional groups have been developed and utilized in recent years. However, the major issue with the use of nanomaterials is that when nano-cleaners are added to aqueous contaminated water samples, they often pose regeneration and recycling issues and leach from the end-use device or filtration system due to their nano-sizes. In case of potential leaching, it is highly dangerous, the probabilities of inducing nanotoxicity are high which is also knownas 'secondary contaminants'. These combined concerns have raised worries among environmentalists and occupational health specialists about the potential nano-toxicity of the compounds that are potentially more dangerous than earlier encountered substances and chemical compounds. This also necessitates the assessment of nanomaterials for their specialized health risks and exposure evaluations to estimate the level of risk to human and environmental health and safety at large. To evaluate the aforementioned shortcoming of nanomaterials, this chapter first evaluates the feasibility of nanomaterials used in water purification, their potential regeneration, and reuse, and the risks associated with the use of novel systems at large. This chapter discusses in detail the use of advanced nanomaterials-stock in water purification and wastewater treatment including assembling nanomaterials in their microstructures in various forms and formats including standalone, blends, composites, gels, aerogels, hydrogels, 3D-sponges, beads, filter cakes, filter beds, foams, and membranes. [10-17]

The demand for membrane materials for emerging applications is ever increasing that are selective to water. These attempts have always been considered to enhance the parameters such as high separation and permeation efficiencies via innovative as well as advanced materials and membrane module design. These membrane materials have now been replaced with a combination of unique and viable functionalities that are obtained with the use of nanomaterials via engineered membrane preparation approaches owing to their high surface reactivity and intractability which facilitate robust permeance with improved pollutant retention. In this direction, many types of hybrid systems, both combinations of membrane materials as well as nanomaterials have been endeavored in order to achieve optimal performance in composite membranes or filters. Among many adopted nanomaterials, nano-sized nanomaterials prepared using carbonaceous precursors such as activated carbon (AC), hydrothermal carbon (HC), carbon nanotubes (CNTs), graphene and their derivatives such as oxide of graphene (GO), and reduced graphene oxides (rGO) have been prominent members. Also, macromolecular nanomaterials originating from biopolymers have been used in either packed bed form or direct membrane in household water treatment as well as purification systems. Therefore, advanced efforts have focused on developing affordable membranes or water filters using nanomaterials of various origins and forms. This chapter however evaluates the merits and demerits of using nanomaterials as advanced and futuristic water treatment systems. [18-21]

8.2 Biopolymer-based Advanced Filters

To address future issues associated with water quality and ever-increasing pollution from various sources. In this situation, freshwater resources are facing huge stress both from domestic end users and industrial production houses. On the other hand, over 90% of used water from industrial sources end up as wastewater and most of it reaches freshwater resources or the open environment creating serious concerns among researchers and technologists. In this direction robust, affordable, green, and sustainable materials stock or easily usable technologies are finding increasing interest. Among many biopolymer-based membranes and advanced nanomaterials have been considered as sustainable futuristic alternatives for conventional methods. These materials have been used both in continuous filter form as well as absorptive materials like conventional adsorbate however with improved separation and pollution retention abilities. In addition to this, polymer-based systems have also been extended to treat relatively severely contaminated wastewater from various sources with improved adsorption capacity and enhanced flux using filters in the form of hybrid membranes or adsorptive bed or cartridge in an engineered module. [22, 23]

However, initial efforts as discussed in one of the chapters, fossil-based synthetic polymers have been utilized to prepare membranes and resins used in various water treatment and purification applications. However, fossil-based polymers lately have shown low sustainability due to their various environmental concerns which restricted their large-scale usage. Also, fossil-based filter materials have increasingly shown their decaying proficiency when used for highly contaminated organic as well as biological contaminant treatments. This makes the overall treatment or purification process costly and induces lack of flexibility for various contaminants by limiting their general adaptability. On the other hand, use of synthetic polymers has drastically enhanced fouling proficiencies in the filter as well as related engineering devices. The ever-increasing environmental concerns related to synthetic polymers and commercial membranes led to the extensive use of alternative filter media in the form of natural polymers that can replace the synthetic polymer-based membranes. Thus, preparation of cost-effective nanocomposite water filters either in membrane or nanomaterial-packed bed form or filter devices using combination of synergic composites made from natural macromolecules in combination with suitable inorganic/ceramic or nano-sized active fillers. These combinations of materials have also been preferred due to their chemical resistance such as oxidation tolerance, high modulus, abundance, cost-effectiveness, easy fabrication, easy processability, recycling prospect, green and renewable nature, which make them biocompatible nanocomposites. All these qualities make natural polymer-based filter materials as frontrunner technology in the separation and purification sector. On the other hand, biopolymers can be retracted from various biomass resources originating from both terrestrial as well as marine ecosystems through simple physicochemical techniques. Upon extraction, biopolymers can bring considerable renewability, non-hazardous character, biocompatibility, biodegradability post-usage or soil-assisted decomposability which are essential to realise general global targets of greener technologies. In this direction, several biopolymers have been tried and tested successfully for water treatment applications involving cellulose, cellulose derivatives like cellulose acetate (CA), cellulose triacetate (CTA), agarose, gelatine, chitosan, sodium alginate (Na-Alga), carboxymethylcellulose (CMC), and so on. In recent advances, several natural fibers such as discarded silk protein fibers, cotton fibers and jute have also been used as alternative substrates for the preparation of functional water filters. Following sections will discuss a series of natural polymer-based composites for their water purification and treatment potentials. [24-30]

Cellulose: Cellulose-based composites are the major stock of materials that have been widely used in water treatment applications. Several decades ago, cellulose acetate was the major first ever polymer used as a membrane for desalination application. Cellulose is a one of the carbohydrate family polymers that naturally occur in abundance in the environment in various sources. Cellulose is popular in water treatment applications mainly due to its great affinity towards water. This is due to the presence of numerous hydroxyl groups which make a natural cellulose-based system in the form of beads, membranes, films, and resins an effective alternative for removal of various pollutant such as heavy metals and hazardous organic pollutants. However, a major disadvantage using cellulose-based macromolecules comes from their limited surface activity and surface area which restricts their direct and wider use in pollutant removal applications. Nonetheless, several surface functionalization processes have been designed via blending, compositing, and using active nanomaterials as one precursor components. Following these functionalization processes, cellulose-based macromolecules can be surface activated to retain targeted pollutants from contaminated water and wastewater. [31, 32]

Thus, cellulose-based composites have been thoroughly evaluated which could potentially turn inert polymers into an active adsorbent when used as a filter and optimized pore characteristic in a membrane for high performance water treatment applications. In this direction, researchers (Shu-Ya Tian et al.) have utilized a vacuum-filtered nanocomposite involving natural polymer composite membrane for removal of organic pollutant removal. [33] For this, first cellulose solution was induced with graphene oxide to facilitate the surface and polymeric chain activity via composite functionalized resulting in cellulose composite membranes (CGCMs) which was then used as well-ordered sieves for the removal of organic pollutants from various effluents. This study further discovered that the cellulose/graphene oxide-based filters have successfully retained various dye molecules from simulated wastewater. The main advantage using cellulose/graphene oxide-based systems is that they provide efficient adsorption capacities with high permeate fluxes. For instance, cellulose-composite adsorbents showed a maximum adsorption capacity of 86 mg/g for Rhodamine B from feed stream with high permeate flux indicating their large-scale potentials in removing various organic pollutants. [34]

In another study, Bo Li et al. prepared Fe-modified cellulose having thin-layer morphology involving 2-3 mm in diameter resulting in environmentally benign water filters. [35] Here, researchers focused on utilizing environmentally compatible adsorbent composites having cellulose/Fe composites for organic dye removal as well as retention of heavy metal from various contaminated sources. In this study, cellulose/Fe composites were prepared in the form of filter beads facilitating easy adsorption performance as high as ~1186 and ~150 mg/g for dyes like methylene blue and rhodamine B, respectively. This high adsorption performance has been accounted for by their characteristic porous morphologies and functional architecture having carboxyl active groups on the surface and enhanced mechano-chemical stability. Thus, cellulose-based composites have been effective against various pollutants and can be designed in various forms and shapes. Another major advantage using cellulose nanocomposite is the potential of the adsorbents post-pollutant removal which can beused several times making them economically viable alternatives to conventional polymer-membranes. Interestingly, use of magnetically active Fe nanocomposites also offers easy separation of adsorbent post-pollutant removal. Therefore, cellulose/Fe nanocomposites have been regarded as green and sustainable adsorbents which can be repeatedly used for various types of water treatment applications.

In another study, three-dimensional composites that are mechanically stable and robust have been prepared using cellulose as the main substrate. In this study, Zhang et al. prepared highly porous architecture having large surface area wherein a very effective adsorption of phenolic compounds has been successfully achieved. For this, cellulose has been functionalized using cyclodextrin as an effective functional moiety resulting in cyclodextrin/cellulose nanofibrils (CD-CNF) aerogel. The findings on CD-CNF suggested use of highly porous filter media can result in a high adsorption potential of 148 μmol/g in CD-CNF aerogel. Interestingly, the CD-CNFs have been converted into stable aerogels using optimized concentration of 3% precursor suspension concentration. Further, the CD-CNF aerogel showed outstanding surface regeneration and reusability potential which make them an ecologically friendly and promising adsorbent for the treatment of toxic phenolic pollutants in various wastewater sources. The continuous flow method using aerogel systems indicates that novel filters can be potentially used in industrial wastewater treatment applications under unfavourable circumstances where all other conventional water treatment techniques fail to provide desired results. [36] On the other hand, advanced approaches have also reported the use of several nanomaterials such as graphene oxide, nanocellulose and their synergetic hybrid composites for water treatment due to their specific and targeted adsorption abilities, mechanical robustness, tailor-made functionalities that can easily form coordination with pollutant metal ions, and surface charge which can attract various specific pollutant moieties. In this direction, researchers like Zhu et al. used bio-hybrids having biopolymers with various 2D nanomaterials like graphene oxides. Here, use of graphene oxide sheets in polymer fibrous matrix were vacuum filtered to form self-assembled filter in which 2,2,6,6-tetramethylpiperidine-1-oxyl radical mediated oxidation nanofibers membrane induced with GO (TO/CS-CNF+GO) with uniquely turned into useful filter as shown in Figure 8.2. These hydrolytically stable and easily recyclable bio-hybrids nanocomposites have demonstrated great potential in removing Cu(II) via adsorptive interaction from various contaminated water sources. In this study, high adsorption capacity has been realized via ionic crosslinking between the negatively charged bio-hybrid filter and the pollutant Cu(II). Thus, novel findings in using chitosan bio-hybrid TO/CS-CNF/nano-GO have also demonstrated their surface regeneration and reusability. Also, use of simple vacuum filtration

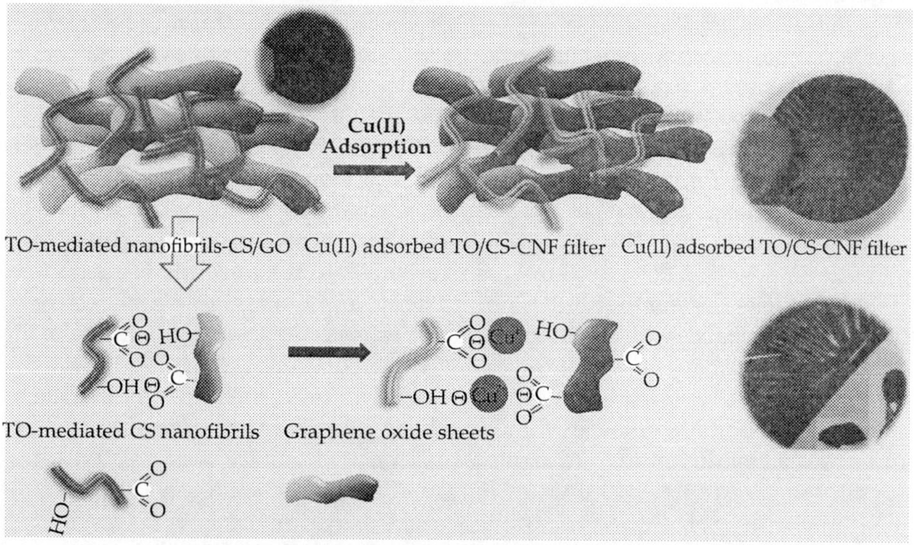

Figure 8.2 Schematic representation of chitosan-based 2,2,6,6-tetramethylpiperidine-1-oxyl radical mediated oxidation nanofibers (TO/CS-CNF+GO) biohybrid filter assembly which is used for the adsorption of Cu(II) on TO/CS-CNF+GO bio-hybrid via hydrogen bond interaction, actual photographs of free standing TO/CS-CNS+GO biohybrid membrane.

technique can also lead to the potential scale-up of the technology which can potentially replace conventional high-pressure membranes with improved flux performance, flexibility, physicochemical stability, and mechanical robustness. [37]

Therefore, cellulose-based absorbents can be an effective alternative with improved separation performance and biopolymer efficiencies can be further improved with the use of nanomaterials in the form of inorganic/organic configuration resulting in unique multifunctional properties. Further, cellulose-based nanocomposites can be potentially engineered as smart and target specific filters for tailor-made and point of use water treatment applications. On the other hand, incorporation of designer nanomaterials can also enhance overall performance of novel ultra-porous filters having improved ductility and Young's modulus. In this direction, researchers like Ru Jiang (2020) et al. adopted acombination of magnetic $NiFe_2O_4$/MWCNT as active functional moieties in cellulose matric (m-$NiFe_2O_4$/MWCNTs/Cellulose) as a promising and recyclable engineering biomaterialfilter. These functionalized cellulose nanocomposites having magnetic properties have been utilized for effective abatement of organic dyes like Congo red (CR) up to 95 mg g^{-1} adsorption capacity from contaminated water. Further, detailed studies have also shown the large scale-up potential of these magnetic adsorbents for a rapid separation of toxic chemicals from wastewater sources as well. [38]

Chitosan: After naturally abundant carbohydrate polymers like cellulose, second abundant natural polysaccharide like chitosan (CS) has been increasingly used for water treatment applications. Advanced studies have categorized chitosan as one of the green and sustainable alternatives for water separation and purification applications. On the other hand, chitosan has been characterized for second most abundance, biocompatibility, non-toxic nature, and adaptability to prepare in various forms and shapes in their end use. Chitosan has also been categorized as a potentially robust biopolymer due its surface chemical functionality in which -NH_2, -OH functional groups have been abundantly found that are useful for retention of series of pollutants like heavy metals, organic substances, and emerging pollutants via their chelating interactions. The incredible functional groups on the chitosan backbone in addition to retention characteristics induce multiple tasks such as hydrophilicity, biocompatibility, non-toxicity, and biodegradability. Active chitosan provides easy chemical modification via their -CH functionality that are useful for metal ion interactions from contaminated water. In this direction, a research team headed by Sankar et al. used metal oxyhydroxide functionalized chitosan meshes that have been used for the removal of toxic ionic species. In this study, Al-oxyhydroxide functionality was created on the chitosan backbone as shown in Figure 8.3. Further, chitosan composite preparation was scaled up via a simple surface activation process to turn into a relatively large packed bed which was fitted to a continuous flow tank to remove commonly available groundwater contaminants such as fluoride, arsenic, and pathogens. This medium size filter cake fitted water filter showed all-in-one type drinking water purifier offers flexible nanomaterial packing efficiencies which can provide relatively larger permeate flux under the influence of gravitational force. One of the major advantages of the chitosan-

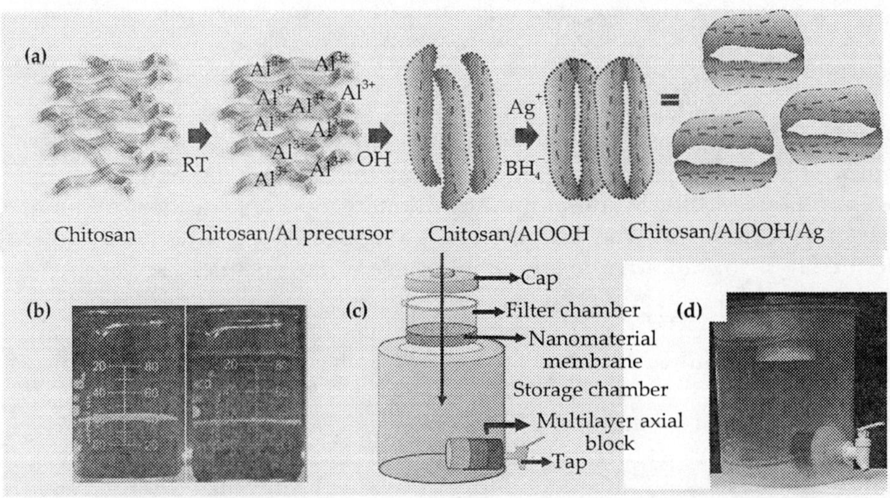

Figure 8.3. Stepwise schematic illustration of preparing (a) aluminium oxyhydroxide-chitosan-based functional granular composites, (b) actual photograph of composite AlOOH functionalized chitosan beads, (c&d) are the detailed schematic of affordable water filter designed to fit portable device and the actual container fitted with AlOOH/chitosan composite filter cakes useful for point-of-use water purification.

based functional nanomaterial system is its flexible packing density which does not require power supply to operate making it one of the suitable water treatment devices useful for remote or point-of-use applications. In this nanomaterial purifier use of metal oxyhydroxide functionalized on chitosan backbone can be prepared on large scale via cost-effective approaches and filter cakes with different packing densities can be engineered to suit the end user application. Also, post-surface modification of Al-functionalized Chitosan using anti-bacterial silver nanoparticles have shown prolonged shelf-life used for ground, surface as well as wastewater samples against various pathogens and organic contaminants. Nonetheless, filters can provide a safe net against microbial passage making it safe water for direct drinking. [39-44]

In a similar attempt, chitosan was further functionalized using ferrihydrite to activate their backbone to retain contaminants like arsenic from groundwater samples. The chitosan/ferrihydrite prepared using a simple functionalization process at room temperature has shown greater affinity towards arsenic species making the overall system an affordable alternative for arsenic-free drinking water production. These devices are useful in circumstances where the conventional water filters do not work or their affordability is an issue. In particular, chitosan-based nanocomposite material having exceptional affinity for the adsorption of toxic ions like arsenic in both As (III) and As(V) valence states make them one of the major players for future treatment media. Therefore, naturally abundant polysaccharide polymer chitosan having low-cost, eco-friendly, and point-of-use features could be an advanced and futuristic alternative in the low and medium water productivity segment. Significantly, these types of chitosan-based nanocomposite water filters are particularly useful for field application in developing and under-developed countries where affordability of large capital-intensive water treatment systems is difficult .

In another approach, chitosan-based ultra-porous filters have been prepared following advanced solid-state materials techniques like sol-gel process. In one such attempt, several combinations of chitosan-based aerogels have been prepared for the removal of a series of pollutants from various water sources. In this advanced approach, highly porous aerogel microarchitectures were achieved via controlled gelation and lyophilisation process. Further, biopolymer surface active groups were utilized to functionalize the large porous network of polymers useful for adsorption of contaminants like metal ions and heavy metals. This is achieved following simple dope-solution-composite formation technique in which the precursor polymer solution was loaded with metal-oxide, metal-hydroxide or metal-oxyhydroxide precursors and further crosslinked to achieve optimal gel formation. In this way, chitosan solution was doped with Fe, Al and Fe/Al metal precursors followed by crosslinking of the polymer resulting in composite hydrogel as shown in Figure 8.4. [45-47]

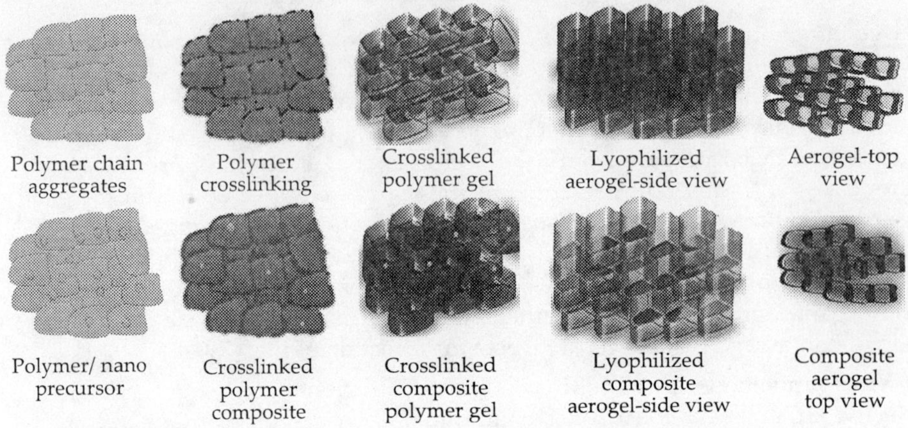

Figure 8.4. Schematic illustration of chitosan-based aerogels and their composite formation as highly porous nanocomposite filters used for the abatement of modern pollutants like fluoride, arsenic, and other emerging pollutants.

This composite hydrogel subsequently frozen and lyophilized to achieve solvent sublimation thereby resulting in highly surface functionalized aerogel water purifiers. When composite aerogels were used as highly porous water filters, aerogel systems showed exceptional pollutant retention efficiency with significantly permeate fluxes in comparison to conventional adsorbents. This characteristic provides aerogel systems' uniqueness in retaining pollutants like arsenic, fluoride, surfactants, active pharmaceutical ingredients and so on, via instant interaction which is absent in conventional filter media. Another major advantages are ultra-light weight and large porous network which makes them ultra-low-density filters creating unique channels useful for faster product water transportation. In addition to this, large porous architecture provides large surface area to facilitate nanomaterials active site to sit via simple functionalization processes. Further, composite having bimetallic oxyhydroxides on within the pores and on the surface of the biopolymer matrix have enhanced pollutant adsorption and purification efficiencies multi-fold. In a study conducted by Nataraj and team, uniquely decorated Al/Fe nanoparticles via a simple biopolymer functionalization process for spontaneous water purification through robust adaptive mechanisms via a cost-effective technique. This study showed use of Al/Fe-functionalized aerogel filters potential in removing pollutants with as high as ˜95% of dye, ˜89% emerging pollutants like surfactants and dye molecules, respectively. Further, retention of fluoride and arsenic to the efficiency up to 93% and 99%, respectively. With this, the overall chitosan-based functional filter showed a remarkable adsorption capacity for a wide range of emerging pollutants. Thus, a chitosan, in general, biopolymer-based functional aerogels can be an economic as well as greener alternative to conventional membrane processes in dealing with real-field ground, surface as well as wastewater treatment for future. Similarly, researchers like Yun Liao et al. also used polydopamine-functionalized attapulgite/chitosan precursor for the preparation of novel three-dimensional aerogel for the effective removal of emerging radioactive pollutants like uranium (VI) from wastewater sources. Further detailed study revealed that most of the adsorption happened in a robust mechanism within 40 min (328k) when feed was optimized at pH 5.0. At this pH chitosan-based functional aerogel showed a maximum adsorption capacity of 175 mg/g for uranium. Another major advantage is recyclability of biopolymer nanocomposites in which simple pH variation can reverse surface activity. In most of the cases, all biopolymer-based nanocomposite aerogels have been reused further for several cycles. However, research is still underway to understand functionalization, surface regeneration as well as adsorption mechanisms which would ease the larger adaptation of biopolymer-based aerogels for real-world applications.

8.3 Ceramic and Inorganic Nanomaterials as Filters

In addition to polymer-based ultra-porous filters, researchers are also focusing on developing loosely packed inorganic active nanomaterials or functionalized ceramics for potential water filtration applications. In this direction, among many metals nanocomposite-based advanced filters have been investigated as alternatives to conventional polymer-based membranes involving advanced inorganic precursors like metal-organic frameworks (MOFs), layered-double hydroxides (LDHs), covalent organic frameworks (COFs) and so on. In particular, hybrid classes of organic–inorganic materials like MOFs have increasingly attracted water technologists and water management innovators due to their potential in retaining pollutants from contaminated water sources. MOFs are prepared using a synthetic building unit that leads to creation of a porous network, tuneable porosity, and topologies. Porosity and the chemical affinity of the MOF can be tuned based on the end use requirements. Following simple techniques like hydrothermal or solvothermal processes the pore aperture and the sizes of pore characteristics can be altered to attract pollutant moieties or contaminant molecules to be separated from contaminated sources. Interestingly, the fixed pore sizes in MOFs have shown great potential to separate pollutants through their size-sieving abilities. On the other hand, MOFs having selective affinity, relatively low hydrophilicity, and fouling resistance characters offer versatile interactive options to have interaction in contaminated sources. Nevertheless, MOFs have been used widely as one of the constituents in nanocomposite membranes for various water treatment applications. In this approach, different types of MOFs have been prepared following various materials preparation techniques followed by their use in membranes. First, MOFs have been dispersed in polymer solution involving either polysulfone, polyvinylidene fluoride, polyacrylonitrile and polyether sulfone to prepare optimized dope solution concentration. In most of the cases, MOFs in polymer matrices have been known to induce optimal pore characteristics such as tuneable porosity, hydrophilicity, morphologies, and surface characteristics useful for handling specific types of contaminated water and wastewater. [48-50]

Therefore, use of MOFs as active pore characteristic altering fillers in polymer membranes have been widely used with varying levels of end results. This is since MOF nano/micro particles in higher concentrations have been known to aggregate in polymer solutions which either induces brittle nature to end polymer or unreliable separation and permeation characteristics to membranes in long run. Interestingly, to overcome the MOF dispersion related issues and further to prevent aggregates, MOFs-based dispersions were uniformly deposited on the support membrane using vacuum filtration technique. This simple vacuum deposited the MOFs on a support porous membrane thus successfully improved the issues associated with poor filler dispersion. Nevertheless, standalone MOFs have also been recently used as an advanced class of products in the treatment of various water types contaminated water. Thus, recent advanced research is adopting innovative filter design and filter preparation techniques to cater to the increasing demand for efficient water treatment techniques.

On the other hand, several unique chemical properties of MOFs were efficiently utilized to achieve pollutant separation or removal applications. Some photo-active MOFs have been used as filler or coating materials due to their self-cleaning and membrane or regeneration of the MOF- based micro- or nanomaterial membranes. In this direction, catalytically active M88A MOF was integrated into the graphene oxide nanosheets (Atian Xie et al.) to achieve an efficient water filter. Further, upon pollutant adsorption or separation using unique surface and pore structures, MOF-based membrane or filter were recycled and reused following photo-Fenton type catalytic regeneration mechanism. In this study, composite nanomaterials-based membranes prepared using the dispersion of graphene oxide/Fe (III)-MOFs showed great potential for utilization in real-world applications. More significantly synergistic characteristic of MOFs both as porous architecture, and potential catalyst that can be packed into an optimally porous filter configuration having tuneable permeability and separation characteristics. Therefore, graphene oxide/Fe (III)-MOFs-based filter showed an improved maximum flux of about ~30 L m^{-2} h^{-1} bar^{-1} with more than 97% a removal efficiency for various contaminants like methylene blue (>98%) and bisphenol A (>87%). On the other hand, MOF-based nanocomposite showed minimal flux decline however, presence of catalytic nature in a membrane helped in degradation of MB and bisphenol A post-surface adsorption. This helped surface regeneration and recycling of membranes up to 12 cycles post-methylene blue (MB) separation. All these proven characteristics of graphene oxide/Fe (III)-MOFs filter make them an important candidate for real-world water purification systems.

In recent efforts, researchers led by Jialin Cao et al. synthesised hydrophilic membranes having anti-fouling characteristics using MOF induced poly(acrylic acid) (PAA) (PAA/MOF) systems. In this attempt, PAA matrix was functionalized by inducing UiO-66-NH_2 MOFs via into the polymer solution, which was further vacuum deposited to obtain in UiO-66-NH_2/PAA filter. Further, composites UiO-66-NH_2/PAA filter was utilized for the selective separation of complex pollutants like oil from contaminated water mixture with as high as >99% rejection with a pure water flux of 2330 L m^{-2} h^{-1} bar^{-1} under optimum operating parameters. Interestingly, UiO-

66-NH_2/PAA filter showed remarkable recycling ability after oil separation from the contaminated mixture retaining its separation ability >80% for flux at the end of 3 cycles of repeated oil/water separation. In another study, vacuum deposited filters having exfoliated manganese oxide filter for organic contaminants water purification. This study showed that exfoliated metal oxide showed a unit adsorption capacitor for dye molecules with a maximum adsorption capacity of >250 mg g^{-1} with remarkable permeate flux of 5736 Lm^{-2} h^{-1} was achieved under vacuum filtration conditions. This study showed a significantly high dye rejection of 99% over a metal oxide filter with remarkable recyclability efficiencies. Here, metal oxide-based filters showed >93% performance retention even after 10 cycles of regeneration and reuse attempts. Overall, several inorganic systems have been used following simple preparation techniques and affordable pollutant removal protocols. Even though traditionally, metal oxides and MOF type inorganic systems have never been considered for their bulk nature, however recent advancements in nanotechnology-assisted nanomaterial preparation have enabled the tuning of surface functionalities and the surface characteristics that are useful for pollutant retention and removal applications from the contaminated sources. [51-56]

8.4 Carbon-based Micro/Nanomaterials as Filters

Recent advances in the carbaneous and graphenaceous materials have opened new avenues for their applications including electronic devices, energy devices, and composite materials and now in water purification and treatment applications. On the other hand, increased environmental concerns associated with emerging and modern pollutants demand for the need to design and develop environmentally benign materials stock for affordable and sustainable water treatment and purification applications. Therefore, recent attempts have been made to devise more innovative, sustainable, and easily scalable materials in filter or adsorbent form. In this direction, use of carbon-based nanomaterials has been prepared in the form of functionalized nanomaterials in various low-dimension form, morphologies, and surface characteristics. This involves activated carbon, carbon nanotubes, and hydrothermal carbon and among graphene materials graphene oxide, reduced graphene oxide, nanocomposites of all prior mentioned materials stock and so on. Another major advantage with these materials is that during the materials possessing properties like exceptionally large surface areas, surface affinity like oleophilicity and hydrophobicity nature can be achieved. These unique characteristics with exceptional morphological patterns and functionalities have been known to be highly suitable for pollutant removal and separation. These combinations of characteristics have drawn tremendous attention from water management and regulators. In addition to this, carbon-based adsorbents and filters have shown unprecedented advantages due to their excellent physicochemical characteristics such as chemical stability, mechanical robustness, affinity towards specific pollutants, and flexibility in design, and so on. Upon functionalization carbon-based nanomaterials provide vastly tuneable structures, properties, and selectivity towards specific target pollutant species. Therefore, functionalized carbon or graphene-based nanomaterials have been considered a vital stock of materials to address the bottleneck situations such as water scarcity due to increasing contamination and environmental pollution which are prevalent in the modern times. On the other hand, functional carbon nanocomposites have been used in the preparation of hybrid membranes for water remediation applications. Conventionally, carbon filters have been utilized as adsorption kit, pre-filter as well as disinfection cartridge in various forms and configurations. Among them, some of the commonly utilized carbon-based devices have been designed or modelled as point-of-use functional nanomaterial filters and devices.

8.5. Carbon Nanotube-based Nanocomposites

Carbon nanotubes (CNTs) are one of the allotropes of carbon which falls between fullerene cages and flat graphene categories having diameters in the range of a nanometre. CNTs are typically termed microscope tubes due to their nanosize; however, CNTs in their nanoscale cylindrical architecture offer remarkable characteristics and physicochemical properties such as flexibility, conductivity, and surface area that are useful in several potential applications. Advanced innovations in electronic, coating applications, actuators, electromagnetic shield, automotive composite parts, mechanochemical and water treatment applications have led to the revolutionary carbon nanotube production targets that recently have exceeded several thousand tons per year. As CNTs carry unique shape at molecule level, porous vertical assemblies can be directly used for preparing uniquely narrow pore diameter filters which can be directly used to produce clean and contaminants free water by allowing only water molecules to pass through the hollow central channel or cylindrical cavity of carbon nanotube molecules simultaneously rejecting pollutant bigger than water molecule and diameter of CNTs pore.

As it is well established, among many issues associated with conventional membranes, controlling pore size and their distribution in polymer matrices is a big challenge. Therefore, detailed understanding of CNTs porous nature can be beneficial for their water treatment applications. With all these potential parameters in place, the evolution of advanced and improved membrane technologies could be a possible scenario with the use of well-ordered controlled pore structure which can be innovatively designed to achieve efficient and cost-effective purification systems. One of the major advantages is, CNTs are extremely adaptable when it comes to tailor made modification or device making via customization process or functionalization protocols. Use of highly ordered and large density pore structure can also drastically reduce the energy consumption with high product flux which can make overall desalination technology affordable and convenient. However, direct use of CNTs in water purification and treatment application may not be feasible, however with slight change in surface chemistry of CNTs can be turned into a useful selective barrier for water purification. As CNTs are relatively hydrophobic in nature which makes it a prerequisite to alter the surface functionality of pristine CNTs. In this direction, several attempts have been successfully made to functionalize the CNTs such as use of amide functionality which helps in turning multi-walled CNTs relatively hydrophilic. This work reported by Fatemeh and partially functionalized CNTs were then introduced into sodium alginate-hydroxyapatite systems to turn the overall system into nanocomposite beads. Further as prepared nanocomposite beads showed several advantages like hydrophilicity and surface functionality due to presence of natural polymer alginate and ceramic constituents making them appropriate for the removal of radioactive pollutants like Co^{+2} ions from contaminated water sources. Interestingly use of functionalized CNT in polymer matrix significantly enhanced the separation and permeation performance of CNTs-based nanocomposite filters. Nevertheless, these composite filters having a surface area of 163.4 $m^2 g^{-1}$ showed optimized adsorption potential of 347.8 mg g^{-1} for radioactive pollutants from contaminated sources. Therefore, CNTs-based nano-systems showed extremely effective, technically viable, and low-cost alternatives, which may be directly used in water treatment contaminated with metal ion pollutants. Also, advanced efforts have been made to design and fabricate special and pollutant specific adsorbents that can quickly be extracted and isolated from the aqueous system. In this direction, novel gelatine-based magnetic adsorbents having -COOH functionalized CNT trapped in the form of gel have been prepared. Here, functional CNTs embedded with magnetic iron oxide nanoparticles (MNPs) prepared by Samaneh et al. have been turned into gel-type adsorbents. As prepared MNPs functionalized magnetic CNTs gel-adsorbent beads were used for the simultaneous removal of organic pollutants like direct red 80 (DR) and methylene blue (MB) dyes with a removal efficiency of 76% and 96%, respectively. These studies revealed the effectiveness of the functionalized CNTs in offering a robust adsorbent that may be useful for wastewater treatment applications.

Further, a similar attempt has been made towards increasing the efficiency of adsorption of CNT-based materials with the help of crosslinked polymer for the removal of metal ions in the contaminated water. In this attempt, amino-group rich-chitosan cross-linked combined with the help of carboxyl group in multi-walled carbon nanotubes (MWCNTs-COOH) as functionalized water treatment system for quick and effective removal of heavy metal ions. This study carried out by Huang et al. effectively prepared crosslinked nanocomposite CNTs-based biopolymer materials were used for the adsorption of chromate ions from a contaminated wastewater having acidic pH and achieved a maximum adsorption capacity of 164 mg/g, within 30 minutes in contact of each other at 313 K. Further, CNT-chitosan-based functionalized nanocomposite material was regenerated for their surface functionality and adsorption activities even after several cycles of effective Cr (VI) adsorption with no or minimum performance decline. Even after 4th cycle regeneration and reuse CNTs-chitosan nanocomposite showed up to 98% performance retention making it an effective reusable adsorbent for heavy metal remediation. Nevertheless, as prepared chitosan-cross-linked MWCNTs-COOH functional nanocomposite have shown great potential with potential to scale up for low to moderate Cr (VI) contaminated industrial wastewater. [56]

8.6. Activated Carbon-based Nanocomposites

Traditionally, activated carbon (AC) has been used for water treatment applications since ancient times as an economical and affordable water remediation and treatment material. Activated carbon having ultra-high surface area even ranging up to ˜1500 m^2/g can be a one best remediation material for various types of contaminated water sources. Activated carbon can be prepared from various precursor materials from natural carbon sources like biomass to synthetic sources such as polymers and organic compounds. Traditionally, biomass-based activated carbon demonstrated excellent abatement capacity for a wide range of organic and inorganic pollutants from water and wastewater sources. Further functionalization assisted with chemical activation of carbon materials can lead to the introduction of exceptionally large surface area of up to 3500 m^2/g with well-organized and developed internal microporosity, and introduction of a wide variety surface functionalities allows it to be used in a wide variety of applications. Composite materials are in high demand for the removal of antibiotics, dyes, and other waste water contaminants because the characteristics of individual components in the composite may be adjusted to suit the needs. Several advanced approaches have also shown the preparation of various forms such as granular activated carbon (GAC) and activated carbon fiber (ACF) for their frequent use as adsorbents due to their sponge-like pore structure and other characteristics. Among several attempts, Doo-Won Kim et al. prepared activated carbon both effective and economical for the removal of cadmium and mercury from contaminated water sources. Here, ACF was modified with NaOH solution (NaACF) to achieve high surface area and functionality simultaneously. Interestingly, NaOH modification of ACF effectively enhanced the rate of heavy metal adsorption (mercury and cadmium ions) with an increased efficiency up to 10 wt%. Besides their direct use as powder in batch-wise adsorption applications, ACFs have also been converted into membranes to be used as continuous flow devices which makes them technologically viable to adopt in real-life applications. [57]

In another study, Radheshyam et al. prepared a sustainable iron oxide functionalized activated carbon and bentonite clay adsorbent which was further encapsulated using eco-friendly biopolymer like sodium alginate. These functional materials have been converted into beads capable of selectively extracting various cationic and anionic pollutants that are hazardous and toxic in nature. In this study, composite activated carbon materials showed a surface area of 433 m^2/g having adsorption capacities of 5.0, 74.2 and 41.3 and mg/g for selective retention of As(V), Pb (II) and Cd(II), respectively. In another study. Hanjing Xu et al. prepared novel activated carbon spheres having a surface area of ˜1491 m^2/g which was derived from greener sources like glucose and sodium dodecyl benzene sulfonate. Further, functionalized activated carbon was used for the removal of chromate ions with 230.15 mg/g adsorption capacity at 25°C. This study showed great potential of activated carbon in removing Cr(VI) contamination from water sources. Recent advances have also shown the utilization of biomass sources for converting it into a high surface area activated carbon through a simple carbonization process. Among many biomass sources, bio-wastes like palm shell, peanut hulls, rice husk, coconut husk, oil-palm trunk fiber, and corncob have been widely utilized for activated carbon preparation. On the other hand, Wara et al. synthesized a composite adsorbent of silver nanoparticles functionalized activated carbon (Ag-AC) originated from bamboo biomass. This activated carbon was then used in a continuous fixed-bed column for the effective removal of formaldehyde from contaminated sources. [58]

8.7. Hydrothermal/Solvothermal-based Nanocomposites

Recent advances in solid-state materials preparation have led to the innovation in hydrothermal, solvothermal and ionothermal process development. These advanced methods result in producing highly functional carbon nano/micro scale materials useful for the water treatment applications. Among these, fabrication of functional carbon has been popularly carried out using hydrothermal/solvothermal carbon (HTC/STC) methods. These methods have been known to induce rich surface functionalities to carbon backbone. On the other hand, hydrophilic functionalities on carbon surface induces exceptional affinity for pollutant capturing and provides anti-fouling characteristics to the overall remediation system. However, HTC has also encountered several limitations such as low surface area, limited degree of functionality and scalability. Nevertheless, use of unique greener solvent-cum-templates in place of water led to the innovation of solvothermal carbon which has considerably improved properties of carbon products with distinct functionality, architecture, porosity, and surface affinity. In one such attempt Nataraj and his team used glucose as carbon precursor and ternary deep eutectic system as solvent as well as structure directing template under solvothermal conditions to yield highly functional carbon microspheres. Further, Fe precursors in the eutectic system helped in Fe doping into solvothermal carbon (Fe-STC) which was then converted into a membrane via vacuum deposition technique and was used as a filter for removal of various contaminants. The Fe-doped STC with a surface area of 26 -175 m^2/g deposited on Whatman filter paper support resulted in scalable potential of the materials. The Fe-STC with optimum surface functionalities showed effective removal efficiency for organic pollutants with relatively high flux rate of 602 $Lm^{-2}h^{-1}$ and rejection over 96% for high molecular weight cationic dyes such as Methylene Blue, Rhodamine B, Methyl Green, and Rh 6B. Further, pharmaceutical active ingredients like dye molecules were rejected up to 60% in a continuous flow method. Therefore, this study showed potential of innovative methodology in removing emerging pollutants from contaminated sources with recyclable functionalities on-demand applications.

Further, a team of researchers led by Nataraj prepared highly oxygenated and AlOOH-functionalized solvothermal carbon (Al-STC) composite in powder as well as deposited membrane form via simple vacuum filtration. Both these approaches have been tested at lab scale to analyse the feasibility of STC for water treatment applications. Here, the use of eutectic systems has facilitated the tuning of morphology as well as surface characteristics that when applied proven a sustainable alternative in dead-end filtration configuration for robust water purification. On the other hand, Al-STCs in membrane form demonstrated excellent retention efficiencies for pharmaceutical waste, organic molecules, and heavy metal ion from contaminated water. In this, functional STC carbon successful retained malachite green and methylene blue dyes over 99.9% and >99.9% rejection for pharmaceutical drugs like ciprofloxacin and 53% rejection for paracetamol. Interestingly, active STC have shown significantly good rejection for hormones oxytocin, surfactants (CTAB), chromate ions showing rejection of 88.6%, 94.9% and >99.9%, respectively from respective contaminated feed solutions. Nonetheless, STC functional materials can be regenerated with simple ethanol washing making it an easy-to-use, reusable, and affordable membrane. These features also offer an alternative and viable technological solution to be a universal water filter. [59] Similarly. several researchers utilized hybrid carbon membranes like GO/CNT, GO/AC, and CNT/AC for the removal of antibiotic like tetracycline hydrochloride and organic dye molecules like methylene blue and rhodamine B from polluted waters. [59]

As discussed in detail, several combinations of hydrothermal/solvothermal carbon have shown great potential in several pollutant retention tasks. Therefore, production of functional carbon composites using either of these techniques have sparked increased attention in using specially designed functional carbon for robust adsorptive removal of both conventional as well as emerging pollutants which can meet the requirements drinking water norms set by international water management and monitoring organizations. In addition to excellent surface area, pore characteristics and surface affinities, feasibility of creating wide varieties of functionalities provide them strong interaction possibilities with contaminants in aqueous media. Therefore, carbon nanotubes, activated carbons, graphene and graphene oxide-based carbon derivative materials have been frequently adopted in adsorptive removal of pollutants from water sources. The several advantages discussed in this previous section have shown the potential of significantly reducing the water treatment duration due to quick contact interaction and excellent adsorptive capabilities.

8.8. Graphene-based Functional Nanomaterials

Conventional composite membranes in the recent past have been prepared using various nano-sized materials including two-dimensional nanomaterials. However, advanced approaches have improved their efficiencies even without the help of dispersion media polymers. These standalone 2D nanomaterials have been widely regarded as next-generation membranes due of their flexibility in stacking, potential in preparing loosely packed self-assemblies to act as loosely packed porous filter, self-supporting film character helps in producing high product flux, ability to retain specific contaminant species, and loosely packed nanomaterials filters resist the trans-membrane pressure which enhances their shelf-life and regeneration efficiencies. Among many two-dimensional materials, graphene and its derivatives have been widely used for preparation of freestanding films that are densely packed suitable to act as filters. [60-65]

Most of the free-standing graphene/GO-based nanocomposite filters or membranes have remarkable molecular separation characteristics in cleaning contaminated water sources. Among many forms, initially self-standing graphene filters were first tested as filters. These filters were prepared using water dispersion filters on a various substrate and extracted thereafter to be a free-standing filter. Interestingly, good dispersibility of graphene-based 2D nanomaterials in aqueous medium make them easy to functionalize and filter to result in membrane fabrication of free-standing nature. Further, use of various types of nanomaterials as one of the components to prepare the composite films and their derivatives have witnessed various applications in different fields other than water treatment like energy generation and storage. In one such attempt, recently metal oxides have been induced into the water dispersions containing graphene sheets further turning them into self-assembled structures. Here, the graphene sheet provided film forming characteristics and metal oxides played as functional active sites. These synergic properties keep graphene and composites stable from aggregation. Also, functionalization of the materials with immobilized nanomaterials like metal oxides in inter-graphene nanosheets layer spaces as inserted active sheets provide them a unique pore and surface activity characteristic. Similarly, various graphene-based composite and hybrid nanomaterials with specific functions have been prepared following self-assembly technique via drop casting, or layer-by-layer deposition, vacuum filtration or spin coating method to obtain economically viable and competitive 2D nanostructure filters. Following one such method, Chao Xu et al. successfully prepared, a novel graphene oxide/TiO_2 nanocomposites with specific surface area of ~489 m^2/g, which is relatively higher than that of bare GO sheets having ~10 m^2/g and TiO_2 of 118.9 m^2/g. Similarly, several GO and metal oxide nanoparticles (GO/NPs) were also designed and stabilized in aqueous media and aqueous composite dispersions were assembled into membrane filters following simple vacuum filtration as shown in Figure 8.5. [66]

Interestingly, these graphene sheets functionalized with TiO_2 have shown broaden the interlayer distance between graphene nanosheets due to the

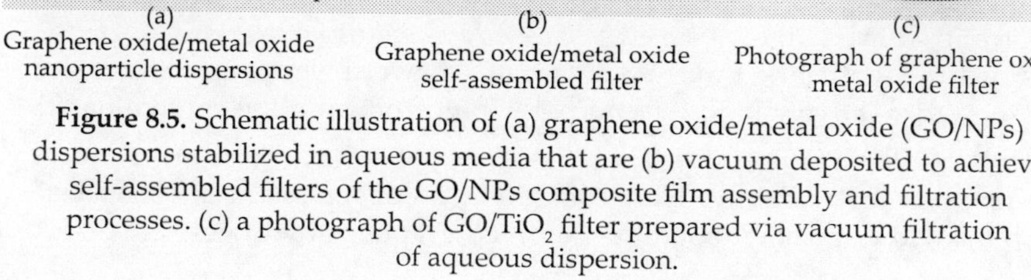

Figure 8.5. Schematic illustration of (a) graphene oxide/metal oxide (GO/NPs) dispersions stabilized in aqueous media that are (b) vacuum deposited to achieve self-assembled filters of the GO/NPs composite film assembly and filtration processes. (c) a photograph of GO/TiO_2 filter prepared via vacuum filtration of aqueous dispersion.

presence of nanoparticles, which led to the formation of highly interconnected pore structure in the free-standing films with unique porous structure arranged in channels like architecture having suitable active sites that facilitate abetment of various pollutants like dye molecules from contaminated sources. This technique has provided a mean pore size of 3.5 nm within the composite filter making it an effective barrier for removal of smaller organic molecules like dyes from contaminated sources. With this pore structure, the membrane provided optimal permeability by removing 100% methyl orange at the initial stage of filtration. However, increase in the concentration of nanoparticles in the dispersion solution results in widening of the pore structures which resulted in the flux and at the same time increased penetration of contaminants through the filter. Nevertheless, graphene oxide-based nanocomposite membranes can be highly effective filters with tuneable capabilities in capturing or selective removal of various types of contaminants at molecular level. Similarly, graphene oxide/MOF based nanocomposites have also been prepared for adsorptive removal of modern pollutants with great potential to replace conventional membranes soon.

In addition to this, several attempts have been made considering increased demand for specialist treatment media or remediation techniques. Also, demand for robust and affordable water filters has been the need of the hour. Conventional water treatment techniques and methods have failed to fulfill this specific demand from the domestic market. Therefore, researchers and innovators have diverted the design of economically viable and robust water filter devices. In this direction, among many modern sets of materials, free-standing nanocomposites with different nature and combinations have shown futuristic potentials. In addition to GO/metal oxides-based nanocomposite filters, Xianfeng et al. also used GO in combination with polymer nanocrystals for the preparation of free standing filters. In this study, GO–chitin nanocrystal (GO/ChNCs) hybrid filter having excellent oleophilicity has been prepared following a vacuum assisted self-deposition technique. These GO/ChNCs self-assembled filters were further used for the selective adsorption of organic pollutants and oil droplets from oil-in-water emulsion mixture. Interestingly, chitin nanocrystals placed in between GO 2D layers have shown favorable permeability factors such as increased interlayer distance due to the presence of uniformly distributed needle-like chitin nanocrystals. The presence of needle-like chitin nanocrystals has provided an interconnected channel network enabling relatively good water flux of around 135 L m^{-2} h^{-1}. While, use of GO:ChNCs in 1:4 volume ratio showed optimal membrane characteristics providing as high as 99.3% rejection for dye contaminants like methylene blue and 97.8% rejection for Congo red. On the other hand, GO:ChNCs in 1:1 volume ratio showed highly packed self-assembly resulting in relatively dense pore structures with low flux of 8.8 Lm^{-2} h^{-1}. Nevertheless, nanocomposite filters showed relatively high rejection of 97% for oil from the contaminated emulsion mixture.

In another such attempt, Hongcai Gao et al. developed a simple facial method to synthesize polydopamine modified graphene hydrogel (PDA/GH) as effective and sustainable adsorbents. In this study, abundant functionality in the polydopamine provides excellent affinity towards toxic pollutants. Further, 2D nanomaterial sheets provide high specific surface area which enables synergic combination to attract pollutants from aqueous media. As prepared PDA/GH gel have been used for the removal of organic as well as inorganic pollutants such as synthetic dyes, aromatic pollutants, and heavy metals from contaminated wastewater. Therefore, PDA/GH composite gel has shown great potential in scale-up for point-of-use water purification and remediation applications. Interestingly, hydrogel-type remediation media in its free standing nature can be a commercially viable media due to their regenerative nature using low-cost reagents. Regeneration of the active surface can also be achieved by altering pH conditions of the ascorbate to retain high adsorption capacity. This study has proven these factors with several repeated adsorption/desorption cycles studies following simple recovery from acidic/basic water after adsorption. [67]

On the other hand, preparation of 3-dimensional graphene oxide hydrogels (3DGOH) obtained with the help of Fe^{+2} ions crosslinking has been attempted by Chen et al. via vacuum filtration for pollution remediation. In this study, crosslinked structure of 3DGOH having interconnected network has shown remarkable pore connectivity resulting in porous material thereby enhanced adsorption capacity. The 3DGOH when used for the wastewater treatment have shown improved water flux of ˜111 L m^{-2} h^{-1} bar^{-1} with >99% retention for methylene blue (MB). During the pollutant retention activity cross-linked GO hydrogel showed remarkable stability in contaminated aqueous medium. As a result, 3DGOH exhibited as a potential alternative for practical application in wastewater treatment and groundwater purification via good penetration which helps in attaining enhanced separation capabilities across a wide range of pH conditions. Feasibility of applying crosslinked GO hydrogel in range of pH and contamination levels helps in successfully eliminating various organic pollutants through a simple filtration process following different separation mechanisms. In another attempt, a group of researchers led by Yiying Chen have prepared 3-dimensional active filter for the retention and abatement of inorganic pollutants like heavy metal ions from contaminated water. In this study, phosphorylethanolamine was used for the functionalization of graphene foam (PNGF). As prepared PNGF was characterized for their superhydrophilicity owing to its surface functionality and hydrophilic polymer nature. Further, PNGF were used for the robust adsorptive removal of Cd(II) and Pb(II) from contaminated water within 10 minutes of constant. Furthermore, PNGF filters were surface regenerated using simple washing solution of low-cost reagents or altering pH condition of analyte or the gel media. In recent advancements, researchers have also attempted to lyophilize the functional hydrogels to achieve ultra-porous filter media. Here, active functionality is also attained on the surface wall of porous structure to attain adsorptive characteristics.

In addition to earlier attempts, recent demand for sustainable and green materials sources have also forced researchers to adopt eco-friendly and naturally abundant precursor resources. In one such attempt, environment-friendly bio-graphene foams were prepared from renewable materials and used as sustainable adsorbents for heavy metal removal applications. In this study, Melinda et al. followed a simple two-step preparation technique to create GO induced biopolymer foam (bGFs) with a significantly porous architecture and an active surface area of 805 $m^2\ g^{-1}$, which was helpful in eco-friendly applications. Here, biopolymer provides substrate networks whereas GO induced active surfaces are to adsorb nearly 245 mg of Cr (VI)/g on bGFs surface. This remarkable adsorption capacity drastically reduced chromate ion concentration in contaminated water bringing below the maximum permissible levels set by World Health Organization (WHO) of 0.050 mg dm^3 which is even safe for domestic intake. Another major advantage with use of porous foam like filter offers medium to large scale usage in a continuous flow system in real-world applications. Also, use of biopolymeric systems inherently induces superhydrophilicity to the overall system which can be further used for the separation of oil-water contaminated wastewater. The hydrophilic nature of bGFs have shown compelling performance in separation of organic solvents such as toluene and oils in particular petroleum residues from contaminated wastewater media reaching 99.1 % and 98.8% separation efficacy respectively. This architecture enables researchers and technologists to devise real-life application filters. [68]

Further, creation of an advanced class of functional free standing water filters have gained increased attention for affordable water filters. In this direction, use of inorganic functionalities on carbon support or graphene sheets to induce adequate activities. This type of functionalization process was carried out to solve the issues associated with traditional 2D nanomaterials in different forms owing to their affinity towards residual pollutants in water. In this direction, design of nanofiltration range membranes have been made with the use of tailor-made compositions of GO and silica nanomaterials. In this study, Ying-Chen Du et al. prepared characteristic graphene oxide/silica (GO/SiO_2) nanocomposite materials with relatively good hydrophilicity (80.7° of contact angle), mechanochemical robustness which helped in obtaining high water flux and high adsorption capacity for contaminants. Interestingly, the in situ hydrolysis method followed for the preparation of NF-range composite membrane in a basic condition has resulted in mean pore diameter of 3.97 nm, total pore volume of 2.84 $cm^3\ g^{-1}$, and specific surface area 2897 $m^2\ g^{-1}$. With this, as prepared, the GO/SiO_2 filter showed as high as 99% rejection efficiency for rhodamine B with a flux of 229.15 L $m^{-2}\ h^{-1}\ bar^{-1}$ [67]. Interestingly, graphene oxide/silica NF range filters showed good self-standing capabilities without the support substrate however, without the support layer nanocomposite filter lacked robustness. [69]

On the other hand, Farooque et al. synthesized graphene oxide/ginger extract (GO/GE)-based thin-layered filter having tuneable permeability was achieved for the removal of organic pollutants like dye molecules following a simple vacuum assisted deposition technique. The vacuum deposition technique for GO/GE dispersion resulted in optimal filter thickness of ˜250 nm which is useful for attaining excellent separation and permeation efficiencies for their potential application in organic pollutants removal. In this study, a combination of GO/GE dispersion resulted in optimal packing density which provided ultrafast permeability of ˜420 $Lm^{-2}h^{-1}bar^{-1}$ for rhodamine B and methylene blue contaminated water. Also, the filter showed 99% and 98% rejection for rhodamine B and methylene blue, respectively. Overall, the permeance of 320 $Lm^{-2}h^{-1}bar^{-1}$ rhodamine B and 210 $Lm^{-2}h^{-1}bar^{-1}$ for methylene blue was obtained during the removal of both the organic moieties. The overall rejection performance of over >98% for both MB and RD dyes molecules with decent flux was highlighted in this study showed enormous potential of GO/GE-based thin-layered filters in various types of water purification applications. [70]

Nonetheless, the majority of 2D nanomaterials including graphene-based hybrid systems in combination with other nanomaterials in colloidal range have shown great potential in providing effective alternatives to conventional water treatment techniques. As a result, advanced efforts have continued to create simple, affordable, and effective techniques for preparing graphene-based hybrid materials for a wide variety of contaminated water treatment applications.

8.9. Nanocomposite-based Devices

Major portion of natural water resources have been contaminated with one or the other type of contaminants. This problem is expected to aggravate due to increased industrial as well as domestic wastewater release into these natural sources. On the other hand, conventional water treatment technologies and methods have been showing their limitations that are ineffective towards modern pollutants like heavy metals, pharmaceutical ingredients, surfactants, hormones, microplastics and so on. Conventional water treatment techniques either have proven cost intensive or ineffective for multiple contaminants. Therefore, to address such limitations, several attempts have been made to combine multiple water treatment techniques to address complex pollutants from both domestic as well as wastewater treatment. In this direction, membrane technologies have been used in hybrid combinations with adsorption, advanced oxidation, electro-oxidation, Fenton and photo-Fenton type systems have been tried and tested with varying levels of success. Among these, hybrid membranes/adsorption-based processes have evolved as all-in-one solutions for water purification and treatment. In this, use of conventional membranes has continued with new configuration and polymer material forms such as nano-materials induced dope solutions. However, involvement of synthetic polymers inherently shows low sustainability due to their fouling proficiency and degradation potentials in extreme operating conditions. These vital limitations make even hybrid water treatment systems restricted their wider usage and constant threat of replacing membrane materials make them cost-intensive. Also, prolonged use of membrane-based hybrid systems lack flexibility for a variety of emerging contaminants. This is forcing innovations in design, development and fabrication of cost-effective nanomaterials-based water treatment devices having a variety of active nanomaterials in combination in filter device, cartridge or membrane or portable column like membranes that are derived from green and sustainable sources. In this direction, nano-materials-based filters have been widely prepared due to their flexible design, easy to prepare and packing feasibilities, hydrophilic surface characteristics that resist in their chemical degradation, easy availability of nanomaterials precursors across the spectrum make them cost-effective that enable easy processability, renewable precursors make them biocompatible, biodegradable after use in water treatment. Nonetheless, many of the tested nanomaterial-based systems have shown excellent recycling prospects which make them the frontrunner and advanced technologies for universal water treatment purposes. On the other hand, wider use of biopolymers in the preparation of functional nanomaterials make them non-hazardous, non-toxic, and biocompatible water purifiers. Therefore, functional carbons such as STC, HTC and AC in nano-size scales have been considered renewable, environmentally benign materials with high specific functional characteristics making them futuristic universal water filters. Nevertheless, biomass systems both from terrestrial as well as seaweed origins have also been used for the preparation of active nanomaterials-based water filters with exceptional separation performance.

8.10. Conclusion

The ever-increasing concerns associated with environmental pollution, in particular water contamination, has led to the series of innovations in designing effective and affordable water treatment technologies. Even though several conventional water treatment technologies have shown potential to improve in their new form and modification, the separation and treatment performances of modified-conventional water treatment techniques have shown limited success. In this regard, advanced efforts have been made to design modern water treatment materials and technologies to overcome the limitations of conventional technologies. Advanced efforts have also adopted novel nanotechnology-based approaches to design nanomaterial-based functional materials having task-specific affinity towards targeted pollutants. To the surprise, nanomaterials-based functional nanomaterials have shown universal adsorption capacities for a series of pollutants from contaminated water and wastewater sources. Therefore, advanced efforts have shown the potential of functional materials in wider applications which can potentially replace conventional treatment techniques in parts or as a whole.

References

1. https://www.ohchr.org/en/press-releases/2022/03/world-must-end-overexploitation-groundwater-says-un-expert.
2. https://www.unwater.org/water-facts/quality-and wastewater/#:~:text=Globally%2C%2080%25%20of%20wastewater%20flows,%2C%20dysentery%2C%20typhoid%20and%20polio.

3. Jamiu O. Eniola, Rajeev Kumar, M.A. Barakat, Jamshaid Rashid, A review on conventional and advanced hybrid technologies for pharmaceutical wastewater treatment, Journal of Cleaner Production, Volume 356, 2022, 131826, ISSN 0959-6526.
4. Priyanka Kumari, Nupur Bahadur, Ludovic F. Dumée, Photo-catalytic membrane reactors for the remediation of persistent organic pollutants – A review, Separation and Purification Technology, Volume 230, 2020, 115878, ISSN 1383-5866.
5. Sijia Li, Jianan Cui, Xia Wu, Xuan Zhang, Qi Hu, Xiaohong Hou, Rapid in situ microwave synthesis of Fe_3O_4@MIL-100(Fe) for aqueous diclofenac sodium removal through integrated adsorption and photodegradation, Journal of Hazardous Materials, Volume 373, 2019, Pages 408-416, ISSN 0304-3894.
6. Danny Banks, Byung-Moon Jun, Jiyong Heo, Namguk Her, Chang Min Park, Yeomin Yoon, Selected advanced water treatment technologies for perfluoroalkyl and polyfluoroalkyl substances: A review, Separation and Purification Technology, Volume 231, 2020, 115929, ISSN 1383-5866.
7. Gaogui Jing, Shuai Ren Stephen Pooley Wei Sun, Przemyslaw B. Kowalczuk and Zhiyong Gao, Electrocoagulation for industrial wastewater treatment: an updated review, Environ. Sci.: Water Res. Technol., 2021, 7, 1177-1196.
8. Sathya, K., Nagarajan, K., Carlin Geor Malar, G. et al. A comprehensive review on comparison among effluent treatment methods and modern methods of treatment of industrial wastewater effluent from different sources. Appl Water Sci 12, 70 (2022).
9. A. Shahedi, A.K. Darban, F. Taghipour, A. Jamshidi-Zanjani, A review on industrial wastewater treatment via electrocoagulation processes, Current Opinion in Electrochemistry, Volume 22, 2020, Pages 154-169, ISSN 2451-9103.
10. Grace A.V. Magalhães-Ghiotto, Alessandra M. de Oliveira, Jean P.S. Natal, Rosângela Bergamasco, Raquel G. Gomes, Green nanoparticles in water treatment: A review of research trends, applications, environmental aspects and large-scale production, Environmental Nanotechnology, Monitoring & Management, Volume 16, 2021, 100526, ISSN 2215-1532.
11. Yaqoob, A.A.; Parveen, T.; Umar, K.; Mohamad Ibrahim, M.N. Role of Nanomaterials in the Treatment of Wastewater: A Review. Water 2020, 12, 495. https://doi.org/10.3390/w12020495.
12. Bais Madhuri, S. P. Singh, R. D. Batra, Nanotechnology in wastewater treatment: A review, Novel Applications in Polymers and Waste Management, 2018, Imprint Apple Academic Press, Pages 10 eBook ISBN9781315365848.
13. Li, J., Wu, Q., Wu, J. (2015). Synthesis of Nanoparticles via Solvothermal and Hydrothermal Methods. In: Aliofkhazraei, M. (eds) Handbook of Nanoparticles. Springer, Cham.
14. Ndlwana L, Raleie N, Dimpe KM, Ogutu HF, Oseghe EO, Motsa MM, Msagati TAM, Mamba BB. Sustainable Hydrothermal and Solvothermal

Synthesis of Advanced Carbon Materials in Multidimensional Applications: A Review. Materials (Basel). 2021 Sep 6; 14(17): 5094. doi: 10.3390/ma14175094. PMID: 34501183; PMCID: PMC8434334.

15. Mruthunjayappa, Manohara Halanur, S. K. Nataraj, and Dibyendu Mondal. New prospects on solvothermal carbonisation assisted by organic solvents, ionic liquids and eutectic mixtures–A critical review. Progress in Materials Science (2022): 100932.
16. Glenita Bridget D'Souza, Ashok Maraddi, Juno Rose, KN Mahadevaprasad, Smitha Kamath, KN Santosh, S. K. Nataraj, Green carbon induced super hydrophilic BioFoam filters for hazardous emerging pollutants removal and their sustainable value addition, Sustainable Materials and Technologies, Available online 27 July 2023, e00676.
17. Glenita D'Souza, Anshu Kumar S. K. Nataraj Designing engineered biopolymer mesh filters for robust sequestration of chromium (VI), fluoride and other emerging pollutants: A sustainable approach. Chemical Engineering Journal, 2022, 443, 136462.
18. Yi Jiang, Pratim Biswas and John D. Fortner, A review of recent developments in graphene-enabled membranes for water treatment, Environ. Sci.: Water Res. Technol., 2016, 2, 915-922.
19. Syeda Mahnoor Zehra Rizvi, Syeda Hijab Zehra Rizvi, Tanzila Mukhtar,Lubna Nawab Tahira Amir, Asia Bibi, A Comprehensive Review on Applications of Graphene in Water Purification, International Journal of Scientific & Engineering Research Volume 12, Issue 3, March 2021.
20. Yongchen Liu 2017 IOP Conf. Ser.: Earth Environ. Sci. 94 012060.
21. Marta Pedrosa, José L. Figueiredo, Adrián M.T. Silva, Graphene-based catalytic membranes for water treatment – A review, Journal of Environmental Chemical Engineering, Volume 9, Issue 1, 2021, 104930, ISSN 2213-3437.
22. Jiang X, Li Y, Tang X, Jiang J, He Q, Xiong Z, Zheng H. Biopolymer-based flocculants: a review of recent technologies. Environ Sci Pollut Res Int. 2021 Sep; 28(34): 46934-46963. doi: 10.1007/s11356-021-15299-y. Epub 2021 Jul 14. PMID: 34263401; PMCID: PMC8279699.
23. Natalia Fijoł, Andrea Aguilar-Sánchez, Aji P. Mathew, 3D-printable biopolymer-based materials for water treatment: A review, Chemical Engineering Journal, Volume 430, Part 3, 2022, 132964, ISSN 1385-8947.
24. Vahid Vatanpour, Mehmet Emin Pasaoglu, Hossein Barzegar, Oğuz Orhun Teber, Recep Kaya, Muhammed Bastug, Alireza Khataee, Ismail Koyuncu, Cellulose acetate in fabrication of polymeric membranes: A review, Chemosphere, Volume 295, 2022, 133914, ISSN 0045-6535.
25. Wenjuan Chen, Yanlei Su, Lili Zheng, Lijun Wang, Zhongyi Jiang, The improved oil/water separation performance of cellulose acetate-graft-polyacrylonitrile membranes, Journal of Membrane Science, Volume 337, Issues 1–2, 2009, Pages 98-105, ISSN 0376-7388.
26. Idress, H., Zaidi, S.Z.J., Sabir, A. et al. Cellulose acetate based Complexation-NF membranes for the removal of Pb(II) from waste water. Sci Rep 11, 1806 (2021).

27. Hani Nasser Abdelhamid and Aji P. Mathew, Cellulose-Based Materials for Water Remediation: Adsorption, Catalysis, and Antifouling, Front. Chem. Eng., Sec. Environmental Chemical Engineering, 02 December 2021.

28. Yaneth A. Bustos-Terrones, Erick R. Bandala, Gabriela E. Moeller-Chávez, Victoria Bustos-Terrones, Enhanced biological wastewater treatment using sodium alginate-immobilized microorganisms in a fluidized bed reactor, Water Science and Engineering, Volume 15, Issue 2, 2022, Pages 125-133, ISSN 1674-2370.

29. Nechita, P., 2017, 'Applications of Chitosan in Wastewater Treatment', in E. A. Shalaby (ed.), Biological Activities and Application of Marine Polysaccharides, IntechOpen, London. 10.5772/65289.

30. Shanta Biswas, Jannatun Fatema, Tonmoy Debnath, and Taslim Ur Rashid, Chitosan–Clay Composites for Wastewater Treatment: A State-of-the-Art Review, ACS ES&T Water 2021, 1(5), 1055-1085.

31. Oluwaseun J. Ajala, A. Khadir, Joshua O. Ighalo, Great C. Umenweke, Chapter 17 - Cellulose-based nano-biosorbents in water purification, Editor(s): Adil Denizli, Nisar Ali, Muhammad Bilal, Adnan Khan, Tuan Anh Nguyen, In Micro and Nano Technologies, Nano-Biosorbents for Decontamination of Water, Air, and Soil Pollution, Elsevier, 2022, Pages 395-415, ISBN 9780323909129.

32. Daud Hussain, Suhail Ayoub Khan, Salman S. Alharthi, Tabrez Alam Khan, Insight into the performance of novel kaolinite-cellulose/cobalt oxide nanocomposite as green adsorbent for liquid phase abatement of heavy metal ions: Modelling and mechanism, Arabian Journal of Chemistry, Volume 15, Issue 7, 2022, 103925, ISSN 1878-5352.

33. Tian, S.-Y., Guo, J.-H., Zhao, C., Peng, Z., Gong, C.-H., Yu, L.-G., Liu, X.-H., & Zhang, J.-W. (2019). Preparation of Cellulose/Graphene Oxide Composite Membranes and Their Application in Removing Organic Contaminants in Wastewater. Journal of Nanoscience and Nanotechnology, 19(4), 2147–2153.

34. Mansoori, S., Davarnejad, R., Matsuura, T., & Ismail, A. F. (2020). Membranes based on non-synthetic (natural) polymers for wastewater treatment. Polymer Testing, 84, 106381.

35. Yu Liu, Bo Li, Wanbo Mao, Wen Hu, Gang Chen, Yingyao Liu, and Zhiqiang Fang, Strong Cellulose-Based Materials by Coupling Sodium Hydroxide–Anthraquinone (NaOH–AQ) Pulping with Hot Pressing from Wood, ACS Omega 2019 4 (4), 7861-7865.

36. Feifei Zhang, Muhammad Shahidul Islam, Richard M. Berry, and Kam Chiu Tam, β-Cyclodextrin-Functionalized Cellulose Nanocrystals and Their Interactions with Surfactants, ACS Omega 2019 4 (1), 2102-2110.

37. Dabum Kim, Youngsang Ko, Goomin Kwon, Ung-Jin Kim, Jong Hun Lee, and Jungmok You, 2,2,6,6-Tetramethylpiperidine-1-oxy-Oxidized Cellulose Nanofiber-Based Nanocomposite Papers for Facile In Situ Surface-Enhanced Raman Scattering Detection, ACS Sustainable Chemistry & Engineering 2019 7 (18), 15640-15647.

38. Ru Jiang, Hua-Yue Zhu, Yong-Qian Fu, En-Min Zong, Sheng-Tao Jiang, Jian-Bing Li, Jian-Qiang Zhu, Ying-Ying Zhu, Magnetic $NiFe_2O_4$/MWCNTs functionalized cellulose bioadsorbent with enhanced adsorption property and rapid separation, Carbohydrate Polymers, Volume 252, 2021, 117158, ISSN 0144-8617.
39. No HK, Meyers SP. Application of chitosan for treatment of wastewaters. Rev Environ Contam Toxicol. 2000;163:1-27. doi: 10.1007/978-1-4757-6429-1_1. PMID: 10771583.
40. Muhammad Zubair, Muhammad Arshad, Aman Ullah, Chapter 25 - Chitosan-based materials for water and wastewater treatment, Editor(s): Sreerag Gopi, Sabu Thomas, Anitha Pius, Handbook of Chitin and Chitosan, Elsevier, 2020, Pages 773-809, ISBN 9780128179666.
41. Jacques Desbrières,Eric Guibal, Chitosan for wastewater treatment, Polymer International, 2017, Volume 67, Issue 1, p. 7-14.
42. Anny Leudjo Taka, Michael J. Klink, Xavier Yangkou Mbianda, Eliazer Bobby Naidoo, Chitosan nanocomposites for water treatment by fixed-bed continuous flow column adsorption: A review, Carbohydrate Polymers, Volume 255, 2021, 117398, ISSN 0144-8617.
43. Anshu Kumar, Parimal Paul, and Sanna Kotrappanavar Nataraj, Bionanomaterial Scaffolds for Effective Removal of Fluoride, Chromium, and Dye, ACS Sustainable Chemistry & Engineering 2017, 5 (1), 895-903.
44. Biopolymer-reinforced synthetic granular nanocomposites for affordable point-of-use water purification, PNAS, Published online before print May 6, 2013.
45. Paola Franco, Stefano Cardea, Antonio Tabernero and Iolanda Dc Marco, Porous Aerogels and Adsorption of Pollutants from Water and Air: A Review, Molecules 2021, 26, 4440.
46. Liu, Xg., Mao, Qs., Jiang, Y. et al. Preparation of Al_2O_3-SiO_2 composite aerogels and their Cu^{2+} absorption properties. Int J Miner Metall Mater 28, 317–324 (2021).
47. Shibing Ye, Yue Liu, and Jiachun Feng, Low-Density, Mechanical Compressible, Water-Induced Self-Recoverable Graphene Aerogels for Water Treatment, ACS Applied Materials & Interfaces 2017, 9(27), 22456-22464.
48. Zewen Zhang, Yueping Bao, Xun Sun, Ke Chen, Mingjiong Zhou, Liu He, Qing Huang, Zhengren Huang, Zhifang Chai, and Yujie Song, Mesoporous Polymer-Derived Ceramic Membranes for Water Purification via a Self-Sacrificed Template, ACS Omega 2020, 5(19), 11100-11105.
49. Zeming He, Zhiyang Lyu, Qilin Gu, Lei Zhang, John Wang, Ceramic-based membranes for water and wastewater treatment, Colloids and Surfaces A: Physicochemical and Engineering Aspects, Volume 578, 2019, 123513, ISSN 0927-7757.
50. S. Lakshmi Sandhya Rani, R. Vinoth Kumar, Insights on applications of low-cost ceramic membranes in wastewater treatment: A mini-review,

Case Studies in Chemical and Environmental Engineering, Volume 4, 2021, 100149, ISSN 2666-0164.

51. Jialin Cao, Yanlei Su, Yanan Liu, Jingyuan Guan, Mingrui He, Runnan Zhang, Zhongyi Jiang, Self-assembled MOF membranes with underwater superoleophobicity for oil/water separation, Journal of Membrane Science, Volume 566, 2018, Pages 268-277, ISSN 0376-7388.
52. A Maraddi, M Halakarni, MM Halanur, SK Nataraj, Fe-MOF induced biopolymer-based sustainable self-cleaning membranes for effective selective separation and wastewater treatment, Sustainable Materials and Technologies, 2023, 35, e00537.
53. Yi-Ming Li, Liumei Cao, Hao Ren, Chun-Yan Ji, Wenjing Li, Lin Cheng. Chiral Polymer-Mediated Pd@MOF-808 for Efficient Sequential Asymmetric Reaction. Catalysis Letters 2022, 455.
54. Kaikai Ma, Megan C. Wasson, Xingjie Wang, Xuan Zhang, Karam B. Idrees, Zhijie Chen, Yufang Wu, Seung-Joon Lee, Ran Cao, Yongwei Chen, Lifeng Yang, Florencia A. Son, Timur Islamoglu, Gregory W. Peterson, John J. Mahle, Omar K. Farha. Near-instantaneous catalytic hydrolysis of organophosphorus nerve agents with zirconium-based MOF/hydrogel composites. Chem Catalysis 2021, 1(3), 721-733.
55. Mohammad Ali Ghasemzadeh, Boshra Mirhosseini-Eshkevari, Poly(acrylic acid)/Fe_3O_4 Supported on MIL-100(Cr) MOF as a Novel and Magnetic Nanocatalyst for the Synthesis of Pyrido[2,3-d]Pyrimidines, Heliyon, 2022, e10022, ISSN 2405-8440.
56. Huang, B. Carbon nanotubes and their polymeric composites: the applications in tissue engineering. Biomanuf Rev 5, 3 (2020).
57. Doo-Won Kim, Jae-Hyung Wee, Cheol-Min Yang, Kap Seung Yang, Efficient removals of Hg and Cd in aqueous solution through NaOH-modified activated carbon fiber, Chemical Engineering Journal, Volume 392, 2020, 123768, ISSN 1385-8947.
58. Radheshyam R. Pawar, Prabuddha Gupta, Lalhmunsiama, Hari C. Bajaj, Seung-Mok Lee, Al-intercalated acid activated bentonite beads for the removal of aqueous phosphate, Science of The Total Environment, Volume 572, 2016, Pages 1222-1230, ISSN 0048-9697.
59. Halanur Mruthunjayappa Manohara, Kanakaraj Aruchamy, Supratim Chakraborty, Nagaraj Radha, Maalige Rajappa Nidhi, Debasis Ghosh, Sanna Kotrappanavar Nataraj, and Dibyendu Mondal, Sustainable Water Purification Using an Engineered Solvothermal Carbon Based Membrane Derived from a Eutectic System, ACS Sustainable Chemistry & Engineering 2019 7 (11), 10143-10153.
60. Manohara Halanur Mruthunjayappa, Vibha T. Sharma, Kalpana Dharmalingam, Nataraj Sanna Kotrappanavar, Dibyendu Mondal. Engineering a Biopolymer-Based Ultrafast Permeable Aerogel Membrane Decorated with Task-Specific Fe–Al Nanocomposites for Robust Water Purification. ACS Applied Bio Materials 2020, 3 (8) , 5233-5243.

61. Smitha V. Kamath, Manohara Halanur Mruthunjayappa, Dibyendu Mondal, Nataraj Sanna Kotrappanavar. Nanocomposite-based high-performance adsorptive water filters: recent advances, limitations, nanotoxicity and environmental implications. Environmental Science: Nano 2022, 9(7) , 2320-2341.
62. Manohara Halanur Mruthunjayappa, Nataraj Sanna Kotrappanavar, Dibyendu Mondal. New prospects on solvothermal carbonisation assisted by organic solvents, ionic liquids and eutectic mixtures – A critical review. Progress in Materials Science 2022, 126 , 100932.
63. Smitha V. Kamath, Halanur M. Manohara, Kanakaraj Aruchamy, Ashok Shrishail Maraddi, Glenita Bridget D'Souza, Kuchangi Naraseeyappa Santhosh, K. N. Mahadevaprasad, S. K. Nataraj. Sorption based easy-to-use low-cost filters derived from invasive weed biomass for dye contaminated water cleanup. RSC Advances 2022, 12(15), 9101-9111.
64. Halanur M. Manohara, Sooraj S. Nayak, Gregory Franklin, Sanna Kotrappanavar Nataraj, Dibyendu Mondal. Progress in marine derived renewable functional materials and biochar for sustainable water purification. Green Chemistry 2021, 23(21), 8305-8331.
65. Manohara Halanur Mruthunjayappa, Dibyendu Mondal, S.K. Nataraj. Polymer aerogels for energy storage and water purification applications. 2022, 31-52, in Polymer-Based Advanced Functional Materials for Energy and Environmental Applications, Edited by Nithin Kundachira Subramani, S. K. Nataraj, Chetankumar Patel, Sachhidananda Shivanna, ISBN 978-981-16-8754-9, Published: 23 December 2021.
66. Taiba Naseem, Tayyiba Durrani, The role of some important metal oxide nanoparticles for wastewater and antibacterial applications: A review, Environmental Chemistry and Ecotoxicology, Volume 3, 2021, Pages 59-75, ISSN 2590-1826.
67. Hongcai Gao, Fei Xiao, Chi Bun Ching, and Hongwei Duan, High-performance asymmetric supercapacitor based on graphene hydrogel and nanostructured MnO_2, ACS Applied Materials & Interfaces, 2012, 4 (5), 2801-2810.
68. Melinda Krebsz, Tibor Pasinszki, Tran Thanh Tung, Md Julker Nine, Dusan Losic, Multiple applications of bio-graphene foam for efficient chromate ion removal and oil-water separation, Chemosphere, Volume 263, 2021, 127790, ISSN 0045-6535.
69. AmrAbdelkhalek, MonaAbd El-Latif , Hesham Ibrahi, Hesham Hamad & Marwa, Showman Controlled synthesis of graphene oxide/silica hybrid nanocomposites for removal of aromatic pollutants in water, Scientifc Reports | (2022) 12: 7060.
70. Rai, S., Bhujel, R., Biswas, J. et al. Biocompatible synthesis of rGO from ginger extract as a green reducing agent and its supercapacitor application. Bull Mater Sci 44, 40 (2021).

Index